MW01040956

Bioremediation and Phytoremediation

Chlorinated and Recalcitrant Compounds

Editors

Godage B. Wickramanayake
Battelle

Robert E. Hinchee
Parsons Engineering Science, Inc.

The First International Conference on Remediation of Chlorinated and Recalcitrant Compounds

Monterey, California, May 18–21, 1998

Library of Congress Cataloging-in-Publication Data

International Conference on Remediation of Chlorinated and Recalcitrant Compounds (1st : 1998 : Monterey, Calif.)
Bioremediation and phytoremediation : chlorinated and recalcitrant compounds / editors, Godage B. Wickramanayake, Robert E. Hinchee : First International Conference on Remediation of Chlorinated and Recalcitrant Compounds, Monterey, California, May 18–21, 1998.
p. cm.
Includes bibliographical references and index.
ISBN 1-57477-059-4 (alk. paper)
1. Organochlorine compounds--Biodegradation--Congresses. 2. Hazardous waste site remediation--Congresses. 3. Bioremediation--Congresses. 4. Phytoremediation--Congresses. I. Wickramanayake, Godage B., 1953– . II. Hinchee, Robert E.

TD1066.073I58 1998
628.5′2—dc21

98-24755
CIP

Printed in the United States of America

Battelle Press
505 King Avenue
Columbus, Ohio 43201, USA
614-424-6393 or 1-800-451-3543
Fax: 1-614-424-3819
Internet: press@battelle.org
Website: www.battelle.org/bookstore

For information on future symposia and conference programs, write to:
Bioremediation Symposium
Battelle
505 King Avenue
Columbus, Ohio 43201-2693
Fax: 614-424-3667

CONTENTS

Aerobic Mechanisms

Biological Removal of 1,2-Dichloroethane and Tetrachloroethene from Contaminated Groundwater (Large Scale). *G. Stucki and M. Thüer* 1

Removal of Di-, Tri-, Tetra-, and Pentachlorophenol Mixtures in a 5-L Continuous Aerobic Packed Column. *L.G. Torres, A. Salinas, B.E. Jiménez, and E.R. Bandala* 7

Modeling of Remediation with ORC™: Transverse Dispersion. *D.J. Wilson and R.D. Norris* 13

Analysis of Remedial Options for Chlorinated VOCs at Harrison Landfill. *H.W. Bentley, J. Tang, S. Smith, D. Samorano, and R.G. Arnold* 21

Enhancing Dissolved Oxygen to Remediate Vinyl Chloride in Groundwater. *I.J. Verhagen, D.W. Wetzstein, D.R. Bruner, and C.M. Hudak* 27

Biological Degradation of Chlorinated Aromatics in a Pilot-Scale Water Treatment Plant. *W. Dott, M. Steiof, and B. Zettler* 33

Modeling the Effect of Nonionic Surfactants on the Biodegradation of Polycyclic Aromatic Hydrocarbons in Soil Slurry Using Respirometric Technique —1. Physicochemical Effect. *J.-S. Park, Y.J. Kim, and I.S. Kim* 39

Bioremediation of Chlorophenol-Contaminated Soil by Composting at Full Scale. *M.M. Laine and K.S. Jørgensen* 45

Enhanced In Situ Mobilization and Biodegradation of Phenanthrene from the Soil by Paraffin Oil/Surfactant. *E. Kim, A. Liu, I. Ahn, L.W. Lion, and M.L. Shuler* 51

Application of Bioremediation Testing Protocol to PAH-Contaminated Aged Soils. *H.H. Tabak, R. Govind, M. Parvatiyar, Q. Song, and J. Guo* 55

Biological Reductive Dechlorination Processes

Reductive Dechlorination of Tetrachloroethene to Ethene Adsorbed on Activated Carbon. *K. Böckle and P. Werner* 65

4-*tert*-Butylphenol Degradation in Anaerobic Conditions. *L. Di Palma, C. Merli, and E. Petrucci* 71

Medium Optimization for the Cultivation of Bacteria Reductively Dechlorinating Trichlorobenzenes. *L. Adrian, U. Szewzyk, and H. Görisch* 77

Numerical Investigation of Factors Promoting TCE Degradation in Porous Media. *N. Singhal, P. Jaffé, and W. Maier* 83

Reductive Dechlorination Versus Adsorption of Tetrachloroethylene in Fluidized-Bed Reactors. *S. Marcoux, J. Nicell, A. Beaubien, and S.R. Guiot* 91

Anaerobic Degradation of PCE and TCE DNAPLs by Groundwater Microorganisms. *R.B. Nielsen and J.D. Keasling* 97

Removing Recalcitrant Volatile Organic Compounds Using Disaccharide and Yeast Extract. *J.H. Honniball, T.A. Delfino, and J.D. Gallinatti* 103

A Combined Anaerobic and Aerobic Microbial System for Complete Degradation of Tetrachloroethylene. *T.H. Lee, M. Ike, and M. Fujita* 109

Effect of Tween™ Surfactants on the Microbial Reductive Dechlorination of Hexachlorobenzene. *D.H. Yeh, K.D. Pennell, and S.G. Pavlostathis* 115

Enhanced In Situ Reductive Dechlorination. *E.S.K. Becvar, A. Fisher, G. Sewell, V. Magar, J. Gossett, and C.M. Vogel* 121

Microbial Reductive Dechlorination of PCE in a Quasi Two-Dimensional Sandbox Model. *O.A. Cirpka and G. Bisch* 129

Continuous On-Line Monitoring of Organochlorine Compounds in a Fixed-Bed Bioreactor by an Optical Infrared Sensor. *V. Acha, M. Meurens, H. Naveau, and S. Agathos* 135

Bioaugmentation and Biomonitoring

Bioremediation of Pentachlorophenol: A Pilot-Scale Study. *H.R. Compton, G. Scogin, W. Johnson, T.F. Miller, M.F. Mohn, and D.G. Crouse* 143

Microcosm Studies of Bioaugmentation of Butane and Propane Utilizers for In Situ Cometabolism of 1,1,1-Trichloroethane. *P. Jitnuyanont, L. Semprini, and L. Sayavedra-Soto* 149

Reductive Dechlorination of *cis*-1,2-Dichloroethene with an Enriched Mixed Culture. *C. Windfuhr, S. Granzow, H. Scholz-Muramatsu, and G. Diekert* 155

Pentachlorophenol Degradation Using Lignin Peroxidase Produced from *Phanerochaete chrysosporium* Immobilized in Polyurethane Foam. *S.H. Choi, E. Song, K.W. Lee, S.-H. Moon, and M.B. Gu* 161

Biotransformation of Hexachlorobenzene by Anaerobic Enriched Cultures. *I.S. Kim, H. Ishii, G.D. Sayles, M.J. Kupferle, and T.L. Huang* 167

Cometabolic Processes

Reducing VOC Concentrations Through Landfill Gas Removal and Cometabolic Degradation. *J.D. Hartley and C.M. Richgels* 175

Cometabolic Biodegradation of *cis*-1,2-Dichloroethene by Ethene-Utilizing Bacteria. *D. Bryniok, P. Koziollek, S. Bauer, and H.-J. Knackmuss* 181

Biotreatability Studies for Remediation of TCE-Contaminated Groundwater. *M. Eguchi, H. Myoga, S. Sasaki, and Y. Miyake* 187

Field Application of In Situ Methanotrophic Treatment for TCE Remediation. *R. Legrand, A.J. Morecraft, J.A. Harju, T.D. Hayes, and T.C. Hazen* 193

Sustained Biodegradation of Trichloroethylene in a Suspended-Growth Gas Treatment Reactor. *S.-B. Lee, J.P. Patton, S.E. Strand, and H.D. Stensel* 199

Trichloroethylene Bioremediation by *Methylosinus trichosporium* OB3b Immobilized in a Fibrous-Bed Bioreactor. *A.L. Kneidel, H. Shim, and S.-T. Yang* 205

Cometabolism of Chlorinated VOCs Downgradient of a Fuel Hydrocarbon Source. *P.I. Dacyk and W.D. Hughes* 215

Cometabolic Biofiltration of TCE Using Bioluminescent Reporter Bacteria. *C.D. Cox, K.G. Robinson, H.-J. Woo, C.L. Wright, and J. Sanseverino* 221

Cometabolic Bioventing of Chlorinated Solvents at a Former Waste Lagoon. *E.E. Cox, T.A. McAlary, D.W. Major, J. Allan, L. Lehmicke, and S.L. Neville* 227

Phytoremediation of Recalcitrant Organic Compounds

Biodegradation of Tetrachloroethylene and Trichloroethylene Using Mixed-Species Microbial Mats. *W. O'Niell, V. Nzengung, J. Noakes, J. Bender, and P. Phillips* 233

Modeling Phytoremediation of Land Contaminated by Hydrocarbons. *M.Y. Corapcioglu, R.L. Rhykerd, C.L. Munster, M.C. Drew, K. Sung, and Y.-Y. Chang* 239

Pilot-Scale Use of Trees to Address VOC Contamination. *H.R. Compton, D.M. Haroski, S.R. Hirsh, and J.G. Wrobel* 245

Phytoremediation of Dissolved-Phase Trichloroethylene Using Mature Vegetation. *W.J. Doucette, B. Bugbee, S. Hayhurst, W.A. Plaehn, D.C. Downey, S.A. Taffinder, and R. Edwards* 251

Evaluation of Tamarisk and Eucalyptus Transpiration for the Application of Phytoremediation. *R.W. Tossell, K. Binard, L. Sangines-Uriarte, M.T. Rafferty, and N.P. Morris* 257

Phreatophyte Influence on Reductive Dechlorination in a Shallow Aquifer Containing TCE. *R.W. Lee, S.A. Jones, E.L. Kuniansky, G.J. Harvey, and S.M. Eberts* 263

Author Index 269

Keyword Index 293

FOREWORD

Bioremediation and phytoremediation have progressed over the past decade from promising ideas to practical remediation approaches, especially with regard to the treatment of hydrocarbon-contaminated sites. Sites contaminated with chlorinated and recalcitrant compounds have proven more resistant to these approaches, but exciting progress is being made both in the laboratory and in the field. *Bioremediation and Phytoremediation: Chlorinated and Recalcitrant Compounds* brings together the latest breakthrough bioremediation research and field applications in chapters that cover cometabolic processes, aerobic and anaerobic mechanisms, biological reductive dechlorination processes, bioaugmentation, biomonitoring, and phytoremediation of recalcitrant organic compounds.

This is one of six volumes published in connection with the First International Conference on Remediation of Chlorinated and Recalcitrant Compounds, held in May 1998 in Monterey, California. The 1998 Conference was the first in a series of biennial conferences focusing on the more problematic substances—chlorinated solvents, pesticides/herbicides, PCBs/dioxins, MTBE, DNAPLs, and explosives residues—in all environmental media. Physical, chemical, biological, thermal, and combined technologies for dealing with these compounds were discussed. Several sessions dealt with natural attenuation, site characterization, and monitoring technologies. Pilot- and field-scale studies were presented, plus the latest research data from the laboratory. Other sessions focused on human health and ecological risk assessment, regulatory issues, technology acceptance, and resource allocation and cost issues. The conference was attended by scientists, engineers, managers, consultants, and other environmental professionals representing universities, government, site management and regulatory agencies, remediation companies, and research and development firms from around the world.

The inspiration for this Conference first came to Karl Nehring of Battelle, who recognized the opportunity to organize an international meeting that would focus on chlorinated and recalcitrant compounds and cover the range of remediation technologies to encompass physical, chemical, thermal, and biological approaches. The Conference would complement Battelle's other biennial remediation meeting, the In Situ and On-Site Bioremediation Symposium. Jeff Means of Battelle championed the idea of the conference and made available the resources to help turn the idea into reality. As plans progressed, a Conference Steering Committee was formed at Battelle to help plan the technical program. Committee members Abe Chen, Tad Fox, Arun Gavaskar, Neeraj Gupta, Phil Jagucki, Dan Janke, Mark Kelley, Victor Magar, Bob Olfenbuttel, and Bruce Sass communicated with potential session chairs to begin the process of soliciting papers and organizing the technical sessions that eventually were presented in Monterey. Throughout the process of organizing the Conference, Carol Young of

Battelle worked tirelessly to keep track of the stream of details, documents, and deadlines involved in an undertaking of this magnitude.

Each section in this and the other five volumes corresponds to a technical session at the Conference. The author of each presentation accepted for the Conference was invited to prepare a short paper formatted according to the specifications provided. Papers were submitted for approximately 60% of the presentations accepted for the conference program. To complete publication shortly after the Conference, no peer review, copy-editing, or typesetting was performed. Thus, the papers within these volumes are printed as submitted by the authors. Because the papers were published as received, differences in national convention and personal style led to variations in such matters as word usage, spelling, abbreviation, the manner in which numbers and measurements are presented, and type style and size.

We would like to thank the Battelle staff who assembled this book and its companion volumes and prepared them for printing. Carol Young, Christina Peterson, Janetta Place, Loretta Bahn, Lynn Copley-Graves, Timothy Lundgren, and Gina Melaragno spent many hours on production tasks. They developed the detailed format specifications sent to each author, tracked papers as received, and examined each to ensure that it met basic page layout requirements, making adjustments when necessary. Then they assembled the volumes, applied headers and page numbers, compiled tables of contents and author and keyword indices, and performed a final page check before submitting the volumes to the publisher. Joseph Sheldrick, manager of Battelle Press, provided valuable production-planning advice and coordinated with the printer; he and Gar Dingess designed the volume covers.

Neither Battelle nor the Conference co-sponsors or supporting organizations reviewed the materials published in these volumes, and their support for the Conference should not be construed as an endorsement of the content.

Godage B. Wickramanayake and Robert E. Hinchee
Conference Chairman and Co-Chairman

BIOLOGICAL REMOVAL OF 1,2-DICHLOROETHANE AND TETRA-CHLOROETHENE FROM CONTAMINATED GROUNDWATER (LARGE-SCALE)

Gerhard Stucki, Markus Thüer (Ciba Specialty Chemicals Inc., Schweizerhalle, Switzerland)

ABSTRACT: The performance of two biological large-scale processes to purify groundwater contaminated with chlorinated hydrocarbons is presented. At the first site in Lübeck (Germany), 1,2-dichloroethane (DCA) is aerobically mineralised by microorganisms bread under laboratory conditions and inoculated into a groundwater purification plant to treat 5 to 20 m^3/h at 8 to 12°C. Average feed concentrations of 15 mg/l were degraded to below 10 µg/l. The initially conventional activated carbon process was biologically modified. As a result, the monthly activated carbon requirement (5 tons) became redundant, and the operation costs fell by a factor of seven.

At the second site in Albstadt (Germany), tetrachloroethene (TCE) is removed from the unsaturated zone by soil vapour extraction, and the contaminated groundwater (2 m^3/h at 12 to 14 °C containing 1.5 mg/l TCE and 25 mg/l nitrate) is pumped and treated in three 1 m^3 reactors run in series using anaerobic and aerobic microorganisms. The contaminated groundwater is amended with methanol before fed to the denitrifying reactor followed by an anaerobic fixed bed and a reactor filled with granular activated carbon. The latter reactor is fed with traces of H_2O_2 required for aerobic conditions. This groundwater treatment began in October 1995 using sludge of a digestor and small volumes of enrichment and several pure cultures able to convert chlorinated ethenes.

THE LÜBECK SITE

Aerobic removal of 1,2-dichloroethane (DCA). In our company, the first full-scale process using bacteria isolated in the laboratory to treat contaminated groundwater started in 1990. The polluted site of a former pharmaceutical production plant was located in Lübeck (Germany), where DCA served as the single solvent for the extraction of pancreatine. Its production lasted from the early fifties until 1987 when the plant was decommissioned. The groundwater was polluted with **DCA as single contaminant**. It was detected in the top aquifer only, ranging from 3 to 15 m below the surface and flowing into surface waters about 350 m east of the production site, passing an area of private homes.

The major sources were removed and a detailed hydrogeological survey of the site was conducted. The pump and treat technology was chosen as remedial action respecting the chemical and physical characteristics of pollutant and environment of the site. Alternatives were regarded as either impractical or less efficient. The groundwater treatment options evaluated were air stripping, activated carbon adsorption, oxidation by UV-ozone/UV-peroxide and biological degradation. The chemical and physical properties of DCA favored the choice of the biological method, since the alternatives were considered unfavorable due to the high solubility, the low sorptive properties (both favoring also the extraction by pump and treat technology) and the chemical structure (DCA has no double bond thus resists attack by ozone or hydroxyl radicals). Microorganisms able to mineralize DCA had been isolated previously and studies supported the applicability of the laboratory strains for the purification of groundwater (Stucki et. al., 1994).

Groundwater treatment plant and its performance. Details of the first 5 years of the pump and treat activity at the Lübeck site have been published previously (Stucki et al., 1995). The groundwater (up to 20 m^3/h) is pumped from a gallery of wells to two parallel sand filters followed by two filters filled with granular activated carbon run in series. The treatment plant was inoculated with DCA degrading microorganisms, and H_2O_2 and nutrients (N- and P-salts) were fed in trace amounts. Two years after start-up, an 8 m^3 rotating biological disc was taken into operation as an additional process unit in front of the sand filters to increase the biological degradation potential. The electrical power requirement for the whole plant is in the range of 5 ± 1 kW. The key operational figures are shown in Tab. 1

The main reason for the **economic success** of the biological purification of the DCA polluted groundwater was that the use of activated carbon as a major contributor to the operation costs became redundant with the biological degradation potential developing: The content of one of the carbon adsorbers had to be replaced 3 weeks after start-up due to the exhausted adsorption capacity. At that time, the biological degradation potential was almost zero. Later, biodegradation rates improved, and the time period until the next activated carbon filter was exhausted increased to several weeks. A third and so far last carbon exchange was required in the following winter season (January 1992). Since six years, no further activated carbon had to be exchanged.

The remediation technology was also an **ecological success**. Almost no energy in addition to the pumping energy to extract the groundwater was used. Furthermore, DCA was completely mineralized *on-site* as indicated by the oxygen consumption, the pH decrease and the chloride ions production during the passage of the groundwater through the bioreactors (Stucki et al., 1995).

The survival of at least one of the strains (*Xanthobacter autotrophicus* GJ10) originally inoculated could be confirmed four years after the start-up of the plant, when a gene probe taken from the plant effluent hybridized positively with the cloned dehalogenase gene from that organism (D.B. Janssen, Groningen, NL, personal communication).

TABLE 1. Operational Key Figures

Parameter	Units	1990/91	1991/92	1992/93	1993/94	1994/95	1995/96	1996/97	Average	Total
Remediation time	years	1	2	3	4	5	6	7		8
Hydraulic load	m3/d	354	281	215	139	129	86	111	188	
Total water pumped + treated	m3/y	56000	103000	79000	50700	47000	28400	38300	57486	402400
Feed concentration	ug/l DCA	2614	4040	4666	9408	11421	3386	3444	5568	
DCA load	kg/d	0.9	1.1	1	1.3	1.3	0.5	0.4	0.93	
Total DCA removed	kg/y	146	415	369	477	500	195	150	322	2252
Activated carbon used	t/y	15	10	0	0	0	0	0		25
Peroxide used	t/y	11.7	20	10.8	9	8.5	4.2	4.3	10	68.5
Electrical power consumed	kWh	60000	20000	40000	41200	47000	43500	46500	42600	298200
Operation costs	DM/m3	3.2	1.01	0.46	0.54	0.68	0.72	0.42	1.00	
Costs saved	DM									883555

Operation costs = costs for energy (including pumping of groundwater), chemicals and sludge disposal, without manpower, maintain groundwater wells, analytics

During the past 9 years (1989 and 1990 observation, 1991 onwards remediation), almost 2.5 t of DCA was extracted and mineralized form the ground water at the Lübeck site. The contamination has moved in two waves through the site, with the top concentrations being measured in 1990 and 1993/94 (Fig. 1).

FIGURE 1. DCA concentrations in 3 observation wells (B 201 at the former production site, and B233/228 on the way towards the extraction well gallery) and in the most contaminated well E 2 of the extraction well gallery.

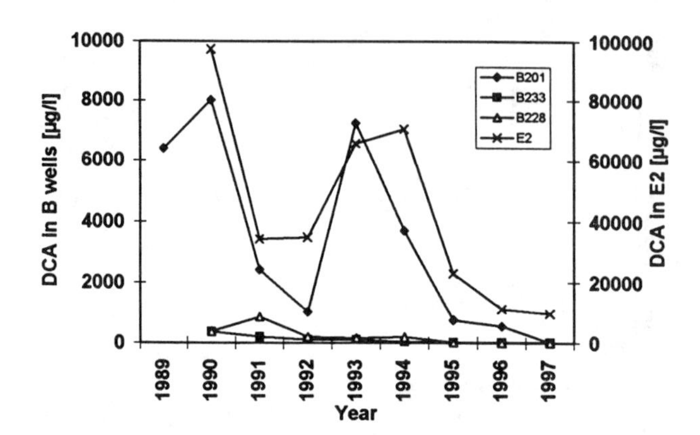

As a result of our remediation activities, the ground water situation at Lübeck has much improved. The DCA concentrations in plume have dropped considerable, and the average DCA concentration in the mostly polluted extraction well E2 has fallen by a factor of 10 (Fig. 1). Only 3 (E1 to E3) out of the 8 extraction wells are still in operation. The others were abandoned because they did not show any contamination for two or more years. The concentrations measured in control wells B 220 and B 221 downstream of the extraction gallery indicate that the spread of the groundwater contamination was successfully prevented. Most of the 13 private wells in the residential area up-stream the groundwater gallery show DCA concentrations of below 5 µg/l for the last 2 years.

***In-situ* observations.** Groundwater samples from both observation and extraction wells show, that DCA may be partially mineralised *in-situ* even though the groundwater was never inoculated with the microorganisms used in the groundwater treatment plant. Groundwater samples showing the highest DCA concentrations also show the highest chloride ion concentrations, the pH is slightly lower than the pH in non-contaminated groundwater, and the oxygen concentration in DCA contaminated samples is depleted. In a few samples fed with additional DCA and kept under aerobic conditions, the chemical was mineralised, too. No DCA degrading microorganisms have been isolated from the Lübeck site so far. Ethene and ethane as potential anaerobic metabolites have not been analysed yet.

THE ALBSTADT SITE

Combined anaerobic-aerobic removal of TCE from groundwater. In a second case, a leaking concrete tank used as inverted siphon to collect and recycle TCE was identified as a source of soil and groundwater contamination. The treatment concept included the cleaning of the soil by soil vapour extraction and the pumping and treating of the groundwater containing up to 2 mg/l TCE using a patented biological process (Thüer et al., 1990), described in parts on laboratory scales by several other authors (e.g. De Bruin et al., 1992). Initially a target concentration of < 10 µg/l for the sum of the chlorinated ethenes was required since the treated water was intended to be re-infiltrated into the groundwater table. However, clogging of the infiltration well was feared due to the high calcium concentrations. Therefore, the environmental office agreed to divert the treated groundwater to the public sewer, where a target concentration of < 100 µg/l had to be reached.

Soil vapour was extracted from the unsaturated zone intermittently using a vacuum pump. It was treated by activated carbon. The condensate containing traces of TCE was fed to the groundwater treatment plant. The groundwater (temperature 12-13 °C, pH 7.5) contained less than 1 mg/l of iron and manganese ions, 150 mg/l of calcium ions, and 25-50 mg/l nitrate. It was pumped at a rate of 2 m^3/h through a series of three reactors each of a volume of 1 m^3 (Fig. 2).

FIGURE 2. Process to remove TCE.

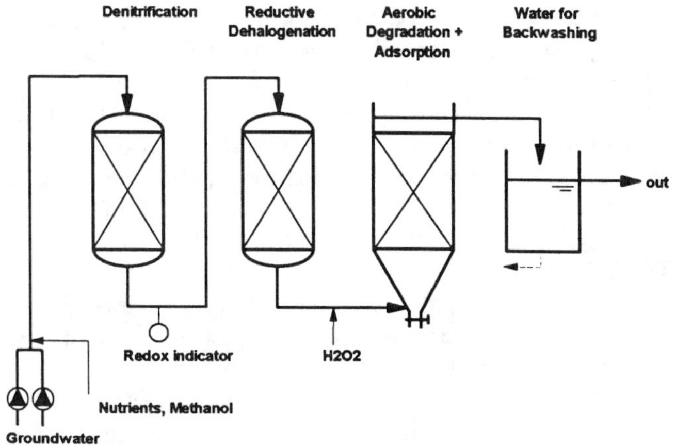

Both the denitrification and the methanogenic reactor were inoculated with sludge of a digestor. The methanogenic reactor was further supplied with 0.3 l sludge obtained from pilot studies and adapted to reduce TCE and trichloroethene, and 0.2 l of a culture supplied by De Bruin (De Bruin et al., 1992). The aerobic activated carbon adsorber was inoculated twice with 2 l of a culture of *Methylosinus trichosporium* Ob3b grown on methane and with 1 l bacteria grown on cumene and isoprene, and 5 l material of a spontaneously operating anaerobic carbon adsorber which converted chlorinated solvents (Strohmeyer et al., 1991).

Anoxic conversion. Nitrate was removed to below 1 mg/l in the first reactor from the beginning of the groundwater treatment. The products built under anaerobic conditions were analysed after the second reactor. The redox potential as a very important process parameter (linked to a telephone alarm) is a reliable process indicator. A **potential of - 300 mV** indicates complete TCE conversion to *cis*-dichloroethene (DCE) via trichloroethene (TRI). About 80 to 90 % of the DCE expected was obtained. It was assumed that the remaining part of the substance might have been reduced further to ethene or ethane, both of which have not been analysed so far. Vinyl chloride was always below the quantification limit (10 µg/l). Traces of methane was detected as degradation product of the methanol fed.

Aerobic conversion. The moving bed activated carbon adsorber was supplied with H_2O_2 and was fed from the bottom (Fig. 2). Methane produced in the preceding steps was available as growth substrate for methanotrophs which have the potential to oxidised DCE. Based on isotherm studies, we calculated a biological DCE removal of about 60 to 90 %. The remaining part of the badly absorbable DCE was held back by the activated carbon. A few times, exhausted activated carbon was **partially exchanged** by virgin carbon added from the top of the reactor, thus keeping some of the grown DCE-oxidizing microorganisms in the system.

Outlook. The combined biological process is still insufficient to make the carbon adsorption redundant. The metabolic bottleneck both of the anaerobic and the aerobic metabolic route remains the DCE conversion. Is it possible to achieve better removal efficiencies if other carbon sources than methanol are fed? Or do we just miss those potent strains with improved catabolic rates?
The time required to maintain the plant amounts to about 30 to 60 min/week. The operation costs including energy for the pumps, methanol and peroxide were in the range of 1.3 DM/m3 when activated carbon was needed, and of about 0.8 DM during the last 15 months. About 35 kg TCE were removed from the site, so far.

REFERENCES

De Bruin, W.P., Kotterman, M.J.J., Posthumus, M.A., Schraa, G., and A.J.B. Zehnder. 1992. "Complete biological reductive transformation of tetrachloroethene to ethane." *Appl. Environ. Microbiol. 58:*1996-2000.

Strohmeyer, S.A., Winkelbauer, W., Kohler, H., Cook, A.M., and T. Leisinger. 1991. "Dichloromethane utilized by an anaerobic mixed culture: acetogenesis and methanogenesis." *Biodegradation 2:*129-137.

Stucki, G., and M. Thüer. 1994. "Increased removal capacity for 1,2-dichloroethane by biological modification of the granular activated carbon process. " *Appl. Microbiol. Biotechnol. 42:*167-172.

Stucki, G., and M. Thüer. 1995. „Experiences of a large-scale application of 1,2-dichloroethane degrading microorganisms for groundwater treatment." *Environ. Sci. Technol. 29*:2339-2345.

Thüer, M., Reisinger, M., and G. Stucki. 1991. "Ground water purification process." *European patent EP 0456 607 A1*, pp 1-4; and *US patent-5180495*.

REMOVAL OF DI-, TRI-, TETRA-, AND PENTACHLOROPHENOL MIXTURES IN A 5 L CONTINUOUS AEROBIC PACKED COLUMN

Luis G. Torres, Alejandro Salinas, Blanca E. Jiménez (Instituto de Ingeniería. Universidad Nacional Autónoma de México. MEXICO), and
Erick R. Bandala (Instituto Mexicano de Tecnología del Agua. MEXICO)

ABSTRACT: A 5 L column packed with *tezontle* (a basaltic scoria) and inoculated with *Pseudomonas fluorescens* was operated at room temperature, under unaerated conditions. Three different chlorophenol mixtures were assessed: 200, 400 and 600 mg/L as total chlorophenols (with proportional concentrations of di-, tri-, tetra-, and pentachlorophenol). Both total (4-aminoantipirine method) and specific chlorophenols (GC/Mass spectrometry system), and total organic carbon (TOC analyzer) removals were measured. Of the total initial chlorophenols for the concentrations of 200, 400 and 600 mg/L, the packed column was able of removing 43-56, 75-80, and 77-86%, respectively. The biodegradation rates for each concentration range were 33-174 mg/L.day (for 200 mg/L), 118-494 mg/L.day (for 400 mg/L), and 179-725 mg/L.day (for 600 mg/L). These variations are due to the hydraulic residence time (HRT), controlled at values of 15.5, 31, and 62 hours. The lowest values correspond most of the time to HRT = 62 hours, and the higher values to HRT=15.5 hours. Total organic carbon TOC removals were around 38-59%. The system demonstrated to be a powerful tool in the bioremediation of waters contaminated with chlorophenols.

INTRODUCTION

Chlorophenols and derivatives have been widely used as pesticides, antifungal agents and herbicides. Due to the high toxicity of this xenobiotics, only a few biological treatment systems are suitable for the bioremediation of contaminated aquifers, water bodies or soils. The use of aerobic submerged filters, where specific microorganisms or mixtures of them have been immobilized, has been reported because of its high potential applications (Arvin, 1991, Seignez *et al.*, 1993, Fava *et al.*, 1996, Torres *et al.*, 1997b). Some of these experiences were carried out in small systems working batchwise. In most of the quoted papers the degradation of simple compounds and never of complex mixtures of them were assessed.

Objective. The aim of this work is to report the successful use of a 5 L continuous aerobic column with a low-cost packing material, for the biodegradation of di, tri, tetra, and pentachlorophenol mixtures in concentrations up to 600 mg/L as total chlorophenols. The relationship between chlorophenol´s removal and the hydraulic retention time was investigated. TOC removals and cell loads in the packed column during the continuous process were also evaluated and discussed.

MATERIALS AND METHODS

Tezontle is a basaltic scoria abundant in the central Valley of México. It is a very cheap, light, resistant, porous and easy to handle construction material. Three different *tezontle* size distributions were characterized in terms of global porosity (%), apparent density (g/ml), and strength indirectly, as the solubilities in HCl and NaOH (%), as well as the suitability of being colonized by *Pseudomonas fluorescens* cells (as $FCU/g_{support}$). These mixtures were termed fine (particles between 0.84-1.19 mm), medium size (particles between 1.19-1.68 mm) and gross portion (particles between 1.68-3.36 mm).

The employed column is a perspex cylinder (95 mm in diameter, 900 mm in height). Active volume is around 3.5 liters according to the porosity of the employed packing material. The cross section is 0.007 m^2. It was packed with 100 mm height of gravel and 710 mm height of a given *tezontle* granulometry. The system was connected as shown on figure 1. A peristaltic pump was used in order to pump CPh solutions and circulate them upwards through the column. The system was started inoculating the *tezontle*. For this purpose, a 15 liters fermentation *of Pseudomonas fluorescens* in YPG (yeast extract, caseine peptone and glucose, 10 mg/L of each one) medium was used. Cell concentrations during the fermentation process were monitored by means of optical density. The rich culture was recirculated through the column for 24 hours and the wasted medium was discharged. Different solutions of chlorophenols in Dapaah medium (Dapaah and Hill, 1992) were fed to the system. The cell load in the column (as $CFU/g_{support}$) was evaluated periodically by platting dishes of YPG and King B (specially designed for fluorescent microorganisms as *P. fluorescens*) media as described by Salinas, 1997. Solutions containing mono-, di-, and trichlorophenols were fed consecutively to the column. The phenols and total organic carbon (TOC) were measured as detailed by Salinas, 1997 . Mixtures consisting of mono-di, mono-di-tri and mono-di-tri-pentachlorophenol were fed to the column. The procedure lasted 3 months. The results of this experimental work were reported elsewhere (Salinas, 1997). After this, di-, tri-, tetra-, and pentachlorophenol mixtures at 200, 400 and 600 mg/L were employed. Every total phenol´s concentration was evaluated at three different hydraulic retention times, HRT, defined in the previous experiences. $FCU/g_{support}$, temperature, total and specific chlorophenol concentrations, and TOC were evaluated periodically. GC/mass techniques were employed in order to determine specific chlorophenols concentration in the effluents, the 4-aminoantipirine method, for the total chlorophenols evaluation, using the appropriate calibration curve. For more details, see Torres *et al.*, 1997a.

RESULTS AND DISCUSSION

The characterization of the packing material with different size distributions is given on table 1. Differences among the three size distributions were observed. Apparent densities were between 0.77 and 0.93 mg/L. Solubility values are an accelerated-test measurement of the decay of the material. The harder mixture is the gross portion, followed by the fine and medium sizes. Regarding the suitability for being colonized by *Pseudomonas*, it is interesting to note that fine and medium portions gave a very good cell load value (around 10^9 $FCU/g_{support}$), but the gross portion showed to be the best one (around 10^{11} FCU/$g_{support}$). Perhaps the most useful parameter is the porosity of the material. Regarding to these values, medium size distribution had the best porosity (65%), followed by the fine (44%) and, at last, the gross portion (21%). In basis of the importance of the packing material porosity for the functioning of the system, it was decided to use the medium size distribution throughout this work.

TABLE 1. Characterization of the *tezontle* mixtures.

Mixture	Particle size	Porosity (%)	App. Densisty (g/ml)	Solubility in HCL (%)	Solubility in NaOH (%)	Cell load (FCU/ $g_{support}$)
Fine	0.84-1.19	44.0	0.9378	3.47	0.897	$8.5x10^{9}$
Medium	1.19-1.68	65.0	0.7700	3.23	1.36	$1.4x10^{9}$
Gross	1.68-3.36	21.3	0.8594	1.88	0.1970	$1.7x10^{11}$

The 15 liters fermentation lasted 24 hours and reached optical densities of around 0.3 (1.08 mg/L, in accord with the dry weight curve). It is interesting to mention that the final YPG total count was about $1x10^{10}$ FCU/$g_{support}$, while the King B was very close ($5x10^{10}$ FCU/$g_{support}$). This means that most of the bacteria present were *P. fluorescens*. The column was packed with the medium size distribution and inoculated as described in the materials and methods section. After the inoculation process, the column reached 10^7 FCU/$g_{support}$ but due to the single CPh´s fed (during three months), the cell load oscillated between 10^5- 10^7 FCU/$g_{support}$ (YPG count). Regarding to the King B count, the FCU/$g_{support}$ value was lower than the YPG count (first two weeks), but rapidly increased up to the same value of the YPGcount. During this period, temperature ranged between 10 and 23°C. 2CPh was not included in the mixture from this point on due to problems with the chlorophenol´s measurement procedure.

The first chlorophenols mixture evaluated was the one containing 400 mg/L (100 mg/L of every chlorophenol), followed by the 600 mg/L one, and the 200 mg/L mixture at the end. Every solution was treated during 2-3 residence times as less. Three samples (morning, noon and afternoon) of the influent and effluent were took in order to determine the chlorophenol and TOC removals.

The CPh´s removal was a function of the HRT for an initial CPh concentration of 200 mg/L. For all the single compounds (except 2,3,4,6TeCPh), the higher the HRT, the lower the CPh removal values. It seems that the 2,3,4,6TeCPh was more degraded at HRT of 62 hours than at 31 hours. The chlorophenol´s removal rate was quite near for the specific compounds and the average for the mixture, except for the 2,4,6TCPh. The correspondent removals for this compound were 36, 5.6 and 5% for the 15.5, 31 and 62 hours of HRT, respectively

In the case of the mixture with 400 mg/L of CPh´s, the removal values were less dependent on the HRT, maybe except for the 2,4,6TCPh. As a general observation the higher removals were found at HRT equivalent to 15.5 and 31 hours for the different compounds. The removal for 2,4,6TCPh was again markedly lower in comparison tothe other compounds or the mixture of total CPh´s. The achieved removals for this mixture were higher (75.6-79.8%) than

TABLE 3. Biodegradation rates for the different chlorophenols, concentrations, and operation systems (in mg_{phenol}/L.day).

Phenol	Conc. mg/L	Batch operation		Continuous operation, mixtures at HRT =		
*	*	Singles**	Mixtures*	62 h	31 h	15 h
2,4DCPh	50	45	9.7	7.7	15.3	35.3
	100	69	34.5	23.8	46	100.1
	150	-	40.4	46.4	100.3	197.8
	200	656	4.6	-	-	-
2,4,6TCPh	50	52	0.7	1.6	20.7	
	100	83	37.3	16.5	29.5	53.6
	150	-	40.1	24.2	83.6	92.3
	200	770	1.7	-	-	-
2,3,4,6TeCPh	50	-	7.6	9.1	19.1	43.7
	100	-	35.7	28	57.1	115
	150	-	38.3	57	108.2	223
	200	-	2.3	-	-	-
PCPh	50	16	7.6	7	13.8	35.7
	100	16	35.4	19.8	41.5	87.9
	150	-	38	50.5	96.7	205.7
	200	204	0	-	-	-
	300	100	-	-	-	-
As total CPh´s	200	-	33.9	24.6	49.8	135.3
	400	-	145.9	88.1	174.1	376.7
	600	-	157	178	367.7	718.9
	800	-	8.6	-	-	-

Notes: *From Torres *et al.*, 1997[a]. ** From Torres *et al.*, 1997b.

column with the same packing material and size distribution, the same immobilized bacteria and the same mineral medium. The only one difference (besides de operation mode, *i.e.* batch *vs.* continuous) was the temperature. The experiments carried out on batch operation were controlled at 28°C. On the contrary, the experiments reported in this work were developed without temperature control (around 17-23°C). As shown, the BDR´s were higher for 2,3,4,TeCPh>PCPh>2,4DCPh>2,4,6TCPh. Regarding to the HRT it is clear that the best BDR´s were obtained at the lower HTR tested (15.5 hours). The behavior in this mixture of chlorophenols is considered quite good: BDR´s between 135-719 mg/L.day were found for the best HTR (15.5 hours). In the case of the 31 HRT, values of 50-367 mg/L.day were achieved. Finally, for the largest HRT (62 hours), the BDR´s were as high as 25-178 mg/L.day. If all these values (specifically those obtained at HRT=15.5 hours) are compared with the values obtained by Torres *et al.*,1997 working the same mixture in batch mode, it is concluded that the BDR´s were increased 1.4-5.8 fold due to the continuous operation. At the other hand, it is necessary to compare this BDR values with those obtained in batch operation, compound by compound (Torres *et al.*,1997b), *i.e.* not mixed. As shown on table 2, the single BDR values were in general lower than those obtained herein. The BDR values for very high CPh´s concentrations are quite high (thousands of those obtained with a 200 mg/L concentration (42.7-58.7%). It is remarkable to compare the behavior of the 2,4,6TCPh for this two mixture concentrations, which is completely different. At the end, the removal values found for the 600 mg/L were slightly higher (77.3-86.6%) than those obtained for the 400 and 200 mg/L experiences. In fact, the trends for the specific phenols and the mixture in the case of 600 mg/L are very similar to those observed for the 400 mg/L test

On table 2, the CPh´s and TOC removals, as well as the cell loads (as $FCU/g_{support}$) and the average temperatures for every biodegradation assessment, are shown. For the 200, 400 and 600 mg/L tests, the TOC removals were 56, 50 and 46%. This fact indicates that the lower the total chlorophenols concentration, the higher the TOC removal values. Except for the case of the 200 mg/L experiments, all the TOC removal values were less than the CPh removal value (1.4-2 fold). These values mean the capability of *Pseudomonas* of mineralizing the phenolic compounds.

TABLE 2. Results of the biodegradation runs, regarding to Total phenols, and TOC degradations. Results are an average of triplicates.

HRT (hours)	Initial total Ph conc. (mg/L)	TOC removal (%)	Total Ph removal (%)	Cell load ($FCU/g_{support}$ x10^{5})	Test temperature (°C)
15.5	200	59.5	58.8	2.97	21.0
31	200	52.1	43.3	2.97	17.0
62	200	57.7	42.7	2.97	17.5
15.5	400	41.3	79.8	2.77	19.0
31	400	52.8	75.6	1.95	18.5
62	400	56.2	76.5	1.95	17.5
15.5	600	45.0	78.0	2.68	20.0
31	600	55.1	86.6	2.68	18.0
62	600	37.9	77.3	2.68	19.2

Regarding to the $FCU/g_{support}$, it is noticeable that the cell load was quite constant ($1.95\text{-}2.97 \times 10^5$ $FCU/g_{support}$). This fact is important when comparing different biodegradation tests. At the end, the average temperature of the column operation was 17-21°C during working hours (temperature was not registered during nights). This is a critical point, since other systems reported in the literature for the treatment of toxic compounds or mixtures are temperature controlled, and that means energy consumption which increases the cost of the waste, superficial or aquifer water treatment.

In order to compare the different removal values obtained in the tests previously described, this values were converted into biodegradation rates BDR´s, taking into account the process time *(i.e.* the HRT) for every case. At table 3, these biodegradation rates for the total phenol concentration and for every compound in the mixtures, are summarized.

For comparison purposes, the BDR´s observed when working with a batch system very similar to that employed in this work (Torres *et al.*,1997a and b). It is important to mention that those experiments were carried out in a 27 mL glass mg/L.day) but in these experiments the specific CPh concentrations ranged between 50 and 150 mg/L.

CONCLUSIONS

The most suitable granulometry for this work´s purposes is the medium one. It showed high porosity (65%), medium apparent density (0.77 g/ml), strength, and suitability for being colonized by *Pseudomonas fluorescens* (around 10^9 FCU/$g_{support}$). The column packed with *tezontle* was able of removing 43-56, 75-80, and 77-86% of the total initial chlorophenols for the concentrations of 200, 400 and 600 mg/L, respectively. The biodegradation rates for every concentration ranges were 33-174 mg/L.day (for 200 mg/L), 118-494 mg/L.day (for 400 mg/L), and 179-725 mg/L.day (for 600 mg/L). These variations are due to the specific HRT, controlled at values of 15.5, 31, and 62 hours. The lowest values correspond most of the times to HRT = 62 hours, and the higher values to HRT=15.5 hours. TOC removals were around 38-59%. The system demonstrated to be a powerful tool in the bioremediation of waters contaminated with chlorophenols.

ACKNOWLEDGEMENTS

This work was financially supported by DGAPA/UNAM (Grant IN503895). The authors wish to acknowledge with thanks the text revision by Manuel de la Torre. The help of Mayra Fernández and Miguel Hernández in the experimental work and analysis is thanked too.

REFERENCES

Arvin E. (1991) Biodegradation kinetics of chlorinated aliphatic hydrocarbons with methane oxidizing bacteria in an aerobic fixed biofilm reactor. Water Research. 25(7): 873-881.

Dapaah S.Y. and G.A. Hill (1992) Biodegradation of chlorinated mixtures by *Pseudomonas putida*. Biotechnology and Bioengineering. 40:1353-1358.

Fava F., F. Baldoni, L. Marchetti and G. Quattroni (19996) A bioreactor system for the mineralization of low-chlorinated biphenyls. Process Biochemistry. 31(7): 659-667.

Salinas A. (1997) Biodegradation of chlorophenol´s mixtures in a packed bed at continuos operation. UNAM. Bachelor´s thesis (In spanish).

Seignez Ch.,V. Mottier, C..Pulgarin, N. Adler and P. Péringer (1993) Biodegradation of xenobiotics in a fixed bed reactor. Environmental Progress. 12(4): 306-311.

Torres L.G., M. Hernández, E.R. Bandala, Y. Pica, V. Albíter and B.E. Jiménez (1997a) Degradation of di-, tri-, tetra-, and pentachlorophenol mixtures in an aerobic biofilter. Applied Microbiology and Biotechnology. Submitted.

Torres L.G., V. Albiter, and B. Jiménez (1997b) Removal of chlorophenols including pentachlorophenol at high concentrations from contaminated waters. *In situ* and on site bioremediation. New Orleans. April-May 1997.

MODELING OF REMEDIATION WITH ORC™: TRANSVERSE DISPERSION

David J. Wilson (ECKENFELDER INC., Nashville, Tennessee)
Robert D. Norris (ECKENFELDER INC., Nashville, Tennessee)

ABSTRACT: Oxygen release compound (ORC™) is used to provide O_2 where costs or inconvenience of other sources make ORC a competitive source of O_2 for biodegradation. One application is the aerobic biodegradation of vinyl chloride and cis-1,2-DCE downgradient of anaerobic zones of natural attenuation or zero-valent metal permeable barriers. Small transverse dispersivities dictate that ORC wells be relatively closely spaced to avoid alternating strips of groundwater containing VOC or dissolved O_2 (D.O.) far downgradient from the wells (pin-striping). A model for ORC cleanups is described which includes transverse and longitudinal dispersion of both O_2 and contaminant. Biodegradation is handled by Monod kinetics. O_2 delivery capacity of a well is limited by Henry's Law and the hydrostatic head. Mineralization is assumed. Inclusion of transverse dispersion for both O_2 and contaminant results in less pin-striping than is modeled with O_2 dispersion alone. This permits somewhat wider spacing of the ORC wells. Dependence of results on model parameters is explored.

INTRODUCTION

Solid MgO_2 mixtures, Oxygen Release Compound (ORC™, Regenesis Bioremediation Products) are used to provide O_2 in bioremediation. The technique is discussed and several field applications have been described (Chapman, et al., 1997; Johnson and Odencrantz, 1997; Koenigsberg, Sandefur, and Cox 1997). Heitkamp (1997) studied degradation of nitrophenols in the presence of sodium percarbonate, MgO_2 (ORC™), and CaO_2 (PermeOx™, FMC Corporation). Regenesis (1997) has discussed use of ORC™ in the remediation of vinyl chloride. R. D. Wilson and Mackay (1997) and Wilson, Mackay, and Cherry (1997) have addressed the concerns associated with transverse mixing and spacing of ORC wells which are considered here.

Most hydrocarbons are degraded to CO_2, water, and biomass by microorganisms in soil if sufficient electron acceptor (usually O_2) is available. The solubility of O_2 is only about 8 mg/L at room temperature if the water is in contact with air at one atm. Hydrocarbons require considerable O_2 for mineralization; 3.13 kg of O_2 per kg of toluene, for example.

The problems and costs associated with the Raymond Process, use of H_2O_2, and biosparging, can in some cases be overcome by use of oxygen-release compounds such as ORC™. The technique uses MgO_2 mixed with materials which control its rate of reaction with water to produce O_2,

$$2\ MgO_2(s) + 2\ H_2O \rightarrow 2\ Mg(OH)_2(s) + O_2(aqueous)$$

The ORC may be placed in "socks", which are then placed in wells. The ORC releases O_2 to the groundwater slowly moving past the well location; this D.O. is available for microbial activity. The O_2 release rate is such that the socks are changed only every six months to a year, so labor costs are low.

In the following we first describe the constraints imposed by stoichiometry and Henry's Law. We then turn to the effects of advection, dispersion, and biodegradation kinetics. The analysis underlying the model is provided elsewhere (Norris, Wilson, and Chang, 1998). The last section describes some model results.

STOICHIOMETRY AND HENRY'S LAW CONSTRAINTS

Several assumptions concern reaction stoichiometry and Henry's Law; these lead to an upper bound for the ORC well spacing. They are:

- The VOC is completely mineralized.
- The groundwater capture cross-section of an ORC well is twice its diameter.
- The maximum O_2 concentration in water leaving an ORC well is given by Henry's Law, the gas phase is pure O_2, and the pressure is the sum of the atmospheric and mean hydrostatic pressures.

Then the maximum spacing between wells which provides enough O_2 for complete mineralization, Lx, is given by

$$L_X = \frac{2b_1(1 + b_2h)a}{R_S C_S}$$

where

a	=	diameter of ORC well, m
h	=	depth of well beneath the water table, m
b_1	=	0.047 kg/m^3 (15 °C)
b_2	=	0.048 m^{-1}
R_s	=	weight ratio of O_2 required per unit mass of VOC
C_s	=	contaminant VOC concentration, kg/m^3

For example, assume that vinyl chloride (R_s = 1.26) is present at 10 mg/L, well diameter is 20 cm, and aquifer thickness is 3 m. The above equation then yields 1.70 m as the maximum well spacing which could provide sufficient D.O. If smaller spacing is needed, one could place the wells in two or three staggered lines, rather than in a single row.

ADVECTION, DISPERSION, AND BIODEGRADATION. THE MATHEMATICAL MODEL

We next address the rate phenomena affecting the system: groundwater velocity, transverse and longitudinal dispersion, and biodegradation kinetics. These wells are commonly placed in one or more rows at right angles to the groundwater flow assumed (from left to right). O_2 from the ORC dissolves in the groundwater and moves to the right. Microorganisms then destroy hydrocarbon and O_2, producing biomass, CO_2 and water. This picture is adequate for our purposes; modification to include other nutrients is readily done. As the groundwater moves from the row of wells, the contaminant and O_2 concentrations decrease. Ideally, all the hydrocarbon has been consumed by the time the water has moved 10 to 30 meters downgradient. We model half of the strip of influence of one ORC well. Contaminated water flows in at the left end of the strip, and O_2 from the well enters at the upper left. As water moves to the right O_2 slowly spreads by dispersion, permitting bacterial growth and hydrocarbon destruction. Hydrocarbon also spreads laterally from regions of high concentration to regions in which it is depleted.

To model these processes the half-strip is partitioned into volume elements as shown. We then calculate for each element the following:

- The rates at which oxygen and substrate are carried into and out of the volume element by the movement of the water from left to right (advection); an asymmetrical upwind algorithm is used.
- The fluxes of O_2 and substrate to and from the volume element caused by dispersion, both in the direction of flow and transverse to it.
- The rates at which oxygen and substrate are used up by microbial activity (Monod kinetics).
- The rates of growth (Monod kinetics) and die-off (first-order) of the microorganisms, which are assumed to be immobile.

RESULTS

Figures 1 through 5 represent the half-strip down-gradient from the ORC well, which is in the upper left corner. The x and y axes are drawn to different scales. Contaminant (toluene) concentrations are proportional to the radii of the circles drawn about the grid points. All runs were continued to steady state. The influent water contains no D.O.; the water leaving the well is saturated with O_2. There is no D.O. transport into the aquifer at the water table.

The runs shown describe a single row of ORC wells having well bores of 20 cm and spaced to have capacity to remediate a plume containing 5.0 mg/L of toluene (or 12.4 mg/L of vinyl chloride). The ORC well is at the upper left corner, and contaminated groundwater enters along the left boundary. The mean saturated D.O. concentration is 53.5 mg/L (15°C); the distance between wells (Lx) is 1.367 m. Influent toluene concentrations are 3.0, 4.0, 4.5, 5.0, and 7.5 mg/L (7.5, 9.9, 11.2, 12.4, and 18.6 mg/L of vinyl chloride).

The results are shown in Figures 1 through 5. Half of the strip of influence of one well is shown for 50 m downgradient. In Figure 1 the toluene, initially 3.0 mg/L, is degraded across the full width of the strip by the time the water has moved about 25 m downgradient. The long distance required for mixing of O_2 and toluene is due to the small transverse dispersivities one typically finds. This slow transverse mixing causes the "pin-stripes" noted by Wilson and Mackay (1997), and by Wilson, Mackay, and Cherry (1997), who modeled D.O. dispersion but not dissolved VOC dispersion.

As the influent toluene concentration increases to 4.0 and 4.5 mg/L (Figures 2 and 3), the mixing distance increases to 34 and 42 m, respectively, since as more O_2 is consumed the O_2 concentration gradients are lower. This results in slower transverse dispersion of O_2, so the pin-stripes extend further downgradient. Remediation is satisfactory if the point of compliance is more than 50 m or so from the ORC wells.

Figure 4 describes a run with an influent toluene concentration of 5.0 mg/L, the stoichiometric maximum which can be destroyed. The pin-stripes extend down-gradient off the figure, more than 50 m. Note that the model neglects natural attenuation, the stoichiometry is conservative, and that compliance may not require 100% destruction of the contaminant.

The impact of an influent toluene concentration of 7.5 mg/L is seen in Figure 5. The toluene flux is 50% more than the available O_2 can handle. Toluene is rapidly reduced to low levels a short distance downgradient from the well, but after the D.O. is exhausted (about 25 m downgradient), toluene diffuses from the region of high concentration. Despite the toluene-free domain a short distance down-gradient, this system is overloaded with VOC and complete remediation has not taken place.

Results from this model depend strongly on the transverse dispersivity, given by $v.L_T$, where v is the linear velocity of the groundwater and L_T is the transverse dispersion length. A simplified approach yields the following estimate for the pinstripe length L_y:

$$L_y = L_x2/8L_T$$

Since pinstripe length is inversely proportional to transverse dispersivity, it is essential to select the transverse dispersivity as realistically as possible.

REFERENCES

Chapman, S. W., B. T. Byerly, D. J. Smyth, R. D. Wilson, and D. M. Mackay, 1997, "Semipassive Oxygen Release Barrier for Enhancement of Intrinsic Bioremediation", in In Situ and On-Site Bioremediation: Vol. 4, 4th International In Situ and On Site Bioremediation Symposium, New Orleans, LA, Apr. 28 - May 1, B. C. Alleman and A. Leeson, eds., Battelle Press, Columbus, OH, p. 209.

Heitkamp, M. A., 1997, "Effects of Oxygen-Releasing Materials on Aerobic Bacterial Degradation Processes", Bioremediation Journal, 1(2), 105.

Johnson, J. G., and J. E. Odencrantz, 1997, "Management of a Hydrocarbon Plume Using a Permeable ORC™ Barrier, in In Situ and On-Site Bioremediation: Vol. 4, p. 215.

Koenigsberg, S., C. Sandefur, and W. Cox, 1997, "The Use of Oxygen Release Compound (ORC™) in Bioremediation", in In Situ and On-Site Bioremediation: Vol. 4, p. 247.

Norris, R. D., D. J. Wilson, and E. Chang, 1998, "Bioremediation with Oxygen-Releasing Solids: A Mathematical Model", submitted to Environmental Monitoring & Assessment.

Regenesis Bioremediation Products, 1997, "ORC Technical Bulletin #2.2.2.3, Oxygen Release Compound, ORC™; Remediation of Vinyl Chloride", 27130A Paseo Espada, Suite 1407, San Juan Capistrano, CA 92675.

Wilson, R. D., and D. M. Mackay, 1997, "Arrays of Unpumped Wells for Plume Migration Control or Enhanced Intrinsic Remediation", in In Situ and On-Site Bioremediation: Vol. 4, p. 187.

Wilson, R. D., D. M. Mackay, and J. A. Cherry, 1997, "Arrays of Unpumped Wells for Plume Migration Control by Semi-Passive In Situ Remediation", Ground Water Monitoring and Remediation, 17, 185.

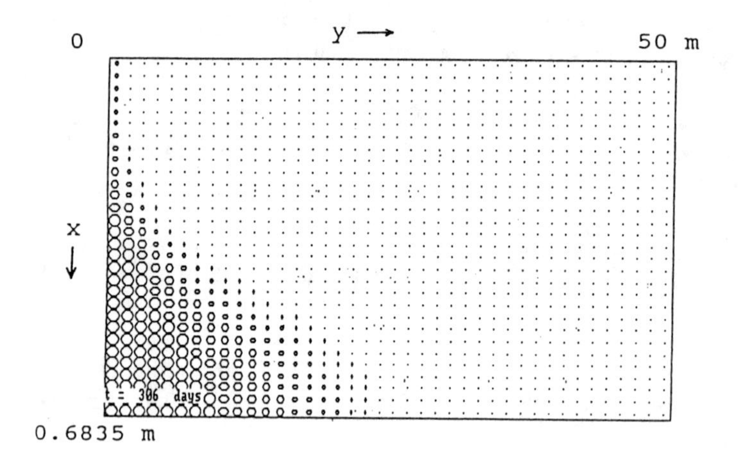

Fig. 1. Plan view of toluene concentration in the half-strip downgradient from an ORC well. Well diameter = 20 cm, influent toluene concentration = 3.0 mg/L, stoichiometric capacity of system = 5.0 mg/L.

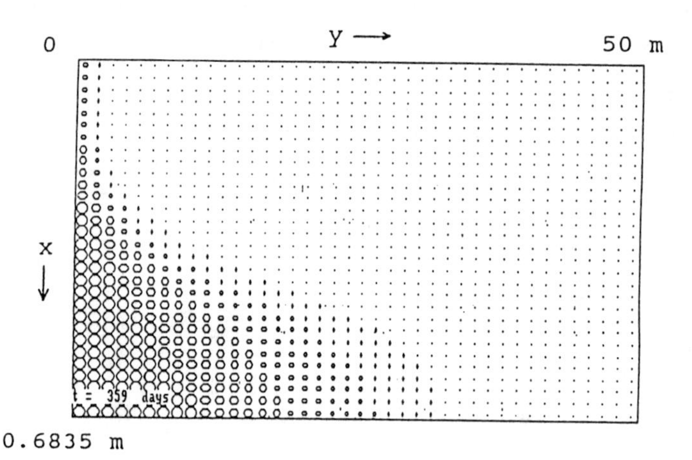

Fig. 2. Plan view of toluene concentration in the half-strip downgradient from an ORC well. Influent toluene concentration = 4.0 mg/L, stoichiometric capacity of system = 5.0 mg/L.

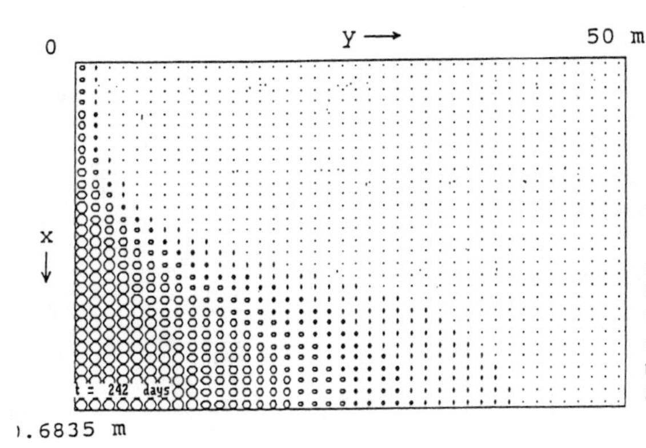

Fig. 3. Plan view of toluene concentration in the half-strip downgradient from an ORC well. Influent toluene concentration = 4.5 mg/L, stoichiometric capacity of system = 5.0 mg/L.

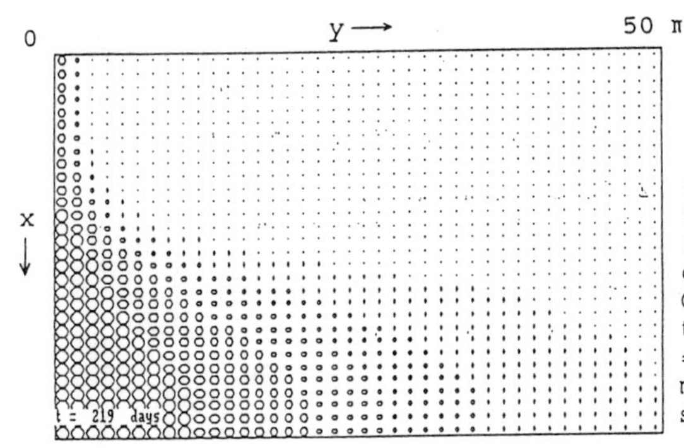

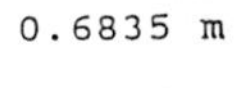

Fig. 4. Plan view of toluene concentration in the half-strip downgradient from an ORC well. Influent toluene concentration = 5.0 mg/L, stoichiometric capacity of system = 5.0 mg/L.

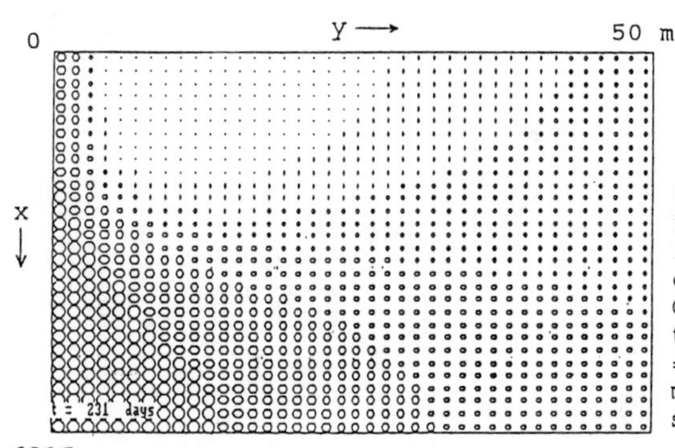

Fig. 5. Plan view of toluene concentration in the half-strip downgradient from an ORC well. Influent toluene concentration = 7.5 mg/L, stoichiometric capacity of system = 5.0 mg/L.

ANALYSIS OF REMEDIAL OPTIONS FOR CHLORINATED VOCS AT HARRISON LANDFILL

Harold W. Bentley, Jinshan Tang, and Stewart Smith
(Hydro Geo Chem, Tucson, Arizona)
Daniel Samorano (City of Tucson, Arizona)
Robert G. Arnold (University of Arizona, Tucson)

ABSTRACT: Preliminary feasibility analyses were conducted for remedial options selected to potentially control and/or cleanup subsurface chlorinated volatile organic compounds (CVOCs) at the closed Harrison Landfill in Tucson. The trial remedial components included (1) in-situ mineralization of groundwater chlorinated ethenes using a combination of anaerobic and aerobic biodegradation; (2) soil vapor extraction (SVE) and treatment of the CVOCs (and methane) in the 250 foot unsaturated zone coupled with active methane withdrawal in and around the landfill; (3) capture and treatment of CVOCs in the proximal, higher concentration part of the groundwater plume, by either an injection-withdrawal recirculating well equipped with an in-well air stripper or (4) a "raining well", a combined pumping and vapor-extraction well that captures groundwater, infiltrates it to the unsaturated zone, and strips its CVOCs by SVE; and (5) "pump-and-treat" control of groundwater at the distal end of the plume. Laboratory studies of site soils and groundwater validated the anaerobic component of biodegradation, but aerobic (heterotrophic or methanotrophic) biodegradation of TCE, DCE, or VC was not observed. Numerical simulation showed that each of other remedial components is potentially feasible but sensitive to hydraulic (or pneumatic) conductivity, leakance, anisotropy, well screen placement and fluid pumping rates. The recirculating well was also found to require a high stripping efficiency and, as a result, to be susceptible to calcite buildup.

INTRODUCTION

Harrison Landfill, in Tucson, Arizona, was an active municipal landfill from 1972 to 1997. The former landfill occupies a footprint of approximately 1800 X 2600 ft and is 100 feet deep. Ongoing remedial investigation has established that chlorinated volatile organic compounds (CVOCs), predominantly tetrachloroethene (PCE) and its reductive dechlorinates trichloroethene (TCE), dichloroethene (DCE) and vinyl chloride (VC) (accompanied by high levels of methane and CO_2) are found in the 250 ft thick unsaturated zone. The upper 50 ft of the underlying aquifer, composed of sands and gravel, contains a 400 ft wide CVOC plume that originates at the landfill and has moved 600 ft down gradient from the landfill boundary. This upper, conductive zone of the aquifer is separated from deeper sands and gravels by a 40 foot thick, lower permeability zone that acts as a confining bed. Groundwater head is approximately 1 foot less in the deeper permeable zone. CVOCs have not been found in the deeper zone. Figure 1 shows a conceptual model of the Harrison landfill site.

Based on the premise that identifying potential remedial options and examining their feasibility is reasonable before embarking on subsequent field investigation, preliminary feasibility analyses were conducted for remedial options selected to potentially control and/or clean up subsurface CVOCs. The results of the study will ultimately be used to identify the most reasonable remedial alternatives in terms of potential effectiveness and cost, the information required to design selected remedial alternatives, and the monitoring data necessary to evaluate the effectiveness of selected remedial alternatives.

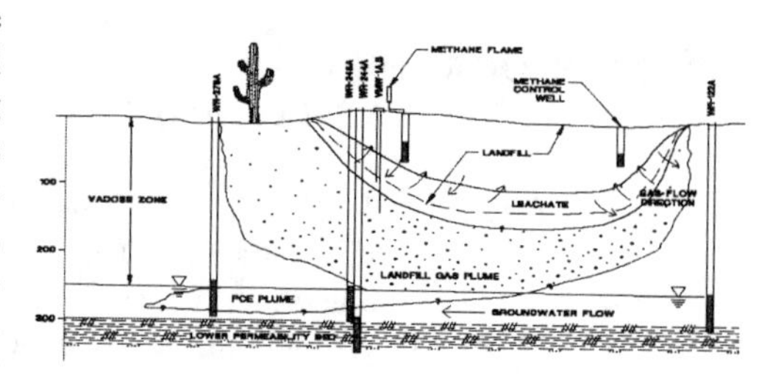

Figure 1. Conceptual cross section of Harrison Landfill's hydrogeology and PCE distribution.

Biodegradation Studies. Laboratory screening tests were conducted to examine the feasibility of in-situ bioremediation of the site's groundwater CVOCs. Because PCE biodegrades only under anaerobic conditions (Vogel at al. 1987) and its partially chlorinated degradation products are regarded as best biodegrading under aerobic conditions, both anaerobic and aerobic microbial processes would be required.

Dissolved oxygen profiles of Well WR-122A, up gradient from the landfill, and WR-245A, down gradient from the landfill (see Figure 1), revealed that the wells were completed, respectively, into aerobic and anaerobic groundwater. Water from well WR-122A and WR-245A was tested for organic and inorganic species and for phospholipid fatty acids (PLFA). Microbes in WR-276A water were counted by acridine orange (AODC) and by direct plate counts for aerobic heterotrophs and methanotrophs. Select cultures that grew on methane were screened for expression of sMMO, a methanotrophic enzyme that co-oxidizes chlorinated ethenes. Soil samples were taken at 95 and 140 feet bls from a borehole used to construct vapor monitoring well VMW-1, completed into unsaturated soils impacted by landfill-origin gases. Microcosms (septum-sealed vials containing soil, water, 5.0 µM/L of various target CVOCs, some with electron donors [benzoate, acetate, lactate, or hydrogen], and some with ammonium phosphate nutrients) were prepared for anaerobic cultures from VMW-1 (140 feet) soils and WR-245A water (taken under anaerobic sampling conditions; and for aerobic cultures from VMW-1 (95 feet) soils and WR-276A (located 300 feet down gradient from well WR-245A) water. Although dissolved oxygen at WR-276A's well head was 6.0 mg/L, later, down-hole DO profiling revealed the well to be anaerobic. The microcosms were sampled and analyzed over a period of 270 days to monitor the degradation of the chlorinated ethenes.

Microbial enumeration of the essentially anaerobic groundwater showed 2.0×10^5 cells/mL (AODC), $6\text{-}7\times10^3$ cells/mL of culturable aerobic heterotrophs (plate count, PTYG medium), and $6\text{-}9\times10^2$ cells/mL of methanotrophs. No sMMO activity was detected. The PLFA results indicated higher microbial populations, higher proportions of Gram negative bacteria in the down gradient well, and favorable conditions for microbial growth. All anaerobic microcosms biodegraded PCE. The addition of nutrients (ammonium, phosphate) and/or electron donors had no observable effect on the degradation rates, about 0.3 μg/L-day, (see Figure 2). In-growth of TCE and DCE was observed in some of the microcosms, but no VC or ethene was found. About 60% of the loss in PCE could not be accounted for by mass balances calculated for these dechlorinates. Regarding the aerobic microcosms, degradation of TCE, DCE, and VC was not observed. We conclude that enough living microorganisms (and physiological variety among them) exist at Harrison Landfill to expect some success in bioremediating its chlorinated ethenes. Reduction of PCE was accomplished in the anaerobic tests, but aerobic (heterotrophic or methanotrophic) degradation of the less chlorinated ethenes was not demonstrated. The micro-organisms necessary for aerobic biodegradation of chlorinated ethenes may not have been present in the microcosm materials taken from anaerobic zones. Further, site-specific studies will be necessary to determine if aerobic CVOC biodegradation is viable at the Harrison Landfill site.

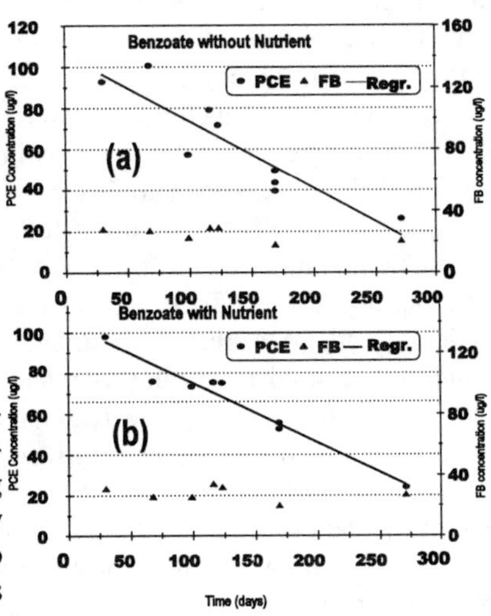

Figure 2. Anaerobic microcosm concentrations of PCE and the internal tracer fluorobenzene with time: a) without and b) with nutrients

Models and Site Parameters Utilized The codes used to perform modeling included the three-dimensional gas flow and transport SUPER 3-D and the three-dimensional water flow and transport ABEL, both proprietary finite-difference codes developed by Hydro Geo Chem (HGC); the two-dimensional particle tracking PATH-2D, also developed by HGC; and the three-dimensional finite difference TRACRN (Travis and Birdsall, 1988) that is capable of both gas and water flow and transport. The site parameters used in the model(s)are listed in Table 1.

Table 1. Parameters of the Feasibility Assessment Models

Saturated Material	**Unsaturated Material**
Horizontal Soil Permeability = 5.6 ft/day	Dispersivity = 25 ft
Vertical Soil Permeability = 0.56-1.4 ft/day	Bubbling Pressure = 1.4×10^{-4} dyne/cm^2
Horizontal Landfill Permeability = 84 ft/day	Pore Size Distribution Index = 2.8
Vertical Landfill Permeability = 8.4 ft/day	Constrictivity Coefficient = 0.2
Soil and Landfill Porosity = 30%	
Soil Specific Yield = 15%	**Chemical (PCE)**
Soil Specific Storage = 1×10^{-4}	Aqueous Diffusion Coeff.= 1×10^{-5}cm^2/s
Aquitard Permeability = 5.5×10^{-3} ft/day	Gaseous Diffusion Coeff. = 0.1 cm^2/s
Aquitard Leakance = 2×10^{-4}	Dimensionless Henry's Coeff. = 1.1
Soil Bulk Density = 1.7 g/cc	K_D = 0.364 mL/g

Modeling SVE Cleanup of Soils Beneath a Methanogenic Landfill. Gas injection or withdrawal near a methane-producing landfill has the potential to introduce air into the landfill, thereby reducing its methane production and waste degradation efficiency and increasing the potential for spontaneous combustion. The objectives of this modeling exercise are to maintain anaerobic conditions in the landfill while reducing PCE to less than MCLs at the groundwater surface. The simulation, performed by SUPER 3-D, used 100,000 layered equidimensional rectangular elements extending 16,000 ft in the X (east) direction, 16,000 ft in the Y (north) direction, and 250 ft (10 layers) in the Z (vertical) direction. Boundaries were ambient; fixed, atmospheric pressure; and zero gas concentration except the lower boundary, specified as no-flow. The landfill, located at the center of the model, was 1,800 ft x 2,600 ft. A regular, rectangular array of 28 methane injection and 34 vapor extraction wells within the landfill served to simulate landfill-methane generation and recovery. The total methane generation of 4,500 standard cubic feet per minute (0.3 mL/ ft^3 landfill/minute) was equal to the specified total vapor extraction rate. A 5-spot array of one air-injection and 4 vapor-extraction wells operated at rates of 250 scfm each was located in soils beneath the landfill. The air injection well, at the center of the landfill, was screened at 175 to 250 ft below land surface (bls). The vapor extraction wells, at the corners of the landfill, were screened at 100 to 175 ft bls.

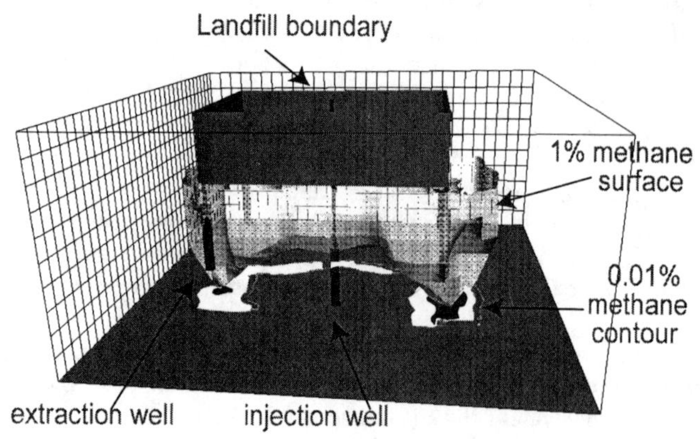

Figure 3. SVE simulation, steady-state methane distribution.

At steady-state, this SVE configuration was found to minimize landfill air intrusion and prevent MCL levels of CVOCs from reaching the water table. Figure 3 shows the 1% residual methane surface and the 0.1 and 0.01% methane contours in the layer contacting the water table. Further field studies will be required to better estimate the model parameters of permeability, porosity, anisotropy, CVOC distribution, and methane distribution and pressure. These better estimates will be required to properly modify this initial SVE model for use in engineering design.

Raining Well Modeling. The "raining" well consisted of a well pumping 30 gpm into concentric infiltration rings centered on the well just beneath the land surface and also served as a SVE well to remove CVOCs from the infiltrating water before reaching the water table. The effects of pumping, infiltration, and SVE were simulated using a 2-step process. The infiltration and SVE operation were simulated using TRACRN. The effect of the infiltrating water on the capture zone of the pumping well was simulated using ABEL.

The TRACRN model consisted of a 2-dimensional cylindrical array of 96 cells in the radial direction and 30 layers. Boundaries were maintained at atmospheric pressure and zero concentration except the left boundary, located at the center of the well installation, which was specified no-flow. This is appropriate because flow in the well would be primarily vertical. No gas flow occurred across the lower boundary, which was maintained at full water saturation.

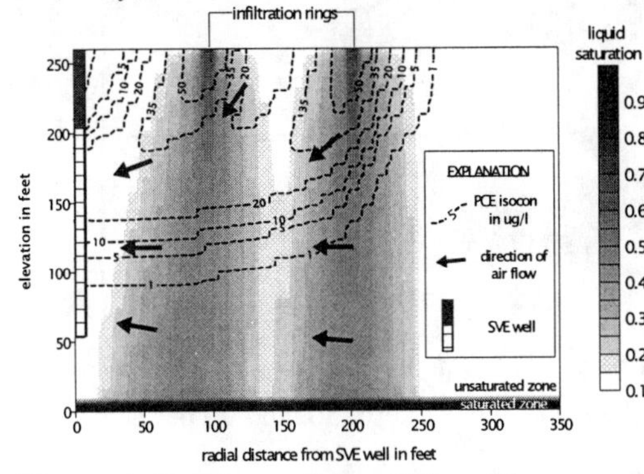

Figure 4. Raining well simulation: 30 gpm groundwater (60 µg/L PCE) and 100 scfm SVE. Isocons show residual PCE distribution.

The SVE well was screened from a depth of approximately 65 ft to 210 ft bls and consisted of a very high permeability material from which gas was extracted at a rate of 100 scfm. Water containing PCE was allowed to infiltrate just beneath the land surface from 2 concentric infiltration rings located at radial distances of 100 and 200 ft from the well. Figure 4 shows the resulting soil water and residual PCE distribution after the simulation reached steady-state.

The ABEL model used a non-equally spaced rectangular array of 241 x 241 elements extending 10,000 ft in the X and 10,000 ft in the Y directions. The left and right boundaries were constant head and the front and back boundaries, no-flow. Heads were specified to result in a groundwater gradient of approximately 0.006 and an initial saturated thickness of 50 ft at the location of the well. The lower boundary had a specified leakance to allow for vertical flow across a lower aquitard. Infiltration rates determined from the TRACRN simulation were imposed as recharge over the proper area surrounding the well location in the ABEL model. The pumping rate of the well in the ABEL model was equivalent to the total recharge rate resulting from injection of the pumping water. PATH-2D calculated a raining well capture zone width of 450 feet using the steady-state head distribution computed by ABEL.

Vertical Circulation Well The vertical circulation well was simulated using ABEL. The model consisted of a three-layer rectangular 73 x 73 array of non-equally spaced elements extending 10,000 ft in both the X and Y directions. All layers were in the upper 50 ft saturated zone. Boundaries were the same as those of the raining well model.

A 500 foot wide VOC plume was generated up gradient of the well by specifying a constant concentration of 60 µg/L at the boundary. The well bisected the width of this generated VOC plume. Water was pumped at 30 gpm from the lower layer, CVOC mass was removed at a specified stripping efficiency, and the water was then injected into the upper layer. The resulting changes in the groundwater plume are displayed in Figure 5. A stripping efficiency greater than

50% was found to reduce the 60µg/L PCE concentration to <10 µg/L, and a 90% efficiency reduced PCE to <5 µg/L. The no-stripping contours provide the interesting insight that the plume's PCE concentration (but not its mass) is reduced by more than 60% simply by dilution. The gas stripping efficiency required for the recirculating well would create operational problems by removing CO_2, causing in-well calcite precipitation from the high calcium-bicarbonate ground waters commonly associated with landfills.

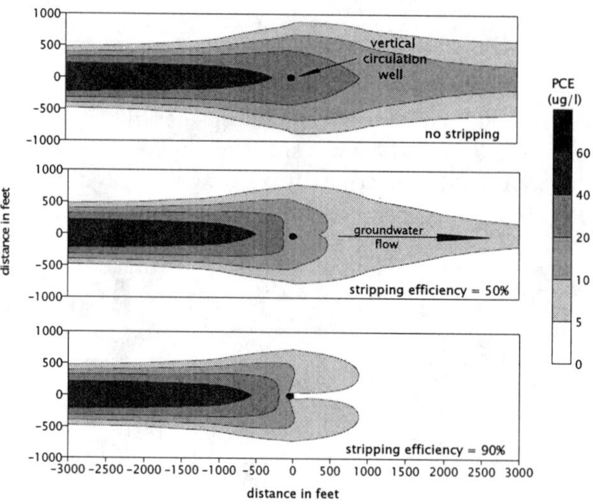

Figure 5. PCE cleanup by recirculating wells with stripping efficiencies of zero, 50, and 90 per cent.

Pump-and Treat Control. The ABEL- and Path 2D-simulated capture zone width of a 30 gpm well was found to be 1200 ft. Grid structures and boundary conditions were the same as those used in the ABEL and PATH-2D raining-well models.

REFERENCES

Smith, S. and G. Walter. 1993. "Numerical Modeling of 'Raining Well' Soil Vapor Extraction Wells for a Hypothetical Alluvial Aquifer." Abstracts with Program: Rocky Mountain Ground Water Conference - Ground Water Technology and Tasks in the 90's, Albuquerque, New Mexico, 1993.

Travis, B. J. and K. H. Birdsell. 1988. TRACRN 1.0: A Model of Flow and Transport In Porous Media for the Yucca Mountain Project - Model Description and User's Manual. Prepared for Los Alamos National Laboratory.

Vogel, T.M., S. S. Criddle and P. L. McCarty. 1987. "Transformations of halogenated aliphatic compounds." Environmental Science and Technology 21(8):722-736.

ENHANCING DISSOLVED OXYGEN TO REMEDIATE VINYL CHLORIDE IN GROUND WATER

Ingrid J. Verhagen and Douglas W. Wetzstein (Minnesota Pollution Control Agency (MPCA), Saint Paul, Minnesota)
D. Roger Bruner and Curtis M. Hudak (Foth & Van Dyke, Eagan, Minnesota)

Abstract: Vinyl chloride (chloroethene) has been shown to aerobically dehalogenate under laboratory conditions, although little data is available to document this process in a field setting. A one year pilot study commenced in the fall of 1996 to document the effectiveness of enhancing dissolved oxygen to degrade chloroethene in-situ. The pilot study was conducted at a former National Priorities List (NPL) site in Northern Minnesota.

The 23 acre (9.3 ha) landfill was constructed on sandy glacial outwash with minor fine-grained lenses. Historical chloroethene concentrations range from below detection outside of the plume to 0.094 milligrams per liter within the plume. The plume is located in the upper 50 feet (15.3 m) of saturated sediments downgradient of the landfill. Oxygen enhancement (OE) wells were installed at the edge of the degradation zone within the plume to facilitate the transfer of dissolved oxygen in the ground water. The monitoring network included monitoring wells MW-6A and -6B, MW-2A and -2B, TEMP-1 and TEMP-2, and P-1 and P-2. Magnesium peroxide socks were installed in the OE wells in May 1997 and removed in December 1997.

Ground water quality was monitored at a 30 minute interval 70 feet (21 m) downgradient of the OE wells with sondes connected to dataloggers. The parameters that were continuously monitored were pH, temperature, conductivity, bromide, and dissolved oxygen. Laboratory analysis of extended parameters was completed monthly. The extended parameters included selected anions, nitrate, nitrite, sulfate, chloride, phosphate, volatile organic compounds, methane, ethane, and ethene. Field analysis of ferrous iron and hydrogen sulfide were also completed monthly. Oxygen was elevated at the water table for one month (July) but was not elevated at the base of the surficial aquifer. Methane concentrations plotted as an inverse bell shaped curve during oxygen enhancement. Near the edge of the plume at the base of the surficial aquifer, chloroethene concentrations significantly dropped for a two month period. Chloroethene concentrations at the water table were inconclusive. The velocity of flow and iron precipitates may have been limiting factors.

INTRODUCTION

A one-year pilot study commenced in the fall of 1996 to document the effectiveness of enhancing dissolved oxygen concentrations to degrade chloroethene in-situ. Prior to the pilot study, a two year bioremediation study that included column studies and batch reactor tests on the microbial flora, e.g. methanotrophs, showed that they were capable of biodegrading chloroethene when dissolved oxygen concentrations were increased (Maier and Tam, 1995). The pilot study was completed at a former NPL site in Northern Minnesota. The initial phase of the pilot study consisted of a bromide tracer test to document the hydrologic connection between the remediation wells and the monitoring wells. Oxygen enhancement commenced May 22, 1997 and was removed from the system December 16, 1997.

The 23 acre (9.3 ha) landfill was constructed on sandy glacial outwash with minor fine-grained lenses. Historical chloroethene concentrations ranged from below detection outside of the plume to 0.094 mg/L within the plume. The cleanup goal for the site was the Maximum Contaminant Level (0.002 mg/L). The plume was located in the upper 50 ft (15.3 m) of saturated sediments downgradient of the landfill. Average linear velocity was greater than 5 ft/day (1.5 m/day).

Objectives. The objective of the study was to enhance the dissolved oxygen in ground water to promote in-situ bioremediation of the chloroethene immediately downgradient of the landfill. Literature sources suggest that dissolved oxygen concentrations must be increased above 10 mg/L for methanotrophs to aerobically degrade chloroethene (Semprini et al, 1991). The chloroethene concentration immediately downgradient of the landfill was an order of magnitude higher than the cleanup goal and a method of supplying dissolved oxygen that was both efficient and low maintenance was sought. With these objectives in mind, air sparging and recirculation were eliminated from consideration. Recirculation did not meet the low maintenance criterion. Air sparging was considered inefficient in the transfer of oxygen and may potentially biofoul the aquifer. Another limiting factor was the concentration of dissolved iron in the aquifer. As such, introduction of dissolved oxygen through the magnesium peroxide was chosen as a promising technology to achieve the site objectives.

MATERIALS AND METHODS

The method chosen to enhance the natural dissolved oxygen of the groundwater system at this site was a commercial product composed primarily of magnesium peroxide, which when hydrated, decomposes to magnesium hydroxide and oxygen. The magnesium peroxide was encased in polypropylene mesh "socks" approximately one-foot long and nominally four-inches (10 cm.) in diameter. The socks were connected to each other via grommets located at each end.

A pilot-scale system consisted of installing six oxygen enhancement (OE) wells constructed with 12 feet of 4-in. (10 cm.) diameter schedule 40 polychloroethene (PVC) riser pipe, flush-threaded to 40 feet of 4-in. (10 cm.) schedule 40 PVC screen with a slot width of 0.02 in. (0.5 cm.). Total depth of each well was 49.5 feet (15.1 m) below ground surface (bgs) and the screened interval extended from above the water table (approximately 12 feet (3.7 m) bgs) to the total depth in an area expected to be contaminated with chloroethene. The wells were aligned approximately perpendicular to groundwater flow and spaced 8 to 10 feet (2.4-3.1 m) apart. Two nested monitoring wells (TEMP-1 and TEMP-2) were constructed 70 feet (21.3 m) downgradient of the six OE wells. Monitoring Well TEMP-1 was constructed with 47 feet (14.3 m) of 2-in. (5 cm.) diameter schedule 40 PVC riser pipe, flush-threaded to 5 feet (1.5 m) of 2-in (5 cm) schedule 40 PVC screen with a slot width of 0.01 in. (0.25 cm.) and monitors the lower sand. TEMP-2 was constructed of similar materials but has 18 ft. (5.5 m) of riser pipe flush-threaded to 10 ft. (3.1 m) of screen and monitors the water table. The wells were installed with a Mobile B-57 drill rig using hollow-stem auger drilling techniques.

An existing side gradient nest of wells MW-6A and MW-6B were used to represent the unimpacted flow system for this study. An existing side-gradient nest of wells MW-2A and MW-2B represent the impacted but un-enhanced (control) flow system for this study. Monitoring wells P-1 and P-2 were used to provide a check of the changing conditions and are located 70 feet (21 m) downgradient of TEMP-1 and TEMP-2. P-1 and P-2 were constructed with a 24 foot (7.3 m) screen from the water table to 46 feet (14 m) bgs. This allowed flexibility in sampling the shallow (P-shallow) and deep (P-deep) portion of the flow system. The pilot study monitoring wells consisted of MW-2A, MW-2B, MW-6A, MW-6B, P-1, P-2, TEMP-1 and TEMP-2.

In-situ monitors (sondes) collected readings of dissolved oxygen, pH, temperature, bromide, and conductivity at 30 minute intervals in TEMP-1 and TEMP-2. The sondes were placed at 46 ft (14 m) below the top of casing in TEMP-1 and 23 ft (7.0 m) below the top of casing in TEMP-2. The contract lab collected and analyzed samples monthly for selected anions, nitrate, nitrite , sulfate, chloride and phosphate by EPA 300.0, 70 volatile organic compounds, which included chloroethene were analyzed by MDH 465E, and field analysis of ferrous iron and hydrogen sulfide by Hack methods were completed monthly. Dissolved methane, ethane, and ethene were also analyzed from samples collected from each monitoring well by HS/GC/FID. Background samples were collected November 16, 1996 from each of the wells installed for this study.

To verify that monitoring wells TEMP-1 and TEMP-2 were down-gradient of the OE wells, a tracer test was conducted. A 3.7 liter solution containing 61 mg/L sodium bromide was introduced into each of the OE wells. Monitoring wells TEMP-1 and TEMP-2 were monitored with both a sonde - Ion Specific Electrode (ISE) and also sampled monthly by the contract laboratory. Both monitoring wells TEMP-1 and TEMP-2 indicated detectable bromide concentrations either by ISE or by the contract lab sampling and analysis.

Magnesium peroxide socks were then placed in each of the OE wells filling the screened length of each well. The socks were placed in the OE wells May 22, 1997 and removed at the end of this study on December 16, 1997. Monthly sampling of the pilot study wells to assess oxygen enhancement began May 14, 1997 and continued through December 16, 1997.

RESULTS AND DISCUSSION

The release of dissolved oxygen resulted in enhancement farther downgradient in the deep system and inconclusive data in the shallow system. However, immediately downgradient of the OE wells (in wells TEMP-1 and TEMP-2) the release of the magnesium peroxide did not lower the chloroethene concentration to meet the cleanup goal and appeared to impact the concentration for only one month (Figure 1). The concentration in the unenhanced well in the plume (2A) of chloroethene ranged from 0.001 to 0.005 mg/L with erratic behavior from May to December. The significant change in chloroethene occurred in P-deep (Figure 1). The concentration of chloroethene at depth dropped from 0.011 mg/L to 0.0013 mg/L for a two month period (September to October). The concentration of chloroethene rebounded back to August levels in November and December. The effect of oxygenation at P-shallow can only be discussed for data from September to December. There was a decline in chloroethene but it is inconclusive without the preceding data.

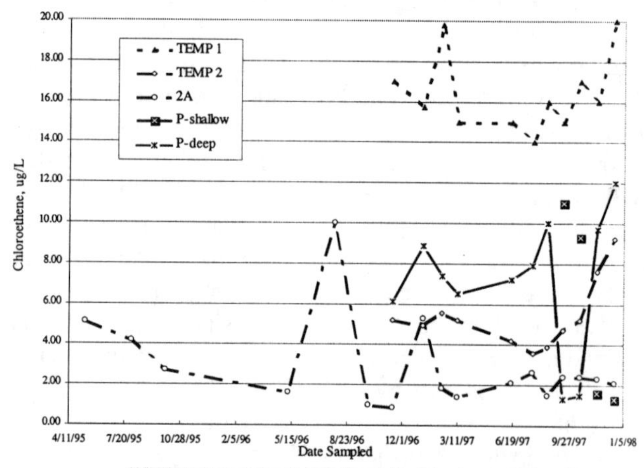

FIGURE 1. Concentration of chloroethene in the pilot system monitoring wells

A plot of methane revealed an inverse bell shaped curve in TEMP-2 with a steady decrease in methane from May to September and then a rebound from September to December. This suggests the magnesium peroxide released dissolved oxygen to the shallow system because in monitoring well 2A methane continued to decrease during this time period. The behavior of chloride was less conclusive, the ambient conditions suggest a high chloride water, it is hard to see significant changes when they occur at the microgram per liter level. Historical chloride data collected at the site indicated the opposite of what was seen during this study; ambient ground water was low in chloride and the plume was high in chloride (Maier and Tam, 1995).

Peaks of oxygen above the background concentration in the plume downgradient of the OE wells at the water table may have also been related to major rainfall events. Major rainfall events produced dissolved oxygen levels of 2 to 6 mg/L within the oxygen depleted plume. These events (rain every day for greater than 2 days) preceded the peaks at well TEMP-2. Continuous monitoring revealed a significant peak of dissolved oxygen at the water table in July. The peak concentration of 20 mg/L detected from July 16 to July 17 can be directly correlated with a major rainfall starting July 13 and lasting until July 17. However, this concentration may have also been enhanced by the magnesium peroxide because the concentration exceeded the range commonly seen after a rainfall event. The magnesium peroxide may have been at its optimum in this system when inundated with a major precipitation event.

Continuous monitoring of dissolved oxygen at the base of the surficial aquifer did not show a change in dissolved oxygen in response to major rainfall events. Deeper portions of the aquifer are naturally depleted in oxygen and may not easily become oxygen enriched.

Ground Water Quality Discussion. The degradation series of tetrachloroethene (PCE), trichloroethene (TCE), 1,1 dichloroethane (DCA), cis 1,2 dichloroethene (DCE), chloroethane (CA), and chloroethene was present at the site (Olsen and Davis, 1990). In monitoring well TEMP-2, TCE was not detected after oxygenation had been in place for 3 months, this left 1,1 DCA and cis 1,2 DCE to produce CA and chloroethene. Farther downgradient of the OE wells, samples collected from P-shallow indicate that TCE was in the reaction during oxygenation. After oxygenation ceased, in December 1997, TCE was not detected and chloroethene continued to drop in concentration. Ethene was detected in the plume downgradient of the OE wells in September. It is a product of anaerobic degradation of chloroethene that appeared in the system only during oxygenation. One quarter mile farther downgradient at well MW-13A, a tie to the degradation series can only be made by examining data from 1991 when PCE, TCE, and chloroethene were present but none of the intermediate products were detected. No detections of chlorinated hydrocarbons were seen in 1995, 1996, 1997. This suggests that the front of the plume has attenuated between P-shallow and 13A.

The deep system reveals that in the unoxygenated plume, TCE was detected in the series in March 1997 after being absent in 1995 and 1996. After oxygenation ceased in December 1997, cis 1,2 DCE, CA, and chloroethene were detected and TCE was not detected. TCE, 1,1 DCA, cis 1,2 DCE, CA, and chloroethene were present prior, during and after oxygenation in monitoring well TEMP 1. The concentrations remained constant except for chloroethene which decreased in June 1997 but then increased significantly in December 1997 after oxygenation had been removed. Continuous monitoring revealed that dissolved oxygen changed within a range from 0.0 to 0.45 mg/L, this change in chloroethene is inconclusive. At P-deep, the same compounds seen in TEMP-1 are duplicated. Despite the presence of both parent and daughter compounds, chloroethene drops significantly (89 percent) in September and October. At P-deep, TCE was not detected and chloroethene increases in

December 1997 in comparison to TEMP-1. Ethene is detected in the deep portion of the plume from June to August. At MW-13B, the entire degradation series is present from PCE to chloroethene. The concentration of chloroethene in samples from MW-13B remained static from May to December 1997.

CONCLUSIONS

The magnesium peroxide socks may have enhanced the degradation of chloroethene at depth farther downgradient of the site and not immediately downgradient of the OE wells. This enhancement was restricted to a two month time period. Several mechanisms may be responsible for the limited time frame of oxygen enhancement. The area of oxygen enhancement was too far up-gradient with parent compounds present that degraded in the presence of oxygen to more of the intermediate products. When the magnesium peroxide socks were removed from the OE wells, some were covered with iron precipitates that may have prevented the generation and release of oxygen. Finally, the rapid flow may have spent the socks within a two month time period and may have been a limiting factor in allowing enough reaction time for a buildup of dissolved oxygen to remediate the plume.

Further study would involve fine tuning the mechanism to deliver the dissolved oxygen. This may involve injecting the magnesium peroxide in a slurry farther downgradient from the reaction zone. The presence of the iron precipitate on some of the magnesium peroxide socks indicates that this must also be studied further. Soil borings could be advanced immediately down-gradient of the OE wells to visually inspect the sediments for iron precipitates, that would indicate that potential aquifer fouling had occurred and if it is a limiting factor to oxygen release and why the socks were rapidly spent.

ACKNOWLEDGMENTS

This pilot study was funded by the MPCA Closed Landfill Program.

REFERENCES

Maier, W.J. and Tam, E.C. 1995. *Bioremediation Study of Leachate Contaminated Soil and Aquifer Materials from Kummer Landfill Site: Executive Summary*. 35 pp.

Olsen, R.L. and Davis, A. 1990. "Predicting the Fate and Transport of Organic Compounds in Groundwater", Part 2, *HMC*, pp. 18-37.

Semprini, L., G. D. Hopkins, P. V. Roberts, D. Grbic-Grbic, and P. L. McCarty. 1991. "A field Evaluation of In-Situ Biodegradation of Chlorinated Ethenes: Part 3, Studies of Competitive Inhibition." *Groundwater* 29(2): 239-250.

BIOLOGICAL DEGRADATION OF CHLORINATED AROMATICS IN A PILOT-SCALE WATER TREATMENT PLANT

Wolfgang Dott (RWTH Aachen, Germany)
Martin Steiof (TU Berlin, Germany)
Berthold Zettler (TU Berlin, Germany)

Abstract: A pilot scale in situ microbiological/hydraulic restoration test was made to examine the possibilty of purifying the groundwater layer underneath a former pesticide production plant. The sitespecific pollutants consisted of chlorobenzenes, chlorophenoles, BTXE-aromatics and hexachlorocyclohexanes. These pollutants are referred to by the (internal company) term „BSS“ (**B**oehringer **S**pecific **S**ubstances"). The chlorobenzenes were found to be the main contaminant group with a mass share of approx. 95 %. The aim of the test was the stimulation of the decomposing indigenous (previous adapted) microorganisms in the subsoil area and in the water treatment plant.

The chlorinated aromatics were decomposed in the single treatment units with high degradation efficiencies (chlorpbenzoles > 95 %, chlorophenoles> 98 % and hexachlorocyclohexanes > 90 % (excl. ε- and β-HCH). The mass-balance of the plant's oxygene consumption indicated that the consumend oxygene was mainly utilized for the oxidation of reduzed iron and manganese and for the oxidation of C_{org}.

INTRODUCTION

The sitespecific pollutants are considered to be partially carcinogenic, mutagenic or teratogenic. These substances endanger the groundwater, and thus potentially the drinking water. (Eisenbrand and Schreier, 1995).

The microbiological degradability of the pollutants was shown under laboratory conditions in numerous degrading tests, however, practical restoration experiences for chlorinated aromatics are still missing. Highly chlorinated aromatics are preferably dechlorinated under anaerobic conditions by reductive dechlorination, and thereby causes the emerging less-chlorinated aromatics to become more easily accessible for aerobic microbiological attack. Thus the in situ operation was subdivided in an initial anaerobic and a continuing aerobic phase (Sander et al., 1991).

Objective. The aim of the test was to clarify that the chosen system of the water treatment plant is suitable for the degradation of the contaminated groundwater. The reconstruction values were 0.1 BSS/L for the groundwater and 0.025 mg BSS/L for the drain of the treatment plant.

Site Description: The Boehringer Company Site in Hamburg (Germany) was selected for this project. From 1953 to 1984 the Boehringer company produced in its Hamburg-Moorfleet factory γ-hexachlorocyclohexane, 2,5-dichloro-4-bromophenol and 2,4,5-trichlorophenoxyacetic acid. Semi finished, by- and end products of the production contaminated wide parts of the site area and the back- and groundwater.

The site has a size of approx. 86,000 m^2. A 6–7.7 m huge superficail artificial layer of rubble, sand and gravel is followed by a 0.75–3.4 m thick clay layer. The groundwater layer consistes of one approx. 9.2 m thick layer of fine sand with a k_f - value of approx. $0.7 \cdot 10^{-4}$ m/s and a following layer of medium sand with a k_f-value of approx. $6 \cdot 10^{-4}$ m/s.

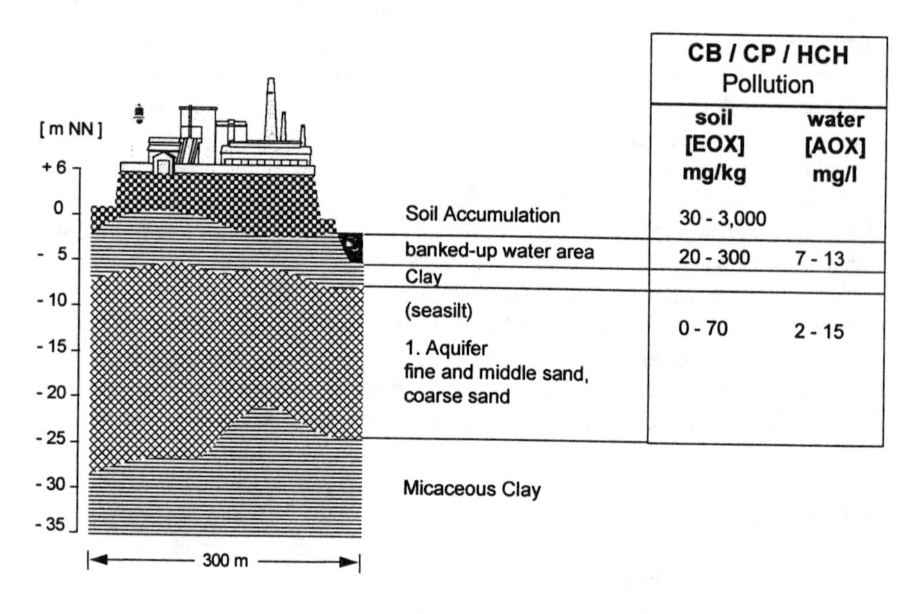

Figure 1. Cross section showing vertical geological profile and the amount of contamination in the soil and in the groundwater

MATERIAL AND METHODS

The contaminated groundwater was taken from the field by 6 wells, pumped through the treatment plant, for microbiological purification, and finally infiltrated into the field (two injection wells). In the water treatment plant the water was aerobisized in the first oxidator unit, and afterwards the iron hydroxide was filtered in the iron removal unit. The microbiological pollutant degradation took place in the fixed-bed reactors of the biological unit. Before the infiltration, the purified water was passed through the activated charcoal filters and once again enriched with oxygene and nutrients (phosphate). The perfomance of the water treatment plant was about 43.2 m/h^3.

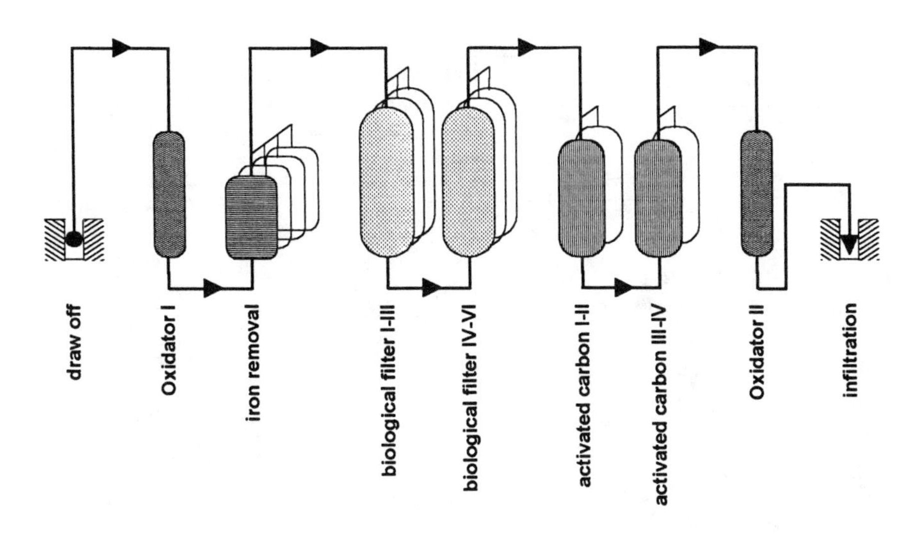

Figure 2. Flow-scheme of the water treatment plant

Water Sampling and Analysis: The water samples were taken in a period of 4 weeks and analyzed for BSS as described elsewhere (Lang et al., 1992).

RESULTS AND DISCUSSION

Lab Scale Degradation. The degradation of the site specific pollutants was demonstrated in svereal test through the use of indigenous bacterias (eidieker, Lang). These results were validated with renewed aerobic degradation tests in batch scale. The results indicated that the chlorobenzenes can be classified in three groups with differing decay rates. The degradatation depends on the number and the position of the chlorinated substitutes.

All chlorobenzenes with chloro-substitutes in meta-position were only slightly decomposed. Monochlorobenzene and the ortho- and meta-substituted di- and trichlorobenzenes were rapidly attacked by the microorganisms. The four- and higher substituted chlorobenzenes were found to be poorly degradable.

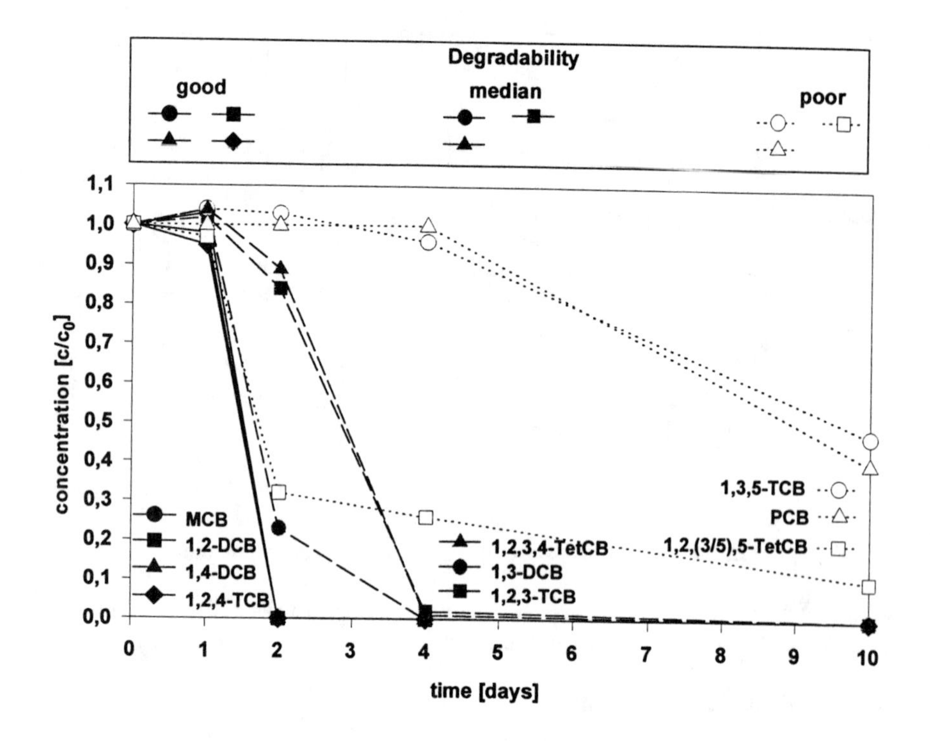

Figure 3. Degradation experiments with chlorobenzenes (initial concentration c_0 see figure 4)

Full Scale Degradation. The pilot scale plant exhibited a similar degradation reaction to that of the lab scale experiments. Highly-substituted chlorbenzenes were less decomposed than lower-substituted contaminants. Notably the chlorobenzenes in meta-position were less degraded than those with ortho- or para sunstitutes (see lab scale experiments.

During the operation time 415,000 m^3 of groundwater were pumped through the treatment plant and thus approx. 1703 kg BSS. 1250 kg BSS in the iron removal unit and 182 kg in the biological unit were allready degraded; 273 kg of the pollutants were adsorbed in the activated carbon filters. In this context it has to be considered that the activated carbon unit was loaded with pollutants, particulary during the first operation months because the microorganisms in the iron removal and biological units were not adequately adapted at that time. After the adaptation period both units showed a degradation efficiency of approx. 99%. As the activated carbon unit showed a high oxygen consumption and had high microbiological densities, it can be assumed that in the test course the pollutants desorbed partially and were degraded afterwards (Voice et al., 1992). Additionally, a negligible amount of pollutants left the treatment plant during backwashing and infiltration.

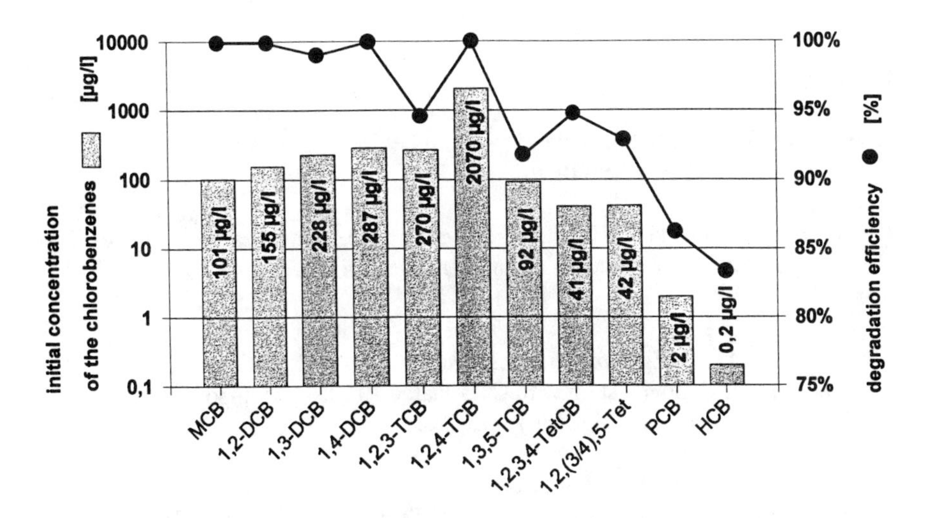

Figure 4. Degradation Perfomance of the „biological unit“ (iron removal and biological filters) for the main contaminant group of the chlorobenzenes

In the iron removal unit the concentrations of the hexachlorocyclohexanes were decreased from 0,1 mg/l to 0,043 mg/l. In the following biological unit this small concentration were further reduced to 0,005 mg HCH/l. The ε-HCH and the β-HCH were only negligibly decomposed. Presumably, the resistence of the β-HCH lies in symmetric structure of the molecul which causes unpolarity (Fiedler et al. 1993).

Table 1. Degradation Perfomance of the „biological unit“ (iron removal and biological filters) for the hexachlorocyclohexanes

Compound	Concentration before treatment (µg/l)	Concentration after treatment (µg/l)	degradation efficieny (%)
α-HCH	49,0	0,7	98,5
β-HCH	3,0	2,7	10,4
γ-HCH	3,0	0,1	97,9
δ-HCH	28,0	2,0	92,9
ε-HCH	6,0	4,5	25,3

REFERENCES

Eisenbrand, G. and P. Schreier. 1995. Römpp Lexikon - Lebensmittelchemie -. 9th ed., Georg Thieme Verlag, Stuttgart New York

Fiedler, H., M. Hub and O. Hutzinger. (1993). Stoffbericht Hexachlorcyclohexan - Hexachlorcyclohexan in Altlasten.; Landesanstalt für Umweltschutz (eds.), Karlsruhe

Lang, E., H. Viedt, J. Egestorff and H.H. Hanert. 1992. „Reaction of the Soil Microflora After Contamination with Chlorinated Aromatic Compounds and HCH". FEMS. Microbiol. Ecol. (86): 275-282

Sander, P., R. M. Wittich, P. Fortnagel, H. Wilkes and W. Francke. 1991. „Degradation of 1,2,4-Trichloro- and 1,2,4,5-Tetrachlorobenzene by Pseudomonas Strains". Appl. Environ. Microbiol. (57): 1430-1440

Voice, T.C., D. Pak,. X. D. Zhao,. J. Shi and R. F. Hickey. 1992. „Biological Activated Carbon in Fluidized Bed Reactors for the Treatment of Groundwater Contaminated with Volatile Aromatic Hydrocarbons". Water. Res. (26): 1389-1401

MODELING THE EFFECT OF NONIONIC SURFACTANTS ON THE BIODEGRADATION OF POLYCYCLIC AROMATIC HYDROCARBON IN SOIL SLURRY USING RESPIROMETRIC TECHNIQUE. — 1. PHYSICOCHEMICAL EFFECT.

Jong-Sup Park, Young J. Kim, and ***In S. Kim*** (Kwangju Institute of Science and Technology, Kangju, Korea)

ABSTRACT: The effects and characteristics of surfactants affecting Polycyclic Aromtic Hydrocarbons(PAHs) solubility and biodegradation were investigated. The critical micelle concentration(CMC) values of surfactants used in this study were 3.50×10^{-5} M (Brij30), 2.50×10^{-5} M (Tween 80), 1.85×10^{-4} M (Triton X-100). The apparent solubility of PAHs was dramatically enhanced in the solutions of surfactants at concentration higher than the CMC. However, near the CMC, the solubility of PAHs was very slightly increased. The solubility of PAH increased as the Hydrophile-Lipophile Balance (HLB) value of surfactant decreases. The micelle-water partition coefficients (Kp) on PAHs of each surfactant were calculated. The range of log Kp investigated in this study was from 4.93 to 6.25.

INTRODUCTION

Polycyclic Aromatic Hydrocarbons (PAHs) are major recalcitrant components in oil contaminants and are known to be toxic to human and other living organisms causing carcinogenecity. These compounds are produced by industrial activities such as oil processing plants and oil storage facilities and are found often in contaminated soil. For the treatment of such contaminated sites, physical, chemical and biological methods can be used. Among these methods, biological treatment method is known to have advantages for environmental and economic reasons. But, because of low aqueous solubility and strong sorption property, soils contaminated with PAHs are not easily treated by biological means. Many studies have been conducted to enhance the biodegradation by increasing the solubility and desorption rate of PAHs from soil particle using surfactants (Aronstein and Alexander, 1993; Bury and Miller, 1993; Volkering et al., 1995).

However, Information is not efficient on the effect of surfactants on the solubility and the biodegradation of PAHs.

Thus, in this study, three phases of experiment are conducted. In the first phase, the physicochemical effects of three nonionic surfactants (Triton X-100, Tween 80, Brij 30) were tested on the solubilities and the desorption rates of PAHs (naphthalene and phenanthrene as model chemicals) in aqueous phase as well as soil slurry phase. In the second phase, biodegradation rates of surfactant with and without soil were determined by measuring the oxygen uptake using respirometer. Finally kinetic experiments for the biodegradation of PAHs experiments with nonionic surfactants and soil were conducted in batch reactor.

MATERIALS AND METHODS

Sand (0.3% organic matter) and clay (0.9% organic matter) was obtained from Kwangju Science Co. (Kwangju, Korea). They were sterilized by autoclaving (121℃and 15psi) for 1hr. No growth was detected when the autoclaved soil was incubated in agar medium at 36.5℃ for 48h. Reagent-grade phenanthrene, naphthalene and nonionic surfactant Brij 30 were purchased from Aldrich (Milwaukee, Wis., USA) and HPLC-grade acetonitrile from Duksan Pure Chemicals (Ansan, Kyungkido, Korea). The non-ionic surfactants, Tween 80 and Triton X-100 were obtained from Sigma Chemical Co (St. Louis, MO, and USA). Hydrophile-Lipopile Balance (HLB) calculated as HLB=% wt EO/5 is 9.7(Brij 30), 13.4(Tween 80), 13.6(Triton X-100) respectively. The critical micelle concentrations (CMC) of surfactants were determined by surface tension measurement at each surfactant concentration.

Well-defined aerobic mixed cultures acclimated to PAH for 2 months were used for batch tests. The nutrient solutions contain NaH_2PO_4 50 mg/L, KH_2PO_4 85 mg/L, K_2HPO_4 165.6 mg/L, NH_4Cl 100 mg/L, $MgSO_47H_2O$ 0.1 mg/L, $FeSO_47H_2O$ 0.1 mg/L, $MnSO_4H2O$ 0.036 mg/L, $ZnSO_47H_2O$ 0.03 mg/L, $CoCl_26H_2O$ 0.01 mg/L, $CaCl_22H_2O$ 0.1 mg/L (Kim and Lee,1996).

Batch tests for the determination of solubility were performed in 250ml flask at 20℃. The surfactant solutions were both above and below the CMC to evaluate the effect on the solubility of phenanthrene and naphthalene. The samples were then centrifuged at 10,000rpm for ten minutes to separate the unsolubilized portion of the PAHs. The PAHs dissolved were determined by injecting samples into a high-performance liquid chromatography (Waters,

Milford, USA) equipped with a C18 column. The eluent was a mixture of acetonitrile and water (75:25). Photodiode Array detector was used for the detection of phenanthrene and naphthalene at 254nm wavelength by measuring at 254nm both phenanthrene and naphthalene.

Biodegradation test of phenanthrene was initiated by transferring acclimated cultures to serum bottles using respirometer (Challenge Environment Sys, AER-200, USA). Oxygen uptake for the degradation reaction was monitored with time. The phenanthrene solution dissolved in methylene chloride was poured into the soil. The methylene chloride was evaporated under a hume hood before experiments. The bottles eventually contain soil with phenanthrene, nutrient solution, buffer solution, and surfactant solution (soil: solution=1:10). Temperature during the entire experiment was maintained at 20℃ constantly.

RESULTS AND DISCUSSION

CMC Measurement. The CMC values were determined from a plot of the surface tension measured by surface tensiometer versus the surfactant solutions of varying concentration. The results are shown in Figure 1.

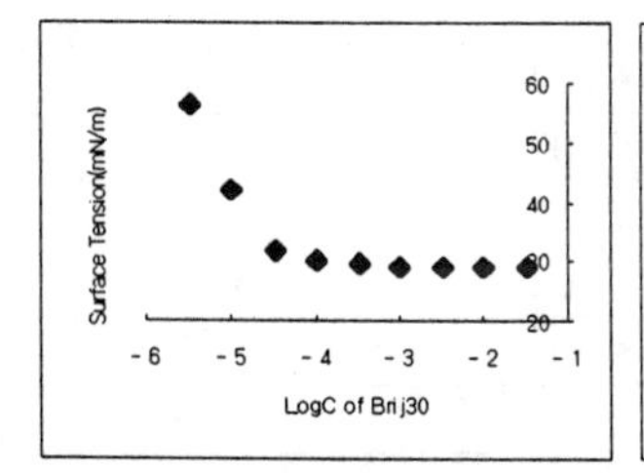

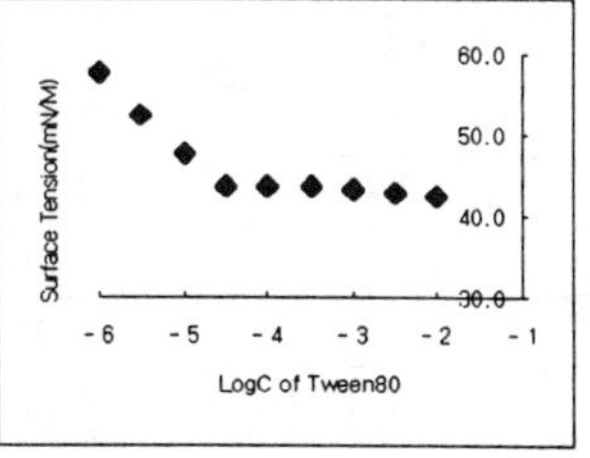

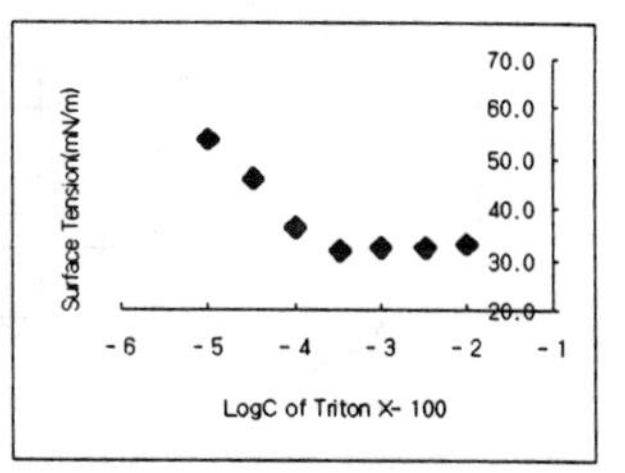

FIGURE 1. Determination of CMC for surfactant solutions from surface tension measurements.

The CMC values of Brij30, Tween 80, Triton X-100 were 3.50×10^{-5}, 2.50×10^{-5}, and 1.85×10^{-4} M respectively.

Solubility of PAHs. The concentration greater than the CMC showed solubility enhancement apparently. However, the solubility of PAHs was slightly increased, when the concentrations were below or near the CMC. The surfactant having lower HLB value produced better solubilization. A molar solubilization ratio

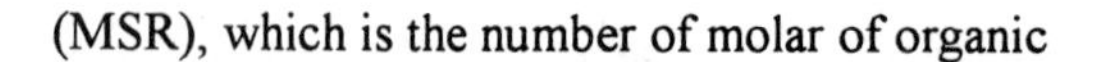

(MSR), which is the number of molar of organic

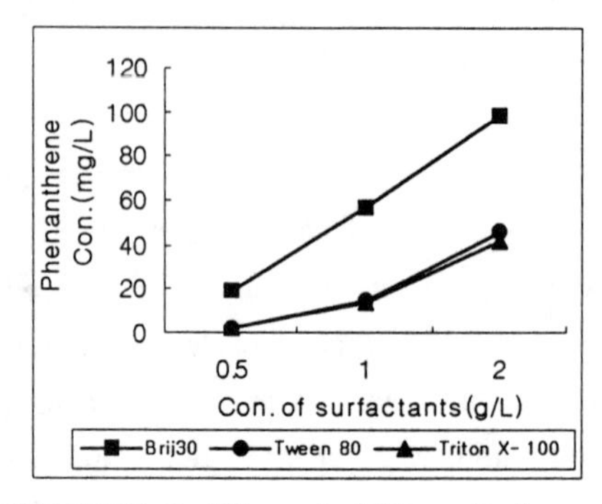

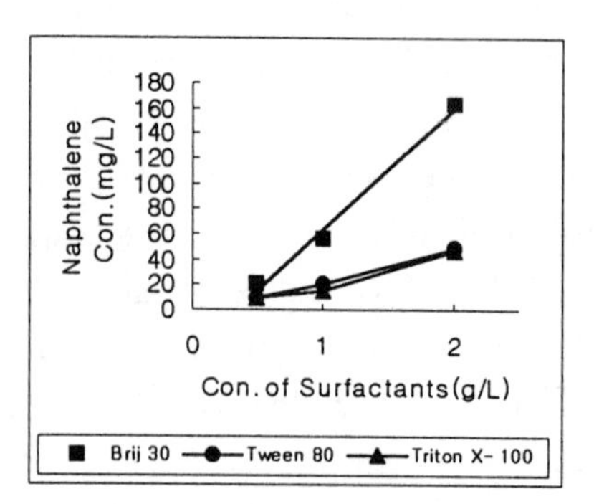

FIGURE 2. The solubility of phenanthrene at greater concentration than CMC.

compound solubilization per mole of surfactant added to solution (Pennell et al., 1997) can be obtained from the solubility curve(Figure 2). MSR values for each surfactants were illustrated Table 1.

TABLE 1. Molar solubility ratio and partition coefficient on phenanthrene of the surfactant solution.

PAHs	Surfactants	MSR	logKp
Phenanthrene	Brij 30	1.06×10^{-1}	5.88
	Tween 80	2.19×10^{-1}	6.25
	Triton X-100	9.78×10^{-2}	5.99
Naphthalene	Brij 30	1.96×10^{-1}	5.81
	Tween 80	1.92×10^{-1}	5.62
	Triton X-100	9.20×10^{-2}	4.93

The micelle-water partition coefficient (Kp) represents the distribution of PAHs between the surfactant micelles and the aqueous phase.

$$Kp = Xm/Xa \tag{1}$$

Where Xm is the mole fraction of PAH in the micelle phase and Xa is the mole fraction of PAH in the aqueous phase. Kp can be derived to equation (2) (Edwards et al. 1991).

$$Kp = (55.56/S_{PAH,CMC}) (MSR/(1+MSR)) \quad (2)$$

Where, $S_{PAH,CMC}$ is the solubility of PAH at the CMC.

REFERENCES

Aronstein, B. N., M. Alexander. 1993. "Effect of a Non-ionic Surfactant added to the Soil Surface on the Biodegradation of Aromatic Hydrocarbons within the Soil." *Appl. Microbiol. Biotechnol.* 39:386-390.

Bury, S. J., and C. A. Miller. 1993. "Effect of Micellar Solubilization on Biodegradation Rates of Hydrocarbons." *Environ. Sci. Techno.* 27(1):104-110.

Edwards, D. A., R. G. Luthy, and Z. Liu. 1991. "Solubilization of Polycyclic Aromatic Hydrocarbons in Micellar Nonionic Surfactant Solutions." *Environ. Sci. Technol.* 25:127-133.

Kim, J. O., and J. J. Lee.1996. "Degradation of Trichloroethylene with Phenol-Oxidizing Microorganisms in Presence of Surfactant." *J. of Korea Solid Wastes Engineering Society.* 1(1):29-37.

Pennell, K. D., A. M. Adinolfi, L. M. Abriola, and M. S. Diallo. 1997. "Solubilization of Dodecane, Tetrachloroethylene, and 1,2-Dichlorobenzene in Micellar Solutions of Ethoxylated Nonionic Surfactants." *Environmental Science and Technology.* 31(5):1382-1389.

Volkering, F., A. M. Breure, J. G. A. Andel, and W. H. Rulkens. 1995. "Influence of Nonionic Surfactants on Bioavailability and Biodegradation of Polycyclic Aromatic Hydrocarbons." *Applied and Environmental Microbiology.* 61(5):1699-1705.

BIOREMEDIATION OF CHLOROPHENOL-CONTAMINATED SOIL BY COMPOSTING IN FULL SCALE

M. Minna Laine and Kirsten S. Jørgensen (Finnish Environment Institute, Helsinki, Finland)

ABSTRACT: A three-year full-scale bioremediation (520 m^3) was performed to test the applicability of using induced straw compost as an inoculum in comparison to addition of bark chips and nutrients to chlorophenol-contaminated soil. The site was a sawmill site, where the commercial chlorophenol mixture, Ky 5, had been used as a wood preservative in a dipping basin. The soil was excavated and treated in the vicinity. The composting was performed in 1995-1997. The composting of chlorophenol-contaminated soil was very efficient during the first year and resulted in 96, 92 and 90 % removal of chlorophenols in 126 days with starting chlorophenol concentrations of 960, 740 and 29 mg (kg dry weight)$^{-1}$, respectively. The compost piles froze completely during winter and during the next 5 months of activity the chlorophenol concentrations decreased with 45, 70 and 50%, respectively. The use of straw compost as inoculum did not enhance the chlorophenol degradation in comparison to addition of bark chips and nutrients. Frequent mixing and control of the nutrient level enhanced the chlorophenol degradation of the indigenous microbes in the contaminated soil. The chlorophenol-degrading microorganisms originating from contaminated soil probably benefited from the enhanced general microbial activity in the piles by cometabolism or synergism.

INTRODUCTION

Chlorophenols are recalcitrant compounds that have been used for decades to impregnate wood, and many residues can be found in the environment long after the use of chlorophenols have been discontinued. Chlorophenols are soluble in water and may leach from contaminated soil to groundwater. Therefore the contaminated sites must be cleaned up to prevent further spread.

Objective. The main objective for this full scale bioremediation (520 m^3) was to test the applicability of using induced straw compost as an inoculum in comparison to addition of bark chips and nutrients to chlorophenol-contaminated soil. The aim was also to produce a well-documented case of composting in biopiles.

The site was a sawmill site, where the commercial chlorophenol mixture, Ky 5, had been used as a wood preservative in a dipping basin. This mixture consisted mainly of 2,4,6-trichlorophenol, 2,3,4,6-tetrachlorophenol and pentachlorophenol (Valo et al. 1984). The soil was excavated and treated in the vicinity.

MATERIALS AND METHODS

Composting of contaminated sawmill soil in full scale. Detailed description of the build-up of the compost piles as well as parameters followed during the composting are presented in Laine et al., 1997b. Briefly, the full-scale composting operated for three summer seasons from 1995 to 1997. Three windrows (piles) were built with the total volume of 520 m^3. All three compost piles contained contaminated sawmill soil, nutrients and lime, but they differed in amendments and the soil type. Piles 1 and 2 contained the same batch of contaminated soil that was sandy soil with partly degraded sawdust. The initial concentration of chlorophenols in piles 1 and 2 was 960 and 740 mg (kg dry wt)$^{-1}$, respectively. The soil in pile 3 was clay, and it had an initial chlorophenol concentration of 30 mg (kg dry wt)$^{-1}$. Pile 1 was amended with bark chips (pine and spruce) as supporting and aerating material. To improve chlorophenol degradation, induced straw compost was added as inoculum to piles 2 and 3. The ratio of the materials was two parts of contaminated soil and one part of straw compost and/or bark by volume. After all the materials were piled, the piles were mixed by turning and covered with tarps. The piles were mixed every third week in 1995 and once a month in 1996 and 1997. Temperature data-loggers were used to follow the temperatures in compost piles on-line.

RESULTS AND DISCUSSION

Chlorophenol removal. The composting of chlorophenol-contaminated soil was very efficient during the first year and resulted in 96, 92 and 90 % removal of chlorophenols in 126 days with starting chlorophenol concentrations of 960, 740 and 29 mg (kg dry weight)$^{-1}$, respectively (Table 1). The compost piles froze completely during winter and during the next 5 months of activity the chlorophenol concentrations decreased with 45, 70 and 50%, respectively. During the third year of composting the rate of degradation continued to decrease. According to the results of earlier laboratory experiments (Laine and Jørgensen 1996) and field test in pilot scale (4 x 13 m^3) (Laine et al. 1997a; Laine and Jørgensen 1997), the mechanism of chlorophenol removal was complete mineralization without harmful side reactions as (bio)methylation or polymerization.

The use of straw compost as inoculum did not enhance the chlorophenol degradation in comparison to addition of bark chips and nutrients. Frequent mixing and control of the nutrient level enhanced the chlorophenol degradation of the indigenous microbes in the contaminated soil (Laine and Jørgensen, 1997). Fast growing microbes responsible for utilization of easily available carbon substrates, measured by respiratory activity and substrate utilization patterns, originated mainly from the added bulking agents, straw compost and bark chips (Laine et al. 1997b). The chlorophenol-degraders originating from contaminated soil probably benefited from the enhanced general microbial activity in the piles by cometabolism or synergism.

TABLE 1. Annual and total chlorophenol removal in % during the composting 1995-1997.

Pile	Chlorophenols at the beginning of composting 1995, (mg (kg dry wt)$^{-1}$) [aver ± SD]	Chlorophenol removal				Chlorophenols at the end of composting 1997, (mg (kg dry wt)$^{-1}$) [aver ± SD]
		1995, %	1996, %	1997, %	Total, %	
Pile 1	960 ± 250	96	45	35	98	15 ± 4
Pile 2	740 ± 380	92	70	0	98	18 ± 10
Pile 3	29 ± 8	90	50	0	93	2 ± 1

Temperature. The average temperature in the compost piles was 35 °C at its highest and it was well above the ambient day temperature (Figure 1). During the winter, the degradation of chlorophenols stopped since the piles were completely frozen.

FIGURE 1. Temperature profiles in compost piles during composting in 1996.

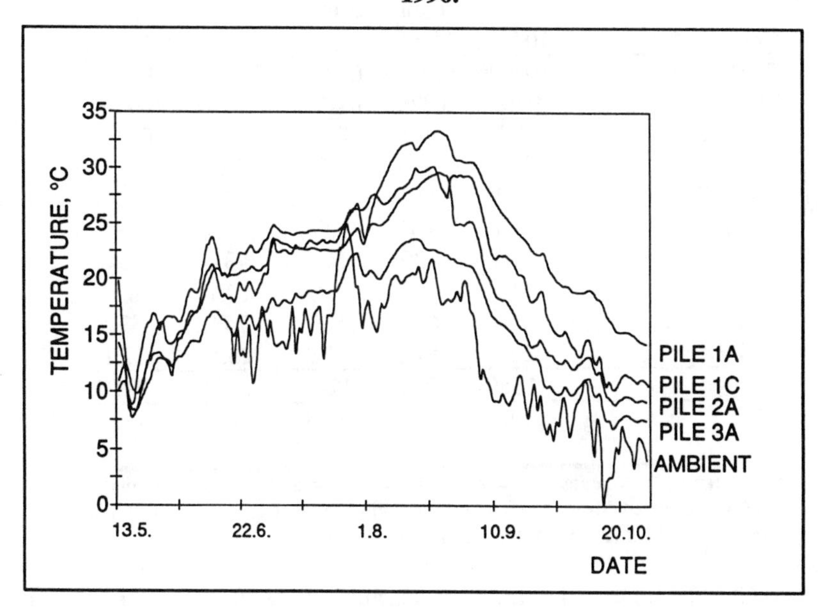

Bioremediation of chlorophenol-contaminated sawmill soil using composting without bioaugmentation is a cheap and feasible method, but outdoor composting is limited to work efficiently only during summer season when the ambient temperature exceeds 10 °C (Laine and Jørgensen, 1997). The drawback of cold climate could be overcome by either constructing the piles indoors (*e.g.* airdome), or supplying extra heating (underground-wires for electric heating).

Mixed contamination of sawmill soil. Chlorophenol-contaminated sawmill soil in Finland is often also contaminated with PCDD/Fs (polychlorinated dibenzo-*p*-dioxins and dibenzofurans), PCPPs (polychlorinated phenoxyphenols), polychlorinated diphenyl ethers and PCBs (polychlorinated biphenyls) since the technical wood preservative called Ky 5 that was produced and used in Finland as well as in other Scandinavian countries contained these dimeric compounds as impurities (Kitunen et al. 1985; Laine et al. 1997a; Vartiainen et al. 1995a). PCDD/F concentrations were followed in pile 2 in 1995 (Table 2). The results suggest that PCDD/Fs were not formed, but not either biodegraded during the composting of contaminated sawmill soil. In order to clean the soil completely, a physical-chemical treatment of the remaining contaminants should follow the biological degradation of the chlorophenols.

TABLE 2. Polychlorinated dibenzo-*p*-dioxins and dibenzofurans in pile 2 during composting in 1995.

	polychlorinated dibenzo-*p*-dioxins and dibenzofurans [µg (kg dry wt) $^{-1}$]				
Congener[a)]	10.7.95	24.7.95	14.8.95	16.10.95	13.11.95
TCDF	0.0906	0.0856	0.0897	0.0883	0.015
TCDD	0.027	0.0262	0.0217	0.0342	0.0254
PCDF	1.1	0.907	0.84	1.15	0.767
PCDD	2.01	1.78	1.63	1.6	1.29
HxCDF	111	113	87.9	125	91.8
HxCDD	9.6	8.57	7.56	8.52	6.87
HpCDF	763	639	565	737	604
HpCDD	11.6	9.83	7.89	10.1	8.9
OCDF	759	623	600	743	531
OCDD	26.6	19.2	12.7	23.5	22.1
Total	1680	1410	1280	1650	1270
I-TEQ[b)]	17.2	16.5	13.3	18.3	13.9

a) T = tetra-, Pe = penta-, Hx = hexa-, Hp = hepta-, and O = octachlorinated. CDF = chlorinated dibenzofuran, CDD = chlorinated dibenzo-*p*-dioxin. b) I-TEQ = International 2,3,7,8-TCDD toxic equivalent.

PCDD/Fs are scored as the most toxic man-made compounds in the world. The toxic mechanism of PCDD/Fs and related compounds is based on their binding to Ah-receptor found in higher organisms. The Ah-receptor-mediated mechanism of action, however, is only one of the several ones suggested as the toxic mechanism of PCDD/Fs (Webster & Commoner, 1994). Researcher's current opinions differ as to how harmful PCDD/F compounds really are, what are their effects on humans, and what is the actual threshold value for the effective dose. Recent studies indicate that chlorophenols rather than PCDD/Fs are the main cause for detrimental health effects to human in contaminated sawmill areas (Vartiainen et al., 1995b). PCDD/Fs are less mobile than chlorophenols since they adsorb to the organic matter in soil. More attention should be paid to study environmental significance of PCPPs, polychlorinated diphenyl ethers and polychlorinated chloroanisoles in contaminated sawmill soils as well.

Survival of the chlorophenol-degrading microorganisms during bioremediation. Chlorophenol degradation was very efficient during the first months of the composting. However, the addition of inorganic nutrients in the form of fertilizer or easily available carbon sources such as fresh bark did not enhance the chlorophenol degradation after winter (Laine et al., 1997b). The revival of chlorophenol-degrading microorganisms after winter was not very efficient. There may be many reasons for that:
1) Cold climate (studies by Laine & Jørgensen (1997) showed no degradation activity below 10 °C),
2) The compost environment had become more suitable and less toxic for other fast-growing microorganisms that may have conquested and colonized over the chlorophenol-degraders,
3) Different microbial species degraded chlorophenols a) in the clay soil pile (Pile 3) compared to organic soil piles (Piles 1 and 2), or b) at different levels of chlorophenol concentrations (bioavailability and tolerance to high concentrations; properties enhancing attachment to chlorophenols in soil).
4) Absence of a possible cosubstrate needed for degradation. The addition of bark chips may have slightly increased the chlorophenol removal (Laine et al., 1997b), maybe because there are phenolic compounds present which may act as cosubstrates for chlorinated phenols, or simply because addition of bark improves the aeration in the compost and chlorophenol degradation seemed to be aerobic.
5) Bioavailability: Too small a fraction of chlorophenols available for degraders, although the chlorophenol concentration in pile 3 decreased from 30 to 2 mg (kg dry wt)$^{-1}$ under similar conditions (Table 1).

ACKNOWLEDGMENTS

We thank Terttu Vartiainen and Teija Strandman at the National Institute of Public Health, Kuopio, Finland, for analyzing and calculating the dioxin congeners.

REFERENCES

Kitunen, V., R. Valo, and M. Salkinoja-Salonen. 1985. "Analysis of chlorinated phenols, phenoxyphenols and dibenzofurans around wood preserving facilities." *Intern. J. Environ. Anal. Chem. 20*: 13-28.

Laine, M. M., J. Ahtiainen, N. Wågman, L. Öberg, and K. S. Jørgensen. 1997a. "Fate and toxicity of chlorophenols, polychlorinated dibenzo-*p*-dioxins and dibenzofurans during composting of contaminated sawmill soil." *Environ. Sci. Technol. 31*: 3244-3250.

Laine, M. M., and K. S. Jørgensen. 1996. "Straw compost and bioremediated soil as inocula for the bioremediation of chlorophenol-contaminated soil." *Appl. Environ. Microbiol. 62*: 1507-1513.

Laine, M. M., and K. S. Jørgensen. 1997. "Effective and safe composting of chlorophenol-contaminated soil in pilot scale." *Environ. Sci. Technol. 31*: 371-378.

Laine, M. M., H. Haario, and K. S. Jørgensen. 1997b. Microbial functional activity during composting of chlorophenol-contaminated sawmill soil. *J. Microbiol. Methods 30*: 21-32.

Valo, R., V. Kitunen, M. Salkinoja-Salonen, and S. Räisänen. 1984. "Chlorinated phenols as contaminants of soil and water in the vicinity of two Finnish sawmills." *Chemosphere 13*: 835-844.

Vartiainen, T., P. Lampi, K. Tolonen, and J. Tuomisto. 1995a. "Polychlorodibenzo-*p*-dioxin and polychlorodibenzofuran concentrations in lake sediments and fish after a ground water pollution with chlorophenols." *Chemosphere 30*: 1439-1451.

Vartiainen, T., P. Lampi, J. T. Tuomisto, and J. Tuomisto. 1995b. "Polychlorodibenzo-*p*-dioxin and polychlorodibenzofuran concentrations in human fat samples in a village after pollution of drinking water with chlorophenols." *Chemosphere 30*: 1429-1438.

Webster, T., and B. Commoner. 1994. "Overview. The Dioxin Debate." In A. Schecter (Ed.), *Dioxins and health.*, pp. 1-50. Plenum Press, New York.

ENHANCED IN-SITU MOBILIZATION AND BIODEGRADATION OF PHENANTHRENE FROM THE SOIL BY PARAFFIN OIL/SURFACTANT

Eunki Kim (Inha University, Inchon, Korea)
Anbo. Liu, Iksung Ahn, L.W. Lion and M.L.Shuler
(Cornell Univ., New York, USA)

ABSTRACT: Mobilization and biodegradation of phenanthrene in the soil was enhanced by using paraffin oil which was stabilized by the addition of surfactant(Brij 30). The ratio of paraffin oil/Brij 30 was determined by measuring the changes in the critical micelle concentration. Stabilized paraffin oil emulsion could solubilize more phenanthrene than control experiments. Column experiment showed increased phenanthrene mobilization from the contaminated soil. Phenanthrene mobilized in the paraffin oil/Brij 30 emulsion could be biodegraded more faster than in water phase or surfactant solution. This result indicates that paraffin oil/surfactant system can be effective and non-toxic for the removal of PAH from the contaminated soil.

INTRODUCTION

Organic solvents, such as alcohols or alkanes, can dissolve high concentrations of PAH. The affinity of the solvent for a given PAH is typically quantified by a distribution coefficient. Among the available solvents, paraffin oil (mineral oil) is safe for human consumption and is used as an oral laxative. Addition of this solvent to water has been shown to increase the biodegradation of PAH (Jimenze and Bartha, 1996). However, paraffin oil is not miscible with water. Therefore remedition of PAH contaminated soils by solvent augmentation with parraffin oil is presently limited to ex-situ treatment processes.

In order to use paraffin oil for in-situ treatment, this solvent must be stabilized by some method, such as the addition of a surfactant. In this work, we used a non-ionic surfactant /paraffin oil emulsion to increase the aqueous concentration of a test PAH (phenanthrene) and to enhance the mobilization of phenanthrene from a test soil.

MATERIALS AND METHODS

Materials: Polyoxyethylene 4 lauryl ether (Brij 30) was used as a non-ionic surfactant. Paraffin oil was purchased from local drug store. Radio-labelled phenanthrene was purchased from Sigma. A bacterial consortium, as a gift of Prof. M. Alexander (Cornell University, USA), was used, which had been isolated from soil.

Determination of Surface tension: The surface tension of the solution was measured by using Fisher Tensiomet 20 (Fisher Scientific, USA). Paraffin oil and surfactant mixture was emulsified by using ultrasonicator for 1 min (Fisher Scientific, USA).

Solubility Enhancement Experiments: In the 20 mL vials with screw caps, 50 mL of mixture of $[^{14}C]$-phenanthrene and non-labeled phenanthrene dissolved in methanol was added. After the methanol was allowed to evaporate, 10 mL sodium azide solution (0.02% with 5mM $CaSO_4$) was added. Total amount of phenanthrene was 10 ppm which was about 8 times of the solubility (1.29 ppm) (Laha and Luthy, 1991). Vials were placed in a shaker at 30 °C for 4 days. Duplicate samples were taken and passed through the 0.1 µm PTFE syringe filters to remove any undissolved phenanthrene. Predetermined amount of sample (4 mLs) was passed through filters before collecting the filtrate to minimize the adsorption during sampling. Filtrate samples were analyzed for radioactivity in a liquid scintillation counter to measure the amount of phenanthrene.

Column Experiments: 10 mL of saturated phenanthrene solution containing 5 mM CaSO4 and 0.005 % sodium azide was added with 5 grams of non-contaminated soil. The soil contains 2.4% organic matter and size of soil ranges between 0.0116 and 0.0087 inch. After equilibrating for 4 days, the soil slurry was packed into 10 cm (1 cm diameter) glass column. The slurry was slowly poured into the column and the effulent from the bottom was collected until one pore volume of the solution was left in the column. The total radioactivity in the collected effulent was measured to calculate the amount of phenanthrene in the soil column. Then the feeding solution was pumped into the column with 10 cm/hr pore velocity. The porosity of the column was determined as 42%. Effulent was collected by using a fraction collector and radioactivity of each fraction was used to measure the amount of eluted phenanthrene.

Mineralization Experiments: Biodegradadation experiments were conducted in 50 mL earlenmyer flasks that contain in their headspace a glass vial in which the CO_2 evolved by mineralization was absorbed into 1 mL of 1N NaOH solution. Silicone rubber stopper wrapped with aluminium foil was used to minimize the loss due to sorption. 50 mL of mixture of [^{14}C]-phenanthrene and non-labeled phenanthrene dissolved in methanol was added. After the methanol was allowed to evaporate, 10 mL of minimal medium was added. Total concentration of phenanthrene was 10 ppm. Minimal medium contained (per Liter) 1 g $(NH_4)_2SO_4$, 3.8 g KH_2PO_4, 0.95 g KOH, 0.1 g NaCl, 0.2 g $MgSO_4$ and 1 mg $FeSO_4.7H_2O$. Cells, grown for 3 days with nonlabeled phenanthrene (1 g/L), was filtered by coarse filter paper (#1 Whattman paper) to remove granular phenanthrene and was inouculated with initial cell density of 0.005 OD 600. During the cultivation at 30C in the shaking incubator, one mL NaOH solution was removed by syringe intermittently and radioactivity was measured to determine the mass of CO_2 and the mineralization ratio.

RESULTS AND DISCUSSION

Determination of the ratio of surfactant and paraffin oil. The oil and surfactant mixture can be stabilized if the surface of the oil drop was surrounded by surfactant molecule. Since the surfactant will locate in the water/air surface after surrounding the oil drop, more surfactant will be required to begin micelle formation (CMC). Therefore, by measuring the difference of CMCs between surfactant-water system and surfactant-oil-water system, the ratio of oil and surfactant could be determined. As shown in Fig.1, with the presence of paraffin oil, more surfactant was needed to achieve the minimum surface tension where the micelle forms. 140 ppm (v/v) of surfactant was required additionaly for

5000ppm paraffin oil (0.5% v/v). Therefore, 2.8 ppm(v/v) of surfactant was required to surround the 100 ppm (v/v) of oil drop.

Solubility of Phenanthrene. Solubility of phenanthrene was determined at various concentrations of surfactant and surfactant/oil. As shown in Fig.2, surfactant/oil mixture can solubilize more phenanthrene than the surfactant alone. This can be attributed to the presence of oil drop in the surfactant/oil mixture. Brij 30/heptane showed similar effects.

Column experiment. In order to compare the stability of surfactant/oil mixture with surfactant only system, column test was done. As shown in Fig.3, addition of surfactant only resulted in poor mobilization of the phenanthrene from soil. However, surfactant/oil mixture could extract phenanthrene significantly more than other solutions. Comparison of the stability of surfactant/oil system can be more clear in the column experiment than in the batch experiment, since only equilibrium data are available in the batch experiment.

Biodegradation of phenanthrene in the surfactant/oil. Phenanthrene washed from the soil by surfactant/oil should be mineralized by microorganisms and surfactant/oil should be recycled to lower the cost of remediation. The effects of surfactant and oil drop on the availability of PAH to the cell were investigated. It was found that bioavailability depended on the structure of surfactant. Also the PAH dissolved in the paraffin oil was mineralized more faster than that in the water. The hydrophobicity of the cell walls of microorganisms was changed during the mineralization.

In order to investigate the bioavailablity of phenanthrene which is dissolved in the oil drop coated by surfactant, mineralization experimentation was performed by measuring the C^{14}-CO_2 produced from the phenanthrene mineralization. As shown in Fig.4, phenanthrene in the surfactant/oil system was mineralized more faster than that in the water. Similar result was observed by other workers (Guha and Jaffe, 1996). For the mineralization, however, adaptation of cells to the surfactants was necessary as shown in the Fig.5. This results indicate that surfactant/oil can be used successfully in the in-situ mobilization of phenannthrene in soil and biodegradation of the phenanthrene with enhanced efficiency.

REFERENCES

Guha,S., and P.R. Jaffe, 1996,"Biodegradation kinetics of phenanthrene partitioned into the micellar phase of nonionic surfactants", 30, 605-611,

Jimenez,I.Y.,and R.Bartha, 1996, "Solvent augmented mineralization of pyrene by a *Mycobacteium* Sp.". Appl. Env. Microbiol. 62(7); 2311-2316

Laha,S. and R.G., Luthy, 1991, "Inhibition of phenanthrene mineralization by nonionic surfactants in soil-water systems". Environ. Sci. Technol., 25(11); 1920-1930

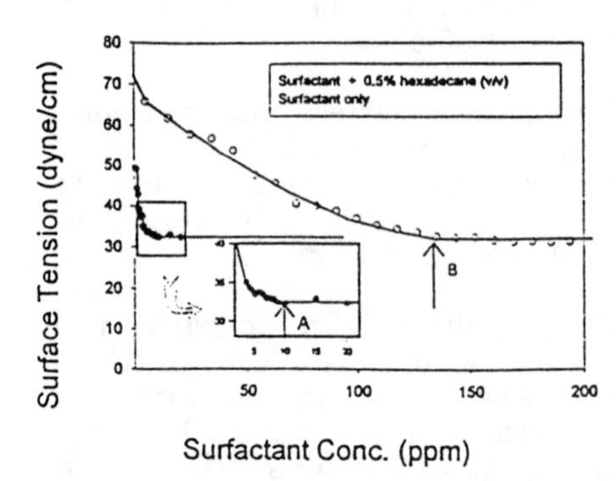

Fig.1 Determination of the ratio between oil and surfactant

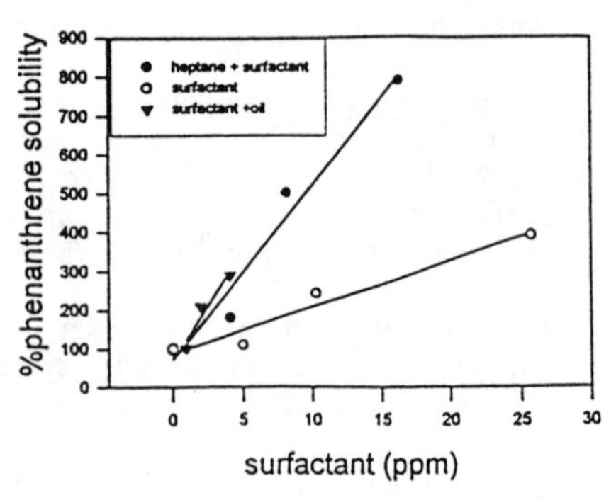

Fig.2 Increased solubility by surfactant-oil system

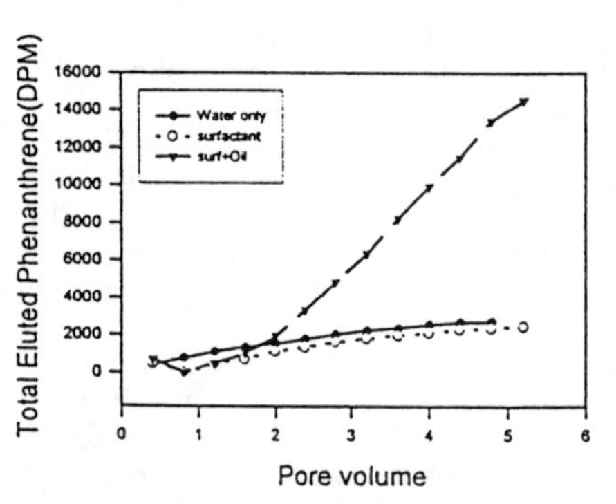

Fig.3 Effects of surfactant and oil washing in the coulmn

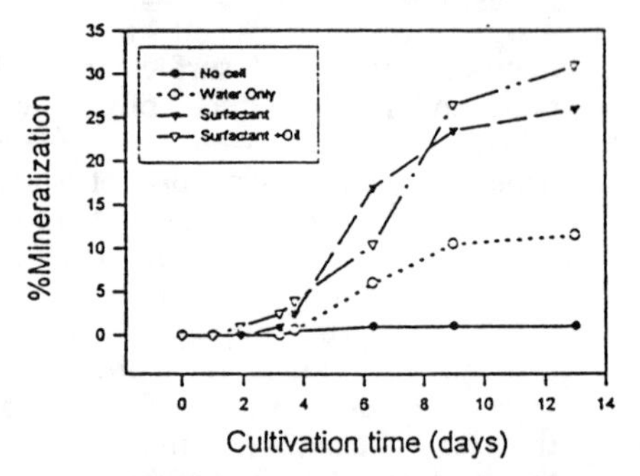

Fig.4 Mineralization of phenanthrene with surfactant and oil

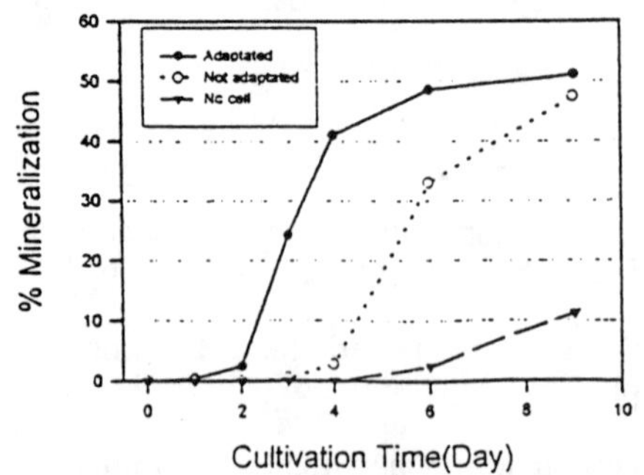

Fig.5 The effects of cell adaptation to surfactant on phenanthrene degradation

APPLICATION OF BIOREMEDIATION TESTING PROTOCOL TO PAH CONTAMINATED AGED SOILS

Henry H. Tabak (U.S. EPA, Cincinnati, Ohio)
Rakesh Govind, M. Parvatiyar, Qi Song, J. Guo
(Department of Chemical Engineering, University of Cincinnati, Ohio)

ABSTRACT: Treatability studies are critical for successful implementation of both *in-situ* and *ex-situ* bioremediation technologies. Few contaminated sites are identical and experience can only be applied within limits. Many variables governing the efficacy of bioremediation processes are a function of environmental conditions, type of contaminants and contaminated media. A systematic protocol has been developed to determine the bioavailability and biokinetics of soil contaminants. Based on this protocol, experimental studies have been conducted using microcosms and soil slurry, wafer and column bioreactors. In this paper, results are presented on the application of soil microcosms to an aged soil, contaminated with several low and high molecular weight polycyclic aromatic hydrocarbons (PAHs), obtained from the Reilly Tar and Chemical Corporation site in St. Louis Park, Minnesota. Six microcosm reactors were set-up to study natural attenuation and the effect of additives (nutrients, surfactants, and inducers).

INTRODUCTION

The determination of the rates of biodegradation of organics in soil systems is important for evaluating the efficiency of *in-situ* bioremediation and approaches for enhancing bioremediation rates. Laboratory studies to determine biodegradation rates can be used as screening tests to determine the rate and extent of bioremediation that might be attained during remediation of polluted soil sites, and to establish design criteria for an optimum bioremediation approach.

From previous studies with uncontaminated soil freshly spiked with individual chemicals [Fu *et al.*, 1996], a systematic approach and modeling equations were developed for quantitative determination of biodegradation rates in intact soil systems. The systematic multi-level protocol (Tabak *et al.,* 1997; Tabak and Govind, 1997) was developed to determine the biokinetic parameters for toxic organic pollutants by the indigenous suspended and immobilized microbiota, the bioavailability of these organics and the transport parameters of the contaminant and oxygen in the soil matrix.

This paper reports on studies with aged contaminated soil from the Reilly Tar and Chemical Corporation site in St. Louis Park, Minnesota. A coal tar refinery and wood-preserving facility formerly operated on the site from 1917 until 1979, after which the site was converted into a city park. The test soil was obtained from the 0.3 m thick asphaltic layer, which is covered with a 0.6 m thick topsoil, and below which is approximately 3.5 m of coarse brown sand.

Polycyclic aromatic hydrocarbons (PAHs) are toxic and hazardous chemicals regulated by the US Environmental Protection Agency as priority pollutants. There is interest in understanding the fate of these compounds in soil-water systems and determining the mechanism and rate of biodegradation. Various studies have identified specific organisms capable of degrading PAH compounds (Heitkamp and Cerniglia, 1989; Cerniglia, 1984, Clover, 1987; Ellis et al., 1991). Information on aerobic pathways is generally limited to two-and three-ring PAH compounds including naphthalene (Fredrickson et al., 1991; Gibson and Subramanian, 1984), acenaphthene (Ellis et al., 1991), and phenanthrene (Fredrickson et al., 1991; Brodkorb and Legge, 1992). Literature on microbial degradation and stability of PAH compounds in soil-water systems has been summarized by Atlas (1981) and Sims and Overcash (1983).

EXPERIMENTAL STUDIES CONDUCTED

A number of microcosms were set up, with and without additives, as given below:

MC1	soil + moisture (6.9%)
MC2	soil + moisture + OECD (Total moisture = 8.7%)
MC3	Soil + OECD + Salicyclic acid (inducer; 100 mg/Kg) (Moisture 8.7%)
MC4	Soil + OECD + Phthalic acid (inducer, 100 mg/Kg) (Moisture 8.7%)
MC5	Soil + OECD + Tween 80 (surfactant, 100 mg/Kg) (Moisture 8.7%)
MC6	soil as received (5.5% moisture content) **(Natural Attenuation)**

MATERIALS AND METHODS

Twenty-two barrels of Reilly Tar soil were collected and shipped to US EPA in Cincinnati, Ohio. The soil was mixed and sieved through 0.5-inch mesh using soil handling equipment. Shaker vial extraction method (Huang et al. 1996) was used to analyze the soil concentration of each PAH. The method involves mixing 4 g of contaminated soil with methylene chloride solvent in 40 mL vials, placing the vials in boxes and attaching the boxes to a circular shaker. The vials are shaken for 18 hours. The extracts are analyzed using a Hewlett Packard Gas Chromatograph equipped with a flame ionization detector (GC-FID).

The soil characteristics were as follows: Loamy Sand; Sand 79%; Silt 13%; Clay 8%; bulk density 1.18 g/mL; % moisture at 1/3 bar: 6.7; % moisture at 15 bar: 5.0; Organic matter: 6.5%; Organic carbon: 3.8%; and total nitrogen: 0.15%. Chemical analysis of the original soil gave the following results for the 18 PAHs identified (in mg/Kg): 2-ring PAHs: Naphthalene (32.7); 2-Methyl Naphthalene (24); 3-ring PAHs: Acenaphthylene (6.6); Acenaphthene (225.7); Dibenzofuran (100); Fluorene (214.1); Phenanthrene (500.7); Anthracene (446.4); 4-ring PAHs:

Fluoranthene (501); Pyrene (343.6); Benzo(a)Anthracene (117.8); Chrysene (140.6); 5-ring PAHs: Benzo(b)Fluoranthene (149.3); Benzo(e)Pyrene (72.8); Benzo(a)Pyrene (75.3); Dibenzo(ah)Anthracene (9.1); 6-ring PAHs: Indeno(123-cd)Pyrene (44.7); Benzo(ghi)Perylene (45.4). The PAH compounds were grouped into the following categories: Sum of 2+3-ring PAHs (1,550.3 mg/Kg); Sum of 4+5+6-ring PAHs (1,499.5 mg/Kg); and Total PAHs (3,049.8 mg/Kg).

A thick-walled sampling bag was placed on a large weighing balance and 10 Kg of contaminated soil was scooped out of the bucket and put into the bag. Required amounts of OECD nutrients and/or additives were added to the sampling bag and the bag was closed with a metal tie. The bag was then placed in a plastic bucket, which has an internal baffle and can be closed with a lid. The bucket was then manually turned on the floor for 30 minutes to mix the soil in the bag.

The plastic bag was removed from the bucket and the soil was scooped out and packed in the microcosm reactor. A microcosm reactor is an airtight rectangular container (20 in. x 12 in. x 12 in.) constructed of glass supported by stainless steel panels. All microcosm reactors were operated at ambient temperature. A controlled flow rate of air was passed through the reactor and the exiting air was vented into the fume hood. The inlet air was humidified by bubbling it through a water trap and a small test tube containing water was also placed inside each microcosm reactor to maintain the moisture content in each microcosm.

Samples taken at time zero were taken from the bulk soil before the microcosms are packed. All subsequent samples were taken from the microcosm. Soil samples were dried using sodium sulfate, extracted, and analyzed by GC to determine the concentrations of the PAHs.

Nutrients, recommended by the Organization for Economic Co-operation and Development (OECD) (OECD, 1981) were used in this study. The amount of OECD nutrients was determined by the nitrogen requirements, which was calculated to give an initial ratio of C:N = 10:1, where C represents the total biodegradable organic carbon (TBOC) present in the experiment. Once the amount of nitrogen (N) was calculated, the amount of ammonium chloride (NH_4Cl) was determined using the ratio of molecular weights. The amounts of other chemicals in the OECD composition were calculated from the following ratios, number in bracket, where the number is the mass of chemical to be added divided by the mass of ammonium chloride needed: KH_2PO_4 (3.4), K_2HPO_4 (8.7), $Na_2HPO_4.2H_2O$ (13.36), NH_4Cl (1), $MgSO_4.7H_2O$ (0.9), $CaCl_2$ (1.1) and $FeCl_3.6H_2O$ (0.01), $MnSO_4.H_2O$ (0.0016), H_3BO_3 (0.0023), $ZnSO_4.7H_2O$ (0.021), $(NH_4)_6Mo_7O_{24}$ (0.0014), $FeCl_3$.EDTA (0.004), and yeast extract (0.006).

RESULTS AND DISCUSSION

Figure 1 shows the natural attenuation of the PAHs in the surface soil of the microcosms. 4-ring PAHs degrade to the maximum percent after 251 days of microcosm operation, while the 2-ring PAHs biodegrade to a substantially lesser

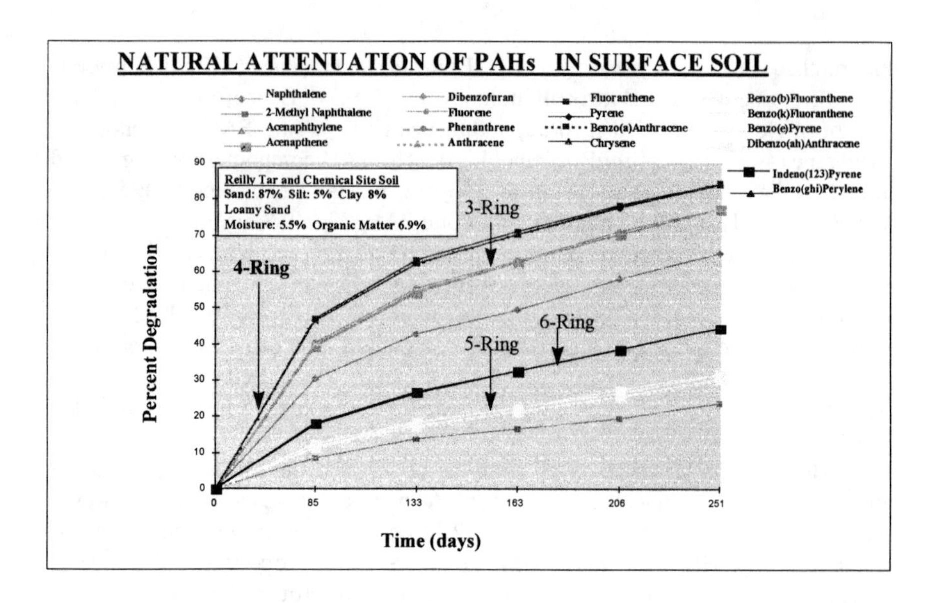

FIGURE 1. Natural Attenuation of PAHs in surface soil (Reilly Tar).

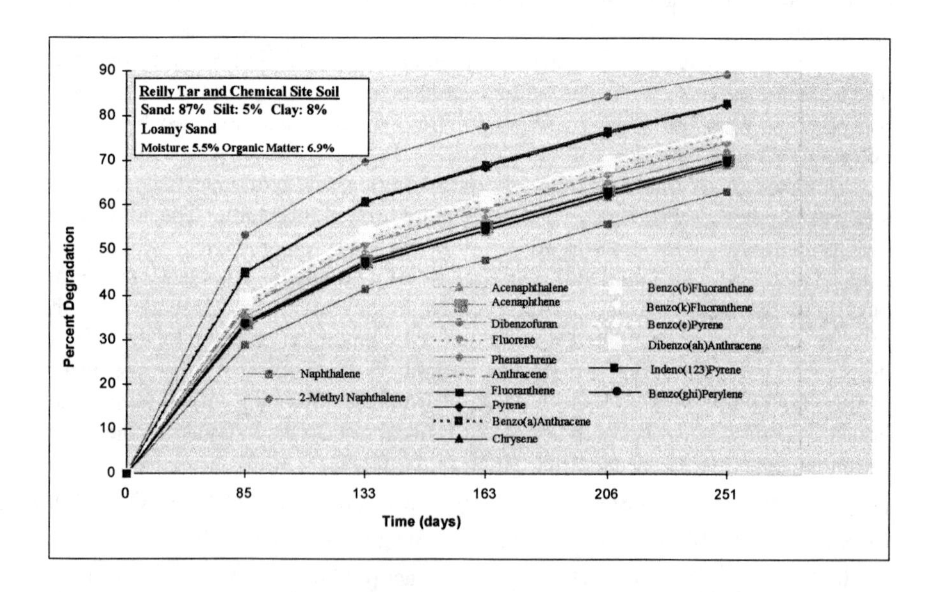

FIGURE 2. Natural Attenuation of PAHs in sub-surface soil (Reilly Tar).

extent. The decreasing order of percent PAH biodegradation as a function of ring size can be summarized as follows:

4-ring > 3-ring > 2-ring > 6-ring > 5-ring

For sub-surface soil (soil at a depth greater than 4 inches) in the microcosm reactors, the percent biodegradation achieved due to natural attenuation, as shown in Figure 2) was different. The decreasing order of the extent of PAH biodegradation, after 251 days, was as follows:

2-Methyl Naphthalene > 4-ring > 5-ring > 3-ring > 6-ring > Naphthalene

Figure 3 shows the percent enhancement of biodegradation for the various PAHs over natural attenuation. The effect of the various additives, nutrients, surfactant, Tween 80 and inducers (Salicyclic acid, Phthalic acid) can be seen in this figure. Inducer, Phthalic acid, produced the largest enhancement for both surface and sub-surface soils. There was a 20% enhancement over natural attenuation due to increase in moisture content from 5.5% to 6.9%. Addition of OECD nutrients produced an enhancement of about 25% over natural attenuation. Natural Reilly Tar soil has a low nitrogen content, and hence addition of ammonia-nitrogen was expected to increase the extent of biodegradation. The enhancement due to the surfactant, Tween 80, was about 30% for both surface and sub-surface soils. The percent enhancement was significantly higher for 2-methyl naphthalene when compared with naphthalene. The differences between surface and sub-surface soils were also significantly higher. The percent enhancements decreased as the PAH ring size increased.

Figure 4 shows the values of the first-order kinetic constant for natural attenuation (Reilly Tar soil at 5.5% moisture content) for both surface and sub-surface soils. The decreasing order of the first-order kinetic constants was as follows:

4-ring > 3-ring > 2-ring > 6-ring > 5-ring

CONCLUSIONS

Several microcosm reactors study was conducted with Reilly Tar soil, contaminated with several PAHs. The PAHs were extracted from the soil samples using a Shaker Vial method and the extracts were analyzed in a gas chromatograph. Natural attenuation rates were obtained for both surface and sub-surface soil samples. The enhancements due to additives - moisture, OECD nutrients, surfactant, Tween 80 and inducers (Salicyclic acid, Phthalic acid) were obtained for both surface and sub-surface soils. Small increases in moisture content (6.9% versus 5.5%) was found to increase the extent of PAH degradation by 20%. Addition of OECD nutrients also increased the extent of PAH biodegradation. The maximum enhancement over natural attenuation was found to occur due to the addition of both OECD nutrients and inducer, Phthalic acid.

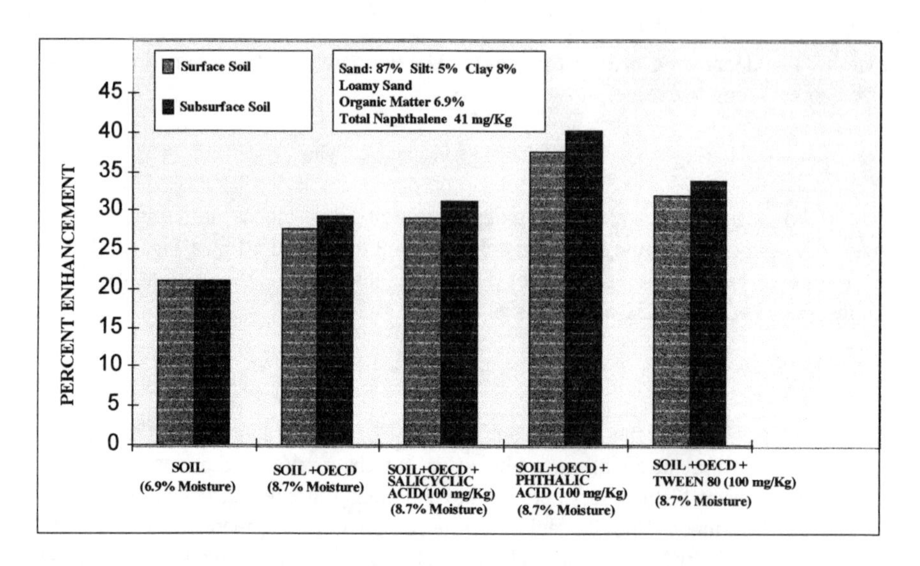

FIGURE 3a. Percent Enhancement of Naphthalene biodegradation in surface and sub-surface soils due to additives.

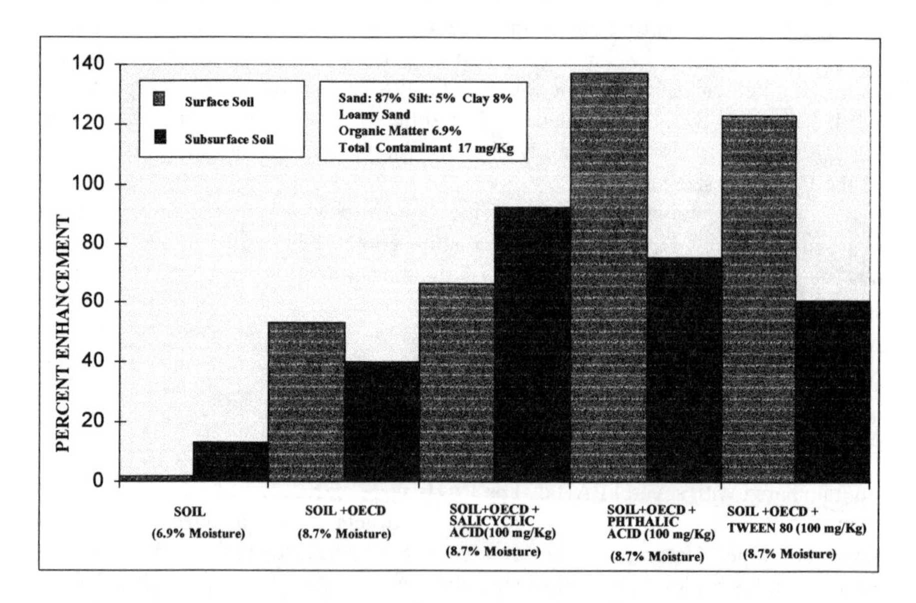

FIGURE 3b. Percent Enhancement of 2-Methyl Naphthalene biodegradation in surface and sub-surface soils due to additives.

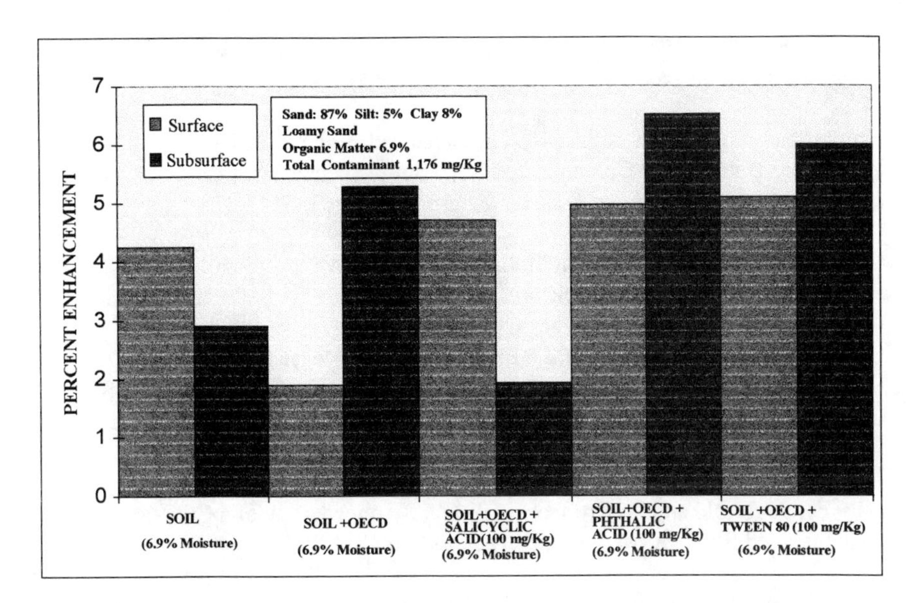

FIGURE 3c. Percent Enhancement of 3-ring PAHs biodegradation in surface and sub-surface soils due to additives.

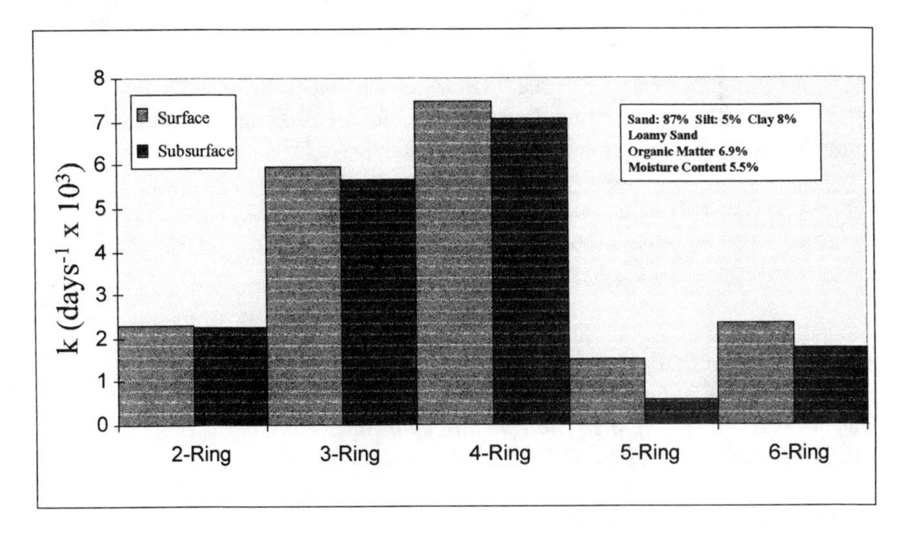

FIGURE 4. First-order kinetic constants for biodegradation of PAHs in surface and sub-surface soils.

REFERENCES

Atlas, R.M. 1981. "Microbial Degradation of Petroleum Hydrocarbons: An Environmental Perspective" *Microbiological Reviews. 45* (1): 180-209.

Brodkorb, T.S. and R.L. Legge. 1992. "Enhanced Biodegradation of Phenanthrene in Oil Tar-Contaminated Soils Supplemented with *Phanerochaete chrysosporium*" *Applied and Environ. Microbiol. 58* (9): 3117-3121.

Cerniglia, C.E. 1984. "Microbial Metabolism of Polycyclic Aromatic Hydrocarbons." *Adv. Appl. Microbiol. 30*: 31-71.

Clover, M.P. 1987. "Studies of the Persistence of Polycyclic Aromatic Hydrocarbons in Unacclimated and Agricultural Soils." M.S. Thesis, Utah State University, Department of Civil and Environmental Engineering, Salt Lake City, UT.

Ellis, B., P. Harold and H. Kronberg. 1991. "Bioremediation of a Creosote Contaminated Site." *Environmental Technology 12*: 447-453.

Fredrickson, J.K., F.J. Brockman, D.J. Workman, S.W. Li and T.O. Stevens. 1991. "Isolation and Characterization of a Subsurface Bacterium Capable of Growth on Toluene, Naphthalene, and Other Aromatic Compounds." *Applied and Environ. Microbiol. 57* (3): 796-803.

Fu, Chunsheng, S. Pfanstiel, Chao Gao, X. Yan, and Rakesh Govind. 1996. "Studies on Contaminant Biodegradation in Slurry, Wafer and Compacted Soil Tube Reactors." *Environ. Sci. and Technol. 30*:743-752.

Gibson, D.T. and V. Subramanian. 1984. "Microbial Degradation of Aromatic Hydrocarbons." In D.T. Gibson (Ed.), *Microbial Degradation of Organic Compounds*, pp. 181-252. Marcel Dekker, New York, NY.

Govind, R., C. Gao, L. Lai, X. Yan, S. Pfanstiel, and H.H. Tabak. 1993. "Development of Methodology for the Determination of Bioavailability and Biodegradation Kinetics of Toxic Organic Pollutant Compounds in Soil." In R.E Hinchee et al. (Eds.), *Applied Biotechnology for Site Remediation,* pp. 229-239, Lewis Publishers, Boca Raton, FL.

Heitkamp, M.A. and C.E. Cerniglia. 1989. "Polycyclic Aromatic Hydrocarbon Degradation by a *Mycobacteriumsp.* in Microcosms Containing Sediment and Water from a Pristine Ecosystem." *Applied and Environm. Microbiol. 55*(8) 1968-1979.

Huang, T., Y. Shan, M. Kuperferle, Q. Zhou, H. Zhu, G.D. Sayles, and C.M. Acheson. 1996. "Extracting PAHs from Soil using a Simple, Effective, Low Cost Shaking Method." *Personal Communication from C.M. Acheson.*

OECD 1981. *OECD Guidelines for Testing of Chemicals,* Section 3, Degradation and Accumulation, Method 301C, Ready Biodegradability: Modified MITI Test (I) adopted May 12, 1981 and Method 302C Inherent Biodegradability: Modified MITI Test (II) adopted May 12, 1981, Director of Information, Organization for Economic Cooperation and Development, Paris, France.

Sims, R.C. and M.R. Overcash. 1983. "Fate of Polynuclear Aromatic Compounds (PNAs) in Soil-Plant Systems. " In F.A. Gunther and J.D. Gunther (Eds.), *Residue Reviews: Residues of Pesticides and Other Contaminants in the Total Environment,* Vol. 88, pp. 1-68. Springer-Verlag, New York, NY.

Tabak, H.H., Rakesh Govind, S. Pfanstiel, Chunsheng Fu, X. Yan, and Chao Gao. 1995. "Protocol Development for Determining Kinetics of In-situ Bioremediation." In R.F. Hinchee et al. (Eds.), *Monitoring and Verification of Bioremediation*, pp. 203-209. Battelle Press, Columbus, OH.

Tabak, H.H., Rakesh Govind, Chunsheng Fu, Qi Song, J. Guo. 1997. "Testing Protocol for Bioavailability, Biokinetics and Treatment End-points." In Bruce C. Alleman and Andrea Leeson (Eds.), *In Situ and On-Site Bioremediation*, pp. 195-203. Battelle Press, Columbus, OH.

Tabak, H.H. and R. Govind. 1997. "Protocol for determining bioavailability and biokinetics of organic pollutants in dispersed, compacted and intact soil systems to enhance *in situ* bioremediation." *J. of Ind. Microbiol & Biotech. 18*: 330-339.

REDUCTIVE DECHLORINATION OF TETRACHLOROETHENE TO ETHENE ADSORBED ON ACTIVATED CARBON

Karin Böckle (DVGW-Technology Center Karlsruhe)
Peter Werner (Technical University of Dresden)

ABSTRACT: A mixed population of bacteria which dechlorinates tetrachloroethylene and less chlorinated metabolites completely to ethene under reductive cometabolic conditions was used to investigate dechlorination kinetics of volatile chlorinated hydrocarbons (VCH) adsorbed on granular activated carbon (GAC). The effect of GAC, VCH-loading and equilibrium concentration on dechlorination kinetics were studied. For all experiments biotransformation kinetics were determined and a mass balance was calculated based on measuring the concentration of the VCH and the chloride released during the biodegradation process. The results showed that a positive effect of GAC on dechlorination activity could be stated. With increasing GAC concentration tetrachloroethylene (PCE) was degraded more completely to ethene even at low equilibrium concentrations. This indicates that the surface of the GAC is of significant importance for the activity of the dechlorinating microorganisms and enhances the biological dechlorination process.

INTRODUCTION

Numerous industries use chlorinated organics such as tetrachloroethene (PCE) and trichloroethene (TCE), as degreasing agents and in textile processing. Under natural conditions partial dechlorination to dichloroethene (DCE) and vinyl chloride (VC) has been reported at many field locations (Major et al., 1991, McCarty et al., 1992). One of the leading technologies for the clean up of VCH contaminated groundwater is GAC-treatment. Due to the low adsorption capacity for the less chlorinated metabolites (Kühn, 1989), the service life of activated carbon filters decreases rapidly and the loaded activated carbon has to be replaced. The used carbon is generally regenerated thermally, which increases the costs of the remediation process.

PCE and its metabolites are known to be biodegradable to ethene under reductive cometabolic conditions (Tandoi et al., 1994, Maymó-Gatell et al., 1995). But up to now only little experience exists on inoculation of GAC with specialized microorganisms and specific stimulation of biodegradation. Thus, the idea of a research project funded by the German Ministry of Research and Development (BMBF) was to remove the adsorbed VCH microbially. In several laboratory studies the dechlorination of PCE was investigated using a mixed enrichment culture which is able to dechlorinate PCE to ethene completely under reductive cometabolic conditions (Böckle & Werner, 1997). The enrichment culture was isolated from a VCH contaminated site and cultivated in the laboratory for a longer period of time (Refae et al., 1992). For these investigations GAC was loaded with PCE, inoculated with the VCH-degrading microorganisms and incubated under reductive cometabolic conditions.

MATERIALS UND METHODS

Loading and Biological Treatment. The GAC (C40/1, particle diameter 1 mm, Carbotech GmbH, Essen) was washed in demineralized water to remove fines and dried at 105°C. Experiments were carried out in Teflon sealed 600-mL bottles, containing 0.2 to 10 g GAC per L. Phosphate buffered mineral medium (pH 7.2) and GAC were sterilized and degased by autoclaving (121°C, 20 min). The bottles were then cooled to room temperature and tetrachloroethylene (Merck, Darmstadt) was added via ethanol as solubilizer. The GAC was preloaded with 6.4 to 320 mg PCE per g. For equilibration the bottles were shaken at 150 rpm for two weeks. After determination of the equilibrium concentrations, the inoculum (material from a chemostate culture that dechlorinates perchloroethylene to ethene), a vitamin solution and sucrose as additional cosubstrate were added. The bottles were then incubated at 30 °C on a rotary shaker (150 rpm) and the biotransformation was determined by measuring the increasing chloride ion and metabolite concentrations in the aqueous phase.

Analytical Methods. VCH and ethene were determined by using a headspace GC system with a flame ionization and an electron capture detector. The samples were filtered through 0.45-µm-pore-size membrane filter to remove fine carbon particles and afterwards transferred to 10-ml-vials. After equilibration at 70°C the analysis was done using a 50-m PONA capillary column (0.2-mm i.d., and a 0.5-µm film). Chloride ions were determined from filtered samples on a Dionex 2010i ion chromatograph and a conductivity detector.

RESULTS AND DISCUSSION

The effect of GAC and equilibrium concentration on dechlorination kinetics was studied in several batch-experiments by using different GAC-concentrations und PCE-loadings.

Figure 1 shows the dechlorination kinetics in the presence of 10 g GAC per L in comparison to an GAC-free control experiment. A positive effect of GAC on the dechlorination activity could be stated. In the presence of activated carbon PCE was degraded more completely to ethene even at low equilibrium concentrations of less than 10 µg PCE per L and less chlorinated metabolites were accumulated. In the absence of GAC no ethene production occured within 50 days. These results suggest that the surface of the GAC is important for the activity of the dechlorinating microorganisms especially the DCE and VC degraders.

Experiments with equal PCE and different GAC addition showed similar kinetics of chloride release, although a limitation by substrate adsorption was expected (figure 2). This indicates that the activity of the dechlorinating microorganisms on the one hand was improved by the adsorbent`s surface, and on the other hand was reduced by the lower substrate availability to a similar extent.

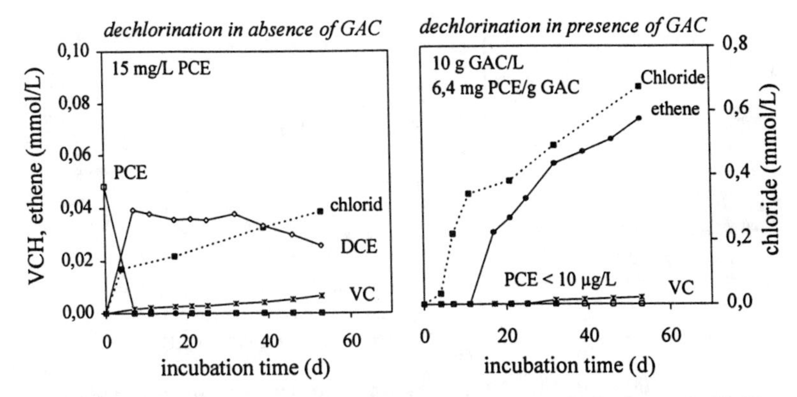

FIGURE 1. Dechlorination kinetics in presence and absence of GAC.

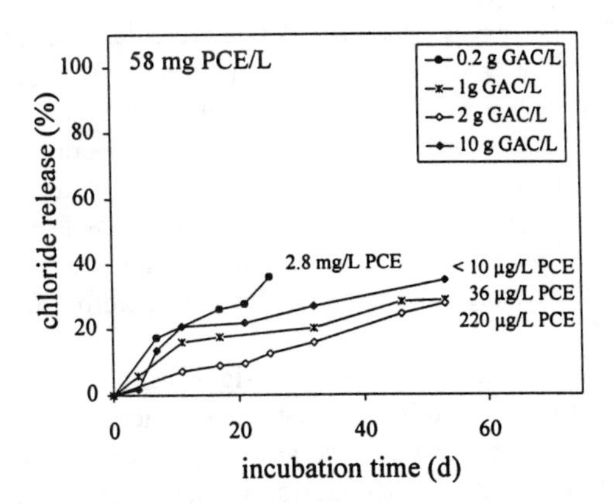

FIGURE 2. Effect of GAC addition on chloride release, equilibrium concentrations < 10 µg/L to 2.8 mg/L PCE.

To verify the influence of substrate adsorption on the dechlorination kinetics further batch experiments were carried out containing equal amounts of GAC (figure 3 a). With increasing PCE loadings and equilibrium concentrations of up to 200 µg PCE per L higher dechlorination rates could be obtained due to the better bioavailability of the adsorbed VCH (figure 3). Even highest PCE-loadings of 320 mg PCE per g GAC did not inhibit the dechlorination process. At equilibrium concentrations > 1 mg PCE per L degradation kinetics were less limited by the desorption rate of the substrate than by the activity of the dechlorinating microorganisms.

Figure 3 b shows the total chloride release and the effect of increasing PCE-loadings on the methane production. With higher substrate availability higher amounts of chloride could be released, whereas the activity of the methanogenic population was already inhibited at low PCE-loadings and equilibrium concentrations > 50 µg PCE per L.

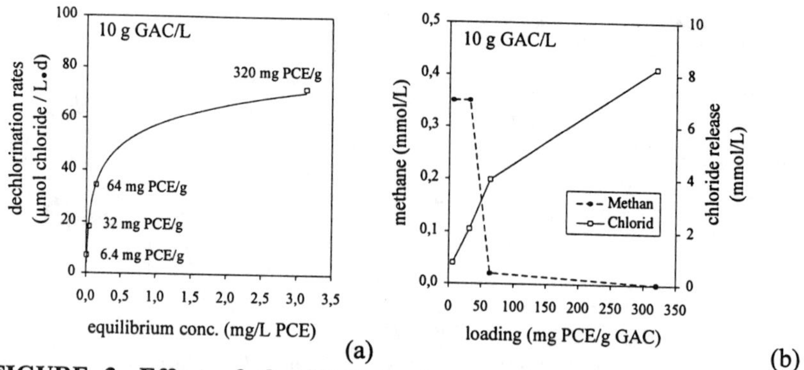

FIGURE 3. Effect of the VCH loading on dechlorination rates (a), chloride release and methane production (b).

In addition the dechlorination kinetics were determined at equal PCE-loading and increasing GAC concentrations (figure 4 a). A linear relationship could be observed between the GAC concentration and the dechlorination rate per L. The dechlorination rates calculated per g GAC were slightly influenced by the GAC concentration. The values increased from 3.8 to 5.6 µmol chloride per g GAC and day with decreasing GAC concentrations from 10 to 1 g GAC per L. This is probably attributed to the increasing solvent/adsorbents rate which enhances the desorption mass transfer and therefore the availability of the substrate.

Methane measurements showed a significant influence of the GAC concentration on the activity of the methanogenic microorganisms (figure 4 b). High methane production was observed at low GAC concentration. This result accords to the high methane concentration of 3.8 mmol per L determined in the GAC-free control. Hence, it is supposed that the methanogenic bacteria are mainly located on the GAC surface, just like the dechlorinating species and are inhibited by the adsorbed PCE.

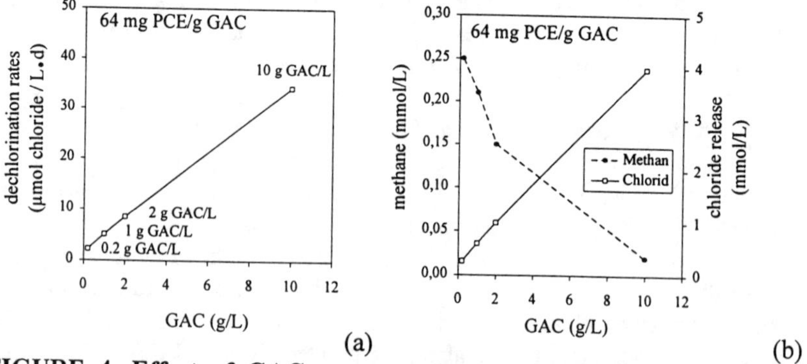

FIGURE 4. Effect of GAC concentration on dechlorination rates (a), chloride release and methane production (b).

CONCLUSIONS

The reductive dechlorination of PCE to ethene could be enhanced by using GAC as adsorbents and surface for the bacterial settlement. Without GAC, PCE was dechlorinated to DCE with high rates, but the kinetic of the further transformation was extremely low and chlorinated metabolites as DCE and VC are recalcitrant over a long period of time. The dechlorination rates were dependent on the PCE-loadings, but even at low substrate concentration high ethene production could be obtained.

Thus, it seems promising to develop a biological GAC system for anaerobic treatment of VCH contaminated water, in which adsorption on GAC and stimulated biodegradation occurs simultaneously.

ACKNOWLEDGEMENTS

This work was carried out at the DVGW-Technology Center, Karlsruhe, Federal Republic of Germany and funded by the German Ministry for Research and Developement (BMBF) (Grant No. 14500736B, August 1993 to June 1996)

REFERENCES

Böckle, K. und P. Werner. 1997. "*Biologischer Abbau der leichtflüchtigen chlorierten Kohlenwasserstoffe (LCKW) Perchlorethylen, Trichlorethylen, cis-1,2-Dichlorethylen und Vinylchlorid*", Abschlußbericht zum Forschungsvorhaben Biologische Regeneration von Aktivkohlen unter Einsatz von Spezialkulturen, FKZ 1450736B.

Böckle, K. and P. Werner. 1997. "*Microbial Regeneration of Activated Carbon Loaded with Volatile Chlorinated Hydrocarbons (VCH)*", in *Fourth International In Situ and On-Site Bioremediation Symposium,* April 1997, New Orleans, Louisiana 4(5):97-101.

Kühn, W. 1989. "Beurteilung und Behandlung von Verschmutzungssituationen mit chlorierten Kohlenwasserstoffen", *DVGW-Schriftenreihe Wasser 205:* 18/1-18/14

Maymó-Gatell, X., V. Tandoi, J. M. Gosset, and S. H. Zinder. 1995. "*Characterization of an H_2--Hutilizing Enrichment Culture that Reductively Dechlorinates Tetrachloroethene to Vinyl Chloride and Ethene in the Absence of Methanogenesis and Acetogenesis*", Appl. Environ. Microbiol. 61:3928-3933.

Major, D. W., W. W. Hodgins, and B. J. Butler. 1991. "*Field and Laboratory Evidence of in Situ Biotransformation of Tetrachloroethylene to Ethene and Ethane at a Chemical Transfer Facility in North Toronto*", in *On-site Bioreclamation.* San Diego, CA: Butterworth-Heinemann Stoneham.

McKay, G. 1996. *Use of Adsorbents for Removal of Pollutants from Wastewater.* CRC Press, Boca Raton.

McCarty, P. L. and J. T. Wilson. 1992. *"Natural Anaerobic Treatment of a TCE Plume, St. Joseph, Michigan, NPL site* ", in *Bioremediation of Hazardous Wastes*, U. S. EPA (EPA/600/R-92/126).

Refae, R. I., O. Meyer, K. Böckle, and P. Werner. 1992. *"Factors Influencing the Biodegradation of Volatile Chlorinated Hydrocarbons and Aromatic Compounds in Soil at the Eppelheim site* ", in *International Symposium of Soil Decontamination Using Biological Processes*, DECHEMA, Frankfurt:797-805.

Tandoi, V., T. D. Distefano, P. A. Bowser, J. M. Gossett, and S. H. Zinder. 1994. *"Reductive Dehalogenation of Chlorinated Ethenes and Halogenated Ethanes by a High-Rate Anaerobic Enrichment Culture* ", Environ. Sci. Technol. 28:973-979

4-TERT- BUTYLPHENOL DEGRADATION IN ANAEROBIC CONDITIONS

*Luca Di Palma, **Carlo Merli**, Elisabetta Petrucci*
(Dip. di Ingegneria Chimica, dei Materiali, delle Materie Prime e Metallurgia
Università di Roma "La Sapienza" via Eudossiana 18, 00184 Roma, Italy)

ABSTRACT: The study involved the simulation of 4-tert-butylphenol degradation in anaerobic conditions. The effect of different toxic concentrations on the removal of a reference substrate was evaluated by means of gas-chromatographic techniques. The simulation was carried out in batch reactors operating at 37°C in which the pH and redox potential were monitored. It was found that, below the methane production inhibitory threshold, the removal of 4-tert-butylphenol involves a first-order kinetic and the constant was calculated.

INTRODUCTION

Phenol compounds are particularly dangerous substances by virtue of their distinguishing characteristics of toxicity, persistence and bio-accumulation. In particular, though not the most toxic compound in the family, 4-tert-butylphenol is remarkably refractory to biodegradation.

The toxicity of 4-tert-butylphenol has been experimentally established in pilot plants simulating the environmental conditions of landfill sites (Di Palma et al., 1995). IC50% (inhibition of biogas production) values of 149.58 mg/l and EC50% (inhibition of methane production) values of 50 mg/l have been obtained.

This study involved simulation of the degradation of 4-tert-butylphenol in anaerobic conditions so as to examine the kinetics and the possible formation of intermediate products. The study was divided into two parts. The first involved establishing the concentrations at which, in an anaerobic environment, the toxic compound would inhibit the production of biogas and methane deriving from the degradation of a synthetic substrate. In the second, we operated within the thresholds established to examine the process and kinetics involved in the degradation of 4-tert-butylphenol. Correlation of the data obtained then make it possible to calculate the kinetic constant characterizing the process.

MATERIALS AND METHODS

Preparation of bacterial culture. The bacteria, derived from a mixed methanogenic consortium obtained from a mixture of landfill percolate effluent from anaerobic digesters, were kept under incubation at 37°C inside airtight vials hermetically sealed of 500 ml in volume (about 350 ml of usable volume).

The biomass was initially acclimated to a nutrient substrate (THBA) consisting of a aqueous solution of THB (Todd Hewitt Broth), the composition of which is given in table 1. 36.4 g of THB and 4.36 g of sodium acetate were dissolved in 1 liter of distilled water. 5 ml of bacterial consortium were inoculated into each vial. This operation was performed in an anaerobic cabinet.

The feed solution. The feed solution (table 2) was prepared by using as organic substrate a solution of sucrose and sodium acetate in distilled water, which acted as a buffer against the acetic acid formed during the digestion process. The organic content of the feed solution was 12,000 mg/l of COD. The relative quantities of sucrose and acetate were calculated on the basis of the oxidation reactions and the buffer effect of the acetate. In order to respect the optimal ratio of nutrients, urea was added to the solution together with mono- and bi-basic phosphates so as to provide the necessary quantities of nitrogen and phosphorus. A multi-vitamin complex enriched with Se, Cu, Zn, Fe, b carotene and methionine was added in order to supply the system with the additional nutrient elements required.

TABLE 1. THB composition

Constituent	mg/l
Beef heart infusion	500
Peptone	20
NaCl	20
$NaHCO_3$	2
Glucose	2
Na_2HPO_4	0.4

TABLE 2. Feed solution

Compound	mg/l
Sucrose	6530
Sodium Acetate	3420
Urea	1065
KH_2PO_4	249.3
K_2HPO_4 3 H_2O	167.3
Multi-vitamine complex	114

Analytical methodology. Methane and carbon dioxide were determined with a TCD gas chromatograph equipped with a Alltech CTR column using helium as carrier gas. In order to calculate the percentage of 4-tert-butylphenol removed, 10 ml samples were taken from the digestion vials. These samples were centrifuged at ambient temperature for 15 minutes and at 14,000 rpm in an Alc 4239 R High Speed Refrigerated Centrifuge. Then, after acidification with sulphuric acid to pH 2 (to avoid the presence of phenol in the aqueous solution) and the addition of 0.25 g of sodium chloride (to avoid the formation of emulsions), extracted by ethyl acetate and analysed with a FID gas-chromatograph equipped with a Supelco SPB 5 column, using helium as carrier gas.

RESULTS AND DISCUSSION

Experimental conditions. An initial series of tests was carried out to tackle the problem of determining the kinetic constant of removal of 4-tert-butylphenol, operating in conditions such as to simulate the behavior of a perfectly mixed continuous system. To this end, 4 digestion reactors (500 ml vials) were prepared, in each of them the level of the toxic compound was kept constant. The concentrations of 4-tert-butylphenol examined were 10, 20 and 40 mg/l, below the toxicity limit of EC20% for the methanogenic biomass, and 80 mg/l, in the field of toxicity between EC50% and EC100%. A fifth vial was used to study the reference solution. The concentrations of 4-tert-butylphenol in solution were measured at weekly intervals and then restored to the initial values by adding a quantity of 4-tert-butylphenol equal to the amount consumed.

Study of concentration over time. Figure 1 shows the results obtained expressed in terms of 4-tert-butylphenol amount removed.

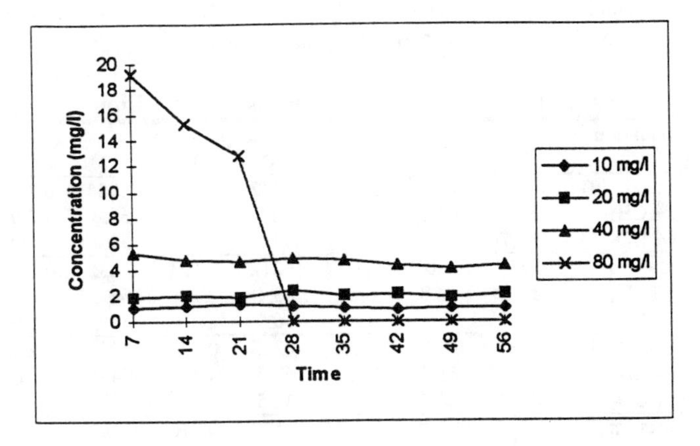

FIGURE 1. Toxic concentration vs. time

The results obtained by processing the data for concentrations below the toxicity threshold lead to the conclusion that the removal of 4-tert-butylphenol, in the experimental conditions, follows a first-order kinetics, or rather that the quantity removed depends on the concentration (T_x) at which the toxic substance is present. After a week of analyses, the variation in concentration (ΔT_x), below the toxicity threshold of EC20%, displays a trend of the following type:

$$\Delta T_x / T_x = K',$$

where, from the data given in the table, K' proves to be equal to 0.11075 d^{-1}, 0.10094 d^{-1} and 0.11665 d^{-1} respectively for the three concentrations studied, and hence to an average of 0.10945 d^{-1} ($\sigma = 0.00794$).

Production of biogas. Table 3 shows the total quantities and the composition of the biogas produced in the digesters in the presence of the toxic compound: biogas production follows a regular trend. Conversely, a constant increase in the production of methane is found in the vials with a concentration of the toxic compound below the threshold of inhibition. This phenomenon is not, however, found when the concentration rises to 80 mg/l, in which conditions there is a sharp decrease in the percentage of methane and a corresponding increase in the percentage of carbon dioxide, thus confirming the inhibiting effect on the methanogenic bacterial population. The data obtained show a peak in gas production corresponding to a value of concentration above which there is a predominance of the toxic effect leading to the inhibition of bacterial activity and hence of biogas production. As the production of biogas, an indicator of the activity of the bacterial consortium, is less influenced than the production of methane by phenol compounds, the conclusion is drawn that the fermentative bacteria have a higher threshold of sensitivity to these compounds.

The measurements of the effluent concentration of organic substances and the percentage of removal with respect to initial concentration provide further confirmation of the satisfactory performance of the process, only small differences being registered between the values of the individual digesters. Redox potential and pH are the variables of the process subjected to daily checking: their values remained in the optimal interval for the anaerobic processes.

TABLE 3. Biogas production and composition

Dig. N.	Parameter	Time (days) 7	14	21	28	35	42	49	56
1	Biogas (ml/l)	0.228	0.251	0.234	0.222	0.171	0.186	0.200	0.222
- 10	% CH_4	56.3	62.1	65.8	62.9	60.7	61.0	62.0	60.0
mg/l	% CO_2	43.7	37.9	34.2	37.1	39.3	39.0	38.0	40.0
2	Biogas (ml/l)	0.177	0.201	0.196	0.212	0.221	0.235	0.225	0.229
- 20	% CH_4	57.1	62.8	65.4	63.1	62.6	62.0	62.1	61.9
mg/l	% CO_2	42.9	37.2	34.6	36.9	37.4	38.0	37.9	38.1
3	Biogas (ml/l)	0.131	0.171	0.220	0.246	0.220	0.228	0.223	0.234
- 40	% CH_4	59.3	65.2	67.2	67.5	64.5	64.0	65.0	66.0
mg/l	% CO_2	40.7	34.8	32.1	32.5	35.5	36.0	35.0	34.0
4	Biogas (ml/l)	0,120	0.114	0.154	0.177	0.710	0.180	0.186	0.171
- 80	% CH_4	42.6	49.2	56.1	56.8	45.3	43.3	42.5	41.8
mg/l	% CO_2	57.4	50.8	43.9	43.2	54.7	56.7	57.5	58.2
5	Biogas (ml/l)	0.224	0.241	0.233	0.235	0.225	0.235	0.231	0.235
blank	% CH_4	60.2	62.1	61.3	63.1	63.1	61.7	62.8	62.6
mg/l	% CO_2	39.8	37.9	38.7	36.9	36.9	38.3	37.2	37.4

Table 4 gives the average weekly values for the concentration of organic matter registered on exiting the digesters during the experiment.

TABLE 4. Effluent Substrate concentration, as mg/l COD

Time (days)	Digester blank	10 mg/l	20 mg/l	40 mg/l	80 mg/l
7	3700	3700	3700	3700	3700
14	3450	3535	3630	3630	3840
21	3410	3400	3500	3450	3400
28	3360	3400	3500	3500	3600
35	3300	3200	3300	3300	3400
42	3150	3300	3300	3400	3500
49	3050	3100	3000	3450	3450
56	3000	3000	3100	3300	3400

The analysis of the data shows that the toxic effect of 4-tert-butylphenol is negligible below the toxicity threshold of EC20%. The kinetics of the degradation of the reference substrate is only slightly disturbed by the presence of the toxic compound. As regards the concentrations of total and volatile suspended solids and the inorganic fraction, measured on a filtered sample, once the process reached stationary conditions these stabilized around values of 750 ± 20 mg/l, 400 ± 15 mg/l and about 350 ± 15 mg/l respectively, without undergoing significant variations in the course of the experiment.

Study of the kinetics of removal of 4-tert-butylphenol. Based on the same initial concentrations of toxic compound, weekly gas-chromatographic analyses were performed in a second series of tests to establish the effective disappearance of the compound under examination. The tests were performed in the same conditions as the previous series, without making up the quantities of 4-tert-butylphenol removed. Table 5 shows the results of the experiment, expressed in terms of the concentration of 4-tert-butylphenol in the individual digesters.

TABLE 5. 4-tert-butylphenol concentration

Time (days)	4-tert-butylphenol concentration in the digesters (mg/l)			
0	10.00	20.00	40.00	80.00
7	8.98	17.89	36.95	60.48
14	8.00	15.17	30.86	44.03
21	7.29	14.45	29.05	28.73
28	6.56	12.63	24.02	16.63
35	4.96	10.69	22.41	16.63
42	4.00	9.03	21.29	16.63
49	3.94	8.41	17.28	16.63
56	3.54	7.92	15.43	16.63

A slow but progressive breakdown of the toxic compound was registered in the digesters with 10, 20 and 40 mg/l, while in the digester with 80 mg/l an initial decease in concentration was followed by stabilization around a value of 16.63 mg/l, which remained the same until the end of the tests. The results shows that in order to study the degradation kinetics of 4-tert-butylphenol it is necessary to consider only the results obtained in reactors with concentrations of the toxic compound below the threshold of inhibition for biogas production, where effective degradation of the compound under examination was registered. The trend registered for the decrease in the concentration of the toxic compound over time, as shown in figure 4, suggested a first-order kinetics of degradation, for which:

$$-\ln\frac{C_A}{C_{A0}} = k \times t\,,$$

where C_{A0} is the initial concentration of 4-tert-butylphenol and C_A the concentration present in the reactor at time t.

Presentation of the experimental data in semi-logarithmic form, provides the results illustrated in figure 2, which confirm the hypothesis of first-order kinetics: the regression performed by means of the least squares method gives a value for of the kinetic constant k of 0.0181 d^{-1}. When compared with the value obtained in the continuous simulation, this value proves to be of the same order of magnitude, albeit slightly higher. If these differences are attributed solely to differences in operating conditions, this provides further confirmation that the hypothesis of first-order kinetics is in fact correct.

CONCLUSIONS

The results obtained during the tests confirm that in order to study the kinetics of the degradation of 4-tert-butylphenol it is necessary to consider solely the results obtained in reactors with concentrations of the toxic compound below the threshold of inhibition for biogas production, where effective degradation of the compound under examination was registered.

The data confirm that the removal of 4-tert-butylphenol for concentrations within the inhibitory threshold of 20% for biogas production takes place in terms of a first-order kinetics only slightly influenced by the quantity of the toxic compound present. In these conditions it was also established that the removal of the reference substrate is not appreciable affected by the toxic effect of the 4-tert-butylphenol.

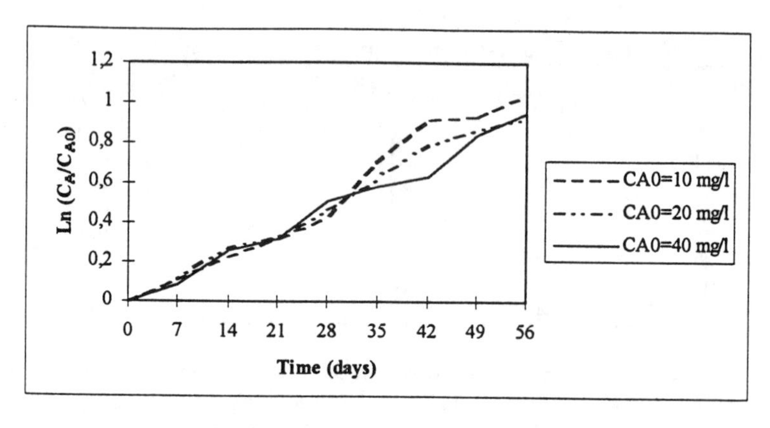

FIGURE 2 - Kinetic constant evaluation

CONTRIBUTIONS

The authors declare that an equal contribution was made by each as regards shaping the work, carrying out the experimental testing, and drafting the paper.

REFERENCES

Di Palma, L., Merli C., and R. Palmieri. 1995. "Confirmation Tests Using Column Percolation on Methanogenic Biomass" In: R.E. Hinchee, C.M. Vogel, F.J. Brockman (Ed.), *Microbial processes for bioremediation*, pp. 335-342. Battelle Press, Columbus, OH.

Juteau, P., Beaudet R., McSween G., Lépine F., Milot S., and G. Bisaillon. 1995. "Anaerobic Biodegradation of Pentachlorophenol by a Methanogenic Consortium." *Appl. Microbiol. Biotechnol. 44*: 218-224.

Kennes, C., Wu W.M., Bhatnagar L., and J. G. Zeikus. 1996. "Anaerobic Dechlorination and Mineralization of Pentachlorophenol and 2,4,6-Trichlorophenol by Methanogenic Pentachlorophenol-Degrading Granules." *Appl. Microbiol. Biotechnol. 44*: 801-806.

O'Connor, O.A., and L.Y. Young. 1996. "Effect of Six Different Functional Groups and Their Position on the Bacterial Metabolism of Monosubstituted Phenols under Anaerobic Digestion." *Environ. Sci. Technol. 30*(5):1419-1428.

Satsangee, R., and P. Ghosh. 1990. "Anaerobic Degradation of Phenol Using an Acclimated Mixed Culture." *Appl. Microbiol. Biotechnol 34*: 27-130.

MEDIUM OPTIMIZATION FOR THE CULTIVATION OF BACTERIA REDUCTIVELY DECHLORINATING TRICHLOROBENZENES

Lorenz Adrian , Ulrich Szewzyk and Helmut Görisch
(Technische Universität Berlin, Germany)

ABSTRACT: Inoculum from a fluidized bed bioreactor, reductively dechlorinating trichlorobenzenes, was transferred to a sulfide reduced synthetic medium. Bromoethanesulfonic acid was used to eliminate the dominant methanogenic population from the consortium resulting in an increased dechlorinating activity. The use of titanium(III) citrate as a reducing agent instead of 1 mM of sulfide led to increased dechlorination. This was partly due to inhibition of dechlorination by sulfide. The addition of cyanocobalamin or the addition of a tungstate-selenite solution to the medium had a positive effect on the reductive dechlorination. A vitamin solution consisting of six different vitamins, however, inhibited the dechlorinating activity.

INTRODUCTION

Chlorobenzenes are widely used chemicals and can be detected within the air, in waste water, ground water, sediments, soil as well as in animal and human bodies (Koch, 1991). They are highly persistent in aerobic and anaerobic environments and enrich within the food chain (Beyer, 1996; Oliver and Nicol, 1982). Chlorobenzenes are toxic for *Bacteria* and *Eucarya* (Holliger et al., 1992; Koch, 1991). Nevertheless, many laboratory studies have shown that reductive dechlorination of chlorobenzenes is a wide spread property that can be found in bacterial consortia from a variety of locations (Adrian et al., 1996; Beurskens et al., 1994; Bosma et al., 1988; Fathepure et al. 1988; Fathepure and Vogel, 1991; Holliger et al., 1992, Middeldorp et al., 1997; Nowak et al., 1996; Ramanand et al., 1993). Studies with other chlorinated aromatics (e.g. Dolfing, 1990) and thermodynamic considerations (Dolfing and Harrison, 1993) have led to the idea that the dechlorination of chlorobenzenes could be linked to bacterial growth. However, no pure culture of chlorobenzene dechlorinating bacteria is currently available to prove this hypothesis.

Objective. The objective of our study was the characterization of a trichlorobenzene dechlorinating mixed culture in a synthetic medium and the determination of the effects of different media components on the dechlorinating activity. The study will allow further attempts to define the physiological role and the phylogenetic position of chlorobenzene dechlorinating bacteria. Furthermore, results from this laboratory study will be used to optimize the performance of a remediation plant where industrial waste water is treated in a pilot scale.

MATERIAL AND METHODS

General. Ti(III) citrate was prepared from Ti(III) chloride and Na_2CO_3 as described by Zehnder and Wuhrmann (1976). Ti(III) chloride was of technical, all other chemicals were of analytical grade. Gases were of 99.999% (N_2) or 99.8% (CO_2) purity and only used after minimizing oxygen contamination by a reduction column (Ochs, Göttingen, Germany).

In situ hybridization was done according to Manz et al., 1998, using one of the following oligonucleotide probes: i) EUB338, specific for *Bacteria.* ii) ARCH915, specific for *Archaea.* iii) EURY499, specific for *Euryarchaeota.* iv) non-EUB338, complementary to EUB338 which was used to test for nonspecific binding (see Amann et al., 1995, for sequences or references).

Inoculum and culture conditions. A polyurethane foam body covered with biomass from a fluidized bed bioreactor was used as the inoculum. The medium was a synthetic, bicarbonate buffered mineral medium, which was reduced prior to inoculation with 1 mM sulfide, or 0.8 mM Ti(III). The basal medium contained minerals, 30 mM bicarbonate buffer, a solution of six vitamins (*p*-aminobenzoate, biotin, nicotinic acid, pantothenic acid, pyridoxine, thiamine), a cyanocobalamin (vitamin B_{12}) solution, a tungstate-selenite solution, resazurin as a redox indicator and trichlorobenzenes. Trichlorobenzene concentrations were 15 μM for 1,2,3-trichlorobenzene and 15 μM for 1,2,4-trichlorobenzene. As a source of carbon and energy 10 mM pyruvate or succinate was added. The exact composition and preparation of this medium have recently been described elsewhere (Adrian et al.,

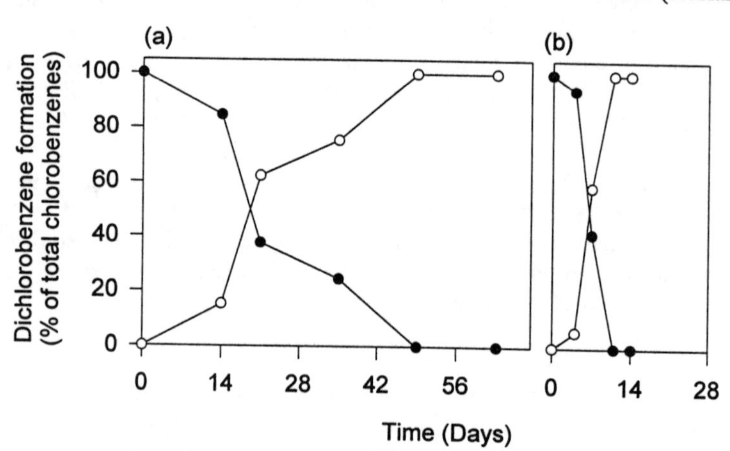

FIGURE 1: Dechlorination of trichlorobenzenes to dichlorobenzenes. (a) In a methanogenic culture transferred from a fluidized bed bioreactor to sulfide reduced medium. (b) In a non-methanogenic culture, established from (a) in a Ti(III) citrate reduced medium. Both cultures contained 10 mM of pyruvate.

Symbols: ●, trichlorobenzenes; ○, dichlorobenzenes.

1998). Cultivation was done in 60 ml serum bottles containing 30 ml of medium. The gas phase was N_2/CO_2 (80%:20%). Each vial was sealed by a Teflon lined butyl rubber septum and inoculated by injection of 0.5 ml inoculum according to Adrian et al. (1998).

Analytical procedures. Chlorobenzenes in aqueous solution were determined by capillary gas chromatography and flame ionization detection after extraction with hexane (Adrian et al., 1998). Methane was analyzed by gas chromatography using a packed column and thermal conductivity detection (Adrian et al. 1998).

RESULTS AND DISCUSSION

From a fluidized bed bioreactor an inoculum was transferred to a bicarbonate buffered, synthetic, sulfide reduced medium containing 1,2,3- and 1,2,4-trichlorobenzene. As a source of carbon and energy 10 mM of pyruvate was added. Initially, the bacterial consortium reductively dechlorinated a mixture of 15 μM of 1,2,3- and 15 μM of 1,2,4-trichlorobenzene completely to dichlorobenzenes within 7 weeks of incubation (Figure 1a).

The consortium contained a high percentage of methanogenic bacteria as determined by in situ hybridization (Figure 2). With the addition of 4 mM bromoethanesulfonic acid (BES), an inhibitor of methanogenic bacteria, an increased dechlorinating activity could be detected within the cultures. The absence of methane as a product within the gas phase, the absence of autofluorescence after excitation at 420 nm, and the absence of hybridization signals after in situ hybridization with the fluorescent labeled oligonucleotide probes EURY499 or ARCH915 indicated that dechlorination was independent from methanogenesis. The transfer of the culture to a Ti(III) citrate reduced medium lead to a stable dechlorinating consortium, in which dechlorination was complete after 10 to 14 days of incubation (Figure 1b). This increase of dechlorinating activity could be caused by the enrichment of the dechlorinating bacteria, adaptation of the microbial consortium to the experimental conditions, an inhibition of trichlorobenzene dechlorination by sulfide when it was used as a reducing agent, a positive effect of the low redox potential of about -500 mV generated by Ti(III) (Zehnder and Wuhrmann, 1976), and/or a positive effect of citrate on the dechlorination activity. An inhibition of reductive dechlorination by sulfide was demonstrated by comparing three cultures containing 1 mM of sulfide with three cultures not

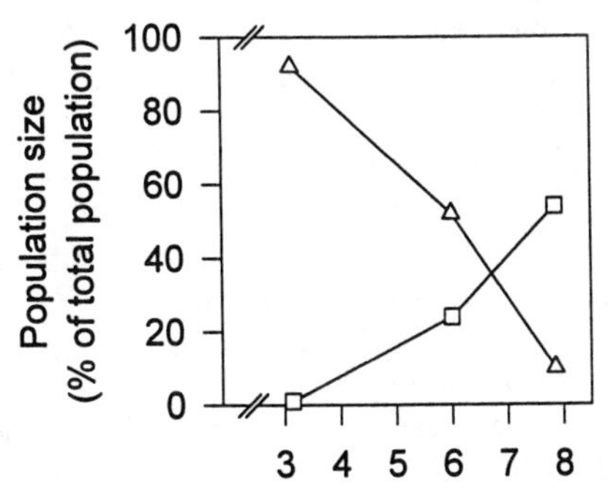

FIGURE 2: Determination of the relative amount of methanogenic bacteria within the orginal culture by in situ hybridization techniques. Symbols: Δ, *Bacteria* detected with probe EUB338; □, *Archaea*, detected with probe EURY499.

containing sulfide. All cultures contained succinate as a source of carbon and energy and were reduced by Ti(III). Without sulfide the extent of dechlorination after 7 days of incubation was 95.8%, ± 5.1. The extend of trichlorobenzene dechlorination in cultures containing 1 mM sulfide was only 3.7% ± 6.4. This result contributes to the understanding of the inhibition of reductive dechlorination in nature as sulfide concentrations of 1 mM or above occur often in anaerobic sediments.

Table 1. Dichlorobenzene formation after 7 days of incubation in cultures containing 10 mM pyruvate, Ti(III) citrate as a reducing agent and different combinations of supplement solutions.

Additions			Dichlorobenzene formation[a]
solution of six vitamins	cyano-cobalamin	tugsten-selenit	
-	+	+	61,2
-	+	-	47,4
-	-	+	44,8
+	+	+	41,3
+	+	-	33,0
-	-	-	32,4
+	-	+	27,1
+	-	-	21,9
+	+	+	0[b]

Symbols: +, present; -, absent

[a] relative amount of dichlorobenzenes formed of total chlorobenzenes detected in the medium

[b] negative control with autoclaved inoculum

Also the effects of other components of the medium on trichlorobenzene dechlorination were evaluated. Table 1 shows the extent of dechlorination in cultures supplemented with different combinations of a vitamin solution, cyanocobalamin solution, and tungstate-selenite solution after 7 days of incubation. The highest dechlorinating activity was found in cultures containing cyanocobalamin and tungstate-selenite but not containing the six-vitamins. A high extend of dechlorination was also determined in cultures containing only cyanocobalamin or tungstate-selenite or containing all three solutions. However, the presence of the six-vitamin solution always led to a lower extend of dechlorination compared to the cultures where the six vitamins were not present. A hypothesis explaining the strong positive effect of cyanocobalamin would be that the dechlorinating bacteria use cyanocobalamin as a cofactor for enzymes within a respiratory electron chain as shown for tetrachloroethene with *Dehalospirillum multivorans* (Neumann et al., 1995). However, further experiments are necessary with a pure culture of trichlorobenzene dechlorinating bacteria to prove this hypothesis. The possibility of an abiotic reduction of clorobenzenes by a strong reductant and cyanocobalamin as demonstrated by Assaf-Anid et al. (1992) was excluded for our cultivation conditions by preparing control cultures that were inoculated with autoclaved inoculum (Table 1). In these cultures no dichlorobenzene formation was detected.

In future work, the effects of different trichlorobenzene concentrations will have to be evaluated. Trichlorobenzene concentrations above 30 to 40 μM inhibit the dechlorinating activity (Adrian et al., 1998; Holliger et al., 1992). However, if the dechlorinating bacteria conserve energy by reductive dechlorination, trichlorobenzenes would promote growth and could be an essential component of the medium. To further understand the ecological role of dechlorinating bacteria in nature or in sewage sludge more information is needed about physiological properties of trichlorobenzene dechlorinating bacteria, about their survival potential in environments containing very low concentrations of chlorinated benzenes, and about their distribution in nature.

ACKNOWLEDGMENTS

The authors thank P. Wendler for expert technical assistance. This work was supported by the Deutsche Forschungsgemeinschaft, Sonderforschungsbereich 193, Biological Treatment of Industrial Wastewaters.

LITERATURE

Adrian, L., W. Manz, U. Szewzyk, and H. Görisch. 1996. "Etablierung einer stabilen, Trichlorbenzol dechlorierenden Mischkultur und deren partielle Populationsbeschreibung mit Hilfe rRNA-gerichteter Oligonukleotidsonden." *gwf Wasser-Abwasser* 137: 612-618.

Adrian, L., W. Manz, U. Szewzyk, and H. Görisch. 1998. "Physiological Characterization of a Bacterial Consortium Reductively Dechlorinating 1,2,3- and 1,2,4-Trichlorobenzene." *Appl. Environ. Microbiol.* 64: 496-503.

Amann, R. I., W. Ludwig, and K.-H. Schleifer. 1995. "Phylogenetic identification and in situ detection of individual microbial cells without cultivation." *Microbiol. Rev.* 59: 143-169.

Assaf-Anid, N., L. Nies, and T. M. Vogel. 1992. "Reductive dechlorination of a polychlorinated biphenyl congener and hexachlorobenzene by vitamin B_{12}." *Appl. Environ. Microbiol.* 58: 1057-1060.

Beurskens, J. E. M., C. G. C. Dekker, H. van den Heuvel, M. Swart, J. de Wolf, and J. Dolfing. 1994. "Dechlorination of chlorinated benzenes by an anaerobic microbial consortium that selectively mediates the thermodynamic most favorable reactions." *Environ. Sci. Technol.* 28: 701-706.

Beyer, W. N. 1996. "Accumulation of chlorinated benzenes in earthworms." *Bull. Environ. Cont. Tox.* 57: 729-736.

Bosma, T. N. P., J. R. van der Meer, G. Schraa, M. E. Tros, and A. J. B. Zehnder. 1988. "Reductive dechlorination of all trichloro- and dichlorobenzene isomers." *FEMS Microbiol. Ecol.* 53: 223-229.

Dolfing, J. 1990. "Reductive dechlorination of 3-chlorobenzoate is coupled to ATP production and growth in an anaerobic bacterium, strain DCB-1." *Arch. Microbiol.* 153: 264-266.

Dolfing, J., and B. K. Harrison. 1993. "Redox and reduction potentials as parameters to predict the degradation pathway of chlorinated benzenes in anaerobic environments." *FEMS Microbiol. Ecol.* 13: 23-29.

Fathepure, B. Z., J. M. Tiedje, and S. A. Boyd. 1988. "Reductive dechlorination of hexachlorobenzene to tri- and dichlorobenzenes in anaerobic sewage sludge." *Appl. Environ. Microbiol.* 54: 327-330.

Fathepure, B. Z., and T. M. Vogel. 1991. "Complete degradation of polychlorinated hydrocarbons by a two-stage biofilm reactor." *Appl. Environ. Microbiol.* 57: 3418-3422.

Holliger, C., G. Schraa, A. J. M. Stams, and A. J. B. Zehnder. 1992. "Enrichment and properties of an anaerobic mixed culture reductively dechlorinating 1,2,3-trichlorobenzene to 1,3-dichlorobenzene." *Appl. Environ. Microbiol.* 58: 1636-1644.

Koch, R. 1991. "*Umweltchemikalien.*" 2nd ed. Verlag Chemie, Weinheim, Germany.

Manz, W., M. Eisenbrecher, T. R. Neu, and U. Szewzyk. 1998. "Abundance and spatial organization of gram-negative sulfate-reducing bacteria in activated sludge investigated by in situ probing with specific 16S rRNA targeted oligonucleotides." *FEMS Microbiol. Ecol.* 25: 43-61.

Middeldorp, P. J. M., J. de Wolf, A. J. B. Zehnder, and G. Schraa. 1997. "Enrichment and properties of a 1,2,4-trichlorobenzene-dechlorinating methanogenic microbial consortium." *Appl. Environ. Microbiol.* 63: 1225-1229.

Neumann, A., G. Wohlfarth, and G. Diekert. 1995. "Properties of tetrachloroethene and trichloroethene dehalogenase of *Dehalospirillum multivorans*." *Arch. Microbiol.* 163: 276-281.

Nowak, J., N.-H. Kirsch, W. Hegemann, and H.-J. Stan. 1996. "Total reductive dechlorination of chlorobenzenes to benzene by a methanogenic mixed culture isolated from Saale river sediment." *Appl. Microbiol. Biotechnol.* 45: 700-709.

Oliver, B. G., and K. D. Nicol. 1982. "Chlorobenzenes in sediments, water, and selected fish from lakes Superior, Huron, Erie, and Ontario." *Environ. Sci. Technol.* 16: 532-536.

Ramanand, K., M. T. Balba, and J. Duffy. 1993. "Reductive dehalogenation of chlorinated benzenes and toluenes under methanogenic conditions." *Appl. Environ. Microbiol.* 59: 3266-3272.

Zehnder, A. J. B., and K. Wuhrmann. 1976. "Titanium(III) citrate as a nontoxic oxidation-reduction buffering system for the culture of obligate anaerobes." *Science* 194: 1165-1166.

NUMERICAL INVESTIGATION OF FACTORS PROMOTING TCE DEGRADATION IN POROUS MEDIA

Naresh Singhal (University of Auckland, Auckland, New Zealand)
Peter Jaffé (Princeton University, Princeton, USA)
Walter Maier (University of Minnesota, Minneapolis, USA)

ABSTRACT: A mathematical model for anaerobic degradation of trichloroethylene (TCE) in soils including gas production, separate phase gas accumulation, anaerobic TCE degradation, growth substrate utilization, and bacterial growth is briefly developed here. The model is used to evaluate the effect of gas production, bacterial distribution, substrate distribution, and TCE volatilization on the total degradation of TCE in soils. It is shown that volatilization is an important mechanism by which TCE may be removed from contaminated water and that excessive gas production can result in decreasing the total amount of TCE degraded in soils. Simulations also show that achieving a uniform active biomass distribution along the entire column can result in increasing the total amount of TCE degraded.

INTRODUCTION

A model for anaerobic degradation of volatile chlorinated compounds in soils or a continuous flow bioreactor must account for substrate uptake, biomass development, biomass transport, gas formation, partitioning of volatile compounds into the gas phase, and dechlorination of chlorinated chemicals. A large number of models have been developed for packed beds, especially in the chemical engineering literature. These models largely ignore the gas phase if the gas flow is small relative to the liquid flow and the system is modeled as a two-phase, solid and liquid, reactor (Hermanowicz and Ganczarczyk, 1985). During anaerobic (methanogenic) degradation of TCE in soils, growth substrate introduced to promote TCE dehalogenation can result in production of gases with limited solubility in water (e.g., methane) and lead to formation of a separate gas phase. Under conditions of stimulated anaerobic activity gas phase is likely to be an important transport mechanism for volatile compounds (such as, TCE) and excessive gas production can lead to significant stripping of TCE from the liquid phase.

The effect of formation of a separate gas phase on the degradation of volatile compounds needs to be studied in order to develop an understanding of the conditions which will maximize the degradation of the targeted volatile contaminant. Singhal (1995) has shown that stimulation of bacterial activity by large additions of growth substrates can result in decreasing the total amount of volatile contaminants degraded, if gas production becomes large. Since the relationships between bacterial buildup in soils, substrate concentration, bacterial activity, gas production and separate phase formation, volatilization, and TCE degradation are complex and non-linear, it is likely that TCE degradation can be influenced by manipulating the distribution of biomass, substrate, and gas production in soils. In this paper a mathematical model for the transport of organic substrate, TCE, biomass, and gas in saturated soils is used to evaluate the effect of bacterial buildup, substrate concentration, gas production, and volatilization on the total amount of TCE degradation and determine conditions under which TCE degradation can be maximized.

MODEL DEVELOPMENT

In this study, the tanks-in-series approach (Hermanowicz and Ganczarczyk, 1985) was used to describe the transport of organic substrate, TCE, biomass, and gas in packed columns. According to the methodology, the column is broken into a number of sections and each section is treated as a completely mixed reactor with uniform concentrations of the various components within it. All reactors receive one inflow stream and originate one outflow stream.

The substrate concentration in the effluent is the same as that within the cell. The biomass in a reactor is divided into attached and suspended biomass. The biomass in the effluent is assumed to be the same as the biomass suspended in the reactor fluid. Additionally, the biomass is composed of active and non-active fractions. Active biomass is the part of total biomass capable of transforming TCE and its daughter products, while the non-active biomass is not able to bring about such transformations. The active biomass is approximated as the biomass utilizing the growth substrate, while the remaining biomass is categorized as non-active biomass. In the equations below, the attached biomass is expressed as *equivalent liquid phase* biomass (i.e., attached biomass concentration in mg/kg is multiplied by $\rho_b/S_w\theta$, where ρ_b is the bulk density of soil, S_w is the water saturation and θ is the soil porosity). The mass balance equations for substrate, bacterial, and dissolved and gas phase TCE transport in columns at steady state, are:

Substrate: $$V_w \frac{dS}{dt} = Q_w S_{in} - Q_w S - \frac{V_L \mu X_v}{Y_{x/s}} = 0 \tag{1}$$

Active Biomass: $$V_w \frac{dX_v}{dt} = Q_w X_v^{in} - Q_w f_v X_v + V_w \mu X_v - V_w b X_v = 0 \tag{2}$$

Non-active Biomass: $$V_w \frac{dX_d}{dt} = Q_w X_d^{in} - Q_w f_d X_d + V_w b X_v - V_w b X_d = 0 \tag{3}$$

Dissolved TCE: $$V_w \frac{dTCE_w}{dt} = Q_w TCE_w^{in} - Q_w TCE_w - V_w r_{TCE_w} + I_g = 0 \tag{4}$$

Gas Phase TCE: $$V_g \frac{dTCE_g}{dt} = Q_g^{in} TCE_g^{in} - (Q_g^{in} + \Delta Q_g) TCE_g - I_g = 0 \tag{5}$$

Where μ = biomass specific growth rate
b = biomass decay rate

f_v, f_d = fraction of active and non-active cells suspended in water
I_g = rate of TCE transferred from the gas phase to the liquid phase
Q_g^{in}, ΔQ_g = influent gas flow rate and the gas produced in the reactor
Q_w = water flow rate
r_{TCE_w} = rate of TCE degraded in the water phase
S, S_{in}= substrate concentration in the reactor and in the reactor influent
TCE_g, TCE_g^{in} = TCE concentration in gas within and that entering a reactor
TCE_w, TCE_w^{in} = TCE concentration in the water within and that entering a reactor
X_d, X_d^{in} = non-active biomass concentration in the reactor and in the reactor influent
X_v, X_v^{in} = active biomass concentration in the reactor and in the reactor influent
V_g, V_w = volume of gas- and liquid- phase in the reactor
$Y_{x/s}$ = yield for biomass formation from substrate.

Degradation of growth substrate was described using a first-order equation. TCE degradation was expressed as a function of substrate degradation, and TCE distribution between water and gas phase was assumed to be at equilibrium. The gas phase is assumed to consist primarily of methane and carbon dioxide. Separate phase gas formation in soils was assumed to be triggered by exceedance of the solubility limit by the dissolved methane in a reactor. Once the separate gas phase formed, carbon dioxide was assumed to distribute between the gas phase and the dissolved carbonate species (i.e., carbonic acid, and carbonate and bicarbonate ions). The volume of pore space occupied by the separate gas phase was estimated using Darcy's Law and the assumption that the capillary pressure was constant along the entire column. Details of the model calibration and verification are presented in Singhal (1995) and Singhal et al. (1997).

FACTORS GOVERNING TCE DEGRADATION IN COLUMNS

Experimental data (adapted from Singhal, 1995) presented in Figure 1 shows that total TCE degraded in sand-filled columns increases with the influent substrate concentration; however, large influent substrate concentrations can lead to reduction in the total TCE degradation. Most of this decrease in TCE degradation was a result of decrease in the conversion of the intermediate TCE daughter products to a non-chlorinated end product; the decrease in the conversion of TCE to an intermediate daughter product was only marginal in the present experiment. Also of some significance is the observation that for low substrate concentrations, TCE degradation was not significantly promoted when the substrate concentration was increased from 500 to 820 mg/L of COD.

TCE degradation in packed columns is governed by the following factors: activity and concentration of the attached biomass, and the concentration of dissolved TCE remaining in the liquid phase. Attached biomass concentration and

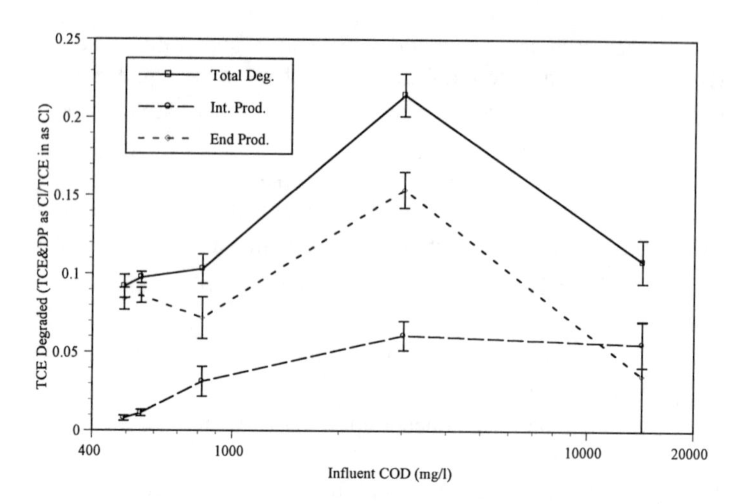

FIGURE 1. Effect of influent substrate concentration on TCE degradation.

activity is a function of substrate degradation, while dissolved TCE degradation is a function of TCE degradation and gas production in the columns (Singhal et al., 1997). Typically, packed are operated by injecting all the substrate at the column inlet, which has the following disadvantages: gas production occurs largely in the column inlet section; volatilization losses are largest in region with the highest active biomass concentration; and, downgradient sections of column with reduced bacterial concentration and activity continue to suffer from high gas flows and volatilization losses. Thus, TCE degradation can be promoted by: minimizing volatilization losses by distributing gas production over the whole column; making bacteria active in a larger fraction of the column; substrate degradation at low substrate concentration; and, introducing substrate at lower concentrations but over a larger fraction of the column to achieve an optimum bacterial distribution in the column. The only direct was of controlling the above factors is by manipulating the pattern of the substrate being injected into the columns. The model developed previously was used to asses the impact of introducing substrate over larger sections of the column on the total amount of TCE degraded.

NUMERICAL SIMULATIONS

The column used in the simulations is 53 cm long with an internal diameter of 5.8 cm. The column is packed with sand having a porosity of 0.306. It is assumed that water content is 0.29 and the pH is 6.72 over the entire column. The yield coefficients for biomass, methane, and carbon dioxide formation due to substrate utilization are 6.8×10^{-5} (mg-protein/mg-COD), 1.41×10^{-5} moles/mg-COD and 6.40×10^{-6} moles/mg-COD, respectively. The influent has a degradable COD of 2089 mg/L. A liquid flow rate of 2 ml/hr is maintained through the column. The column is modeled as 22 completely mixed reactors in series, each with a hydraulic retention time of 0.385 days.

Effect of Substrate Injection Pattern. Simulations were performed for substrate injections to the first reactor, first 3 reactors, first 5 reactors, first 9 reactors, first 14 reactors, and all 22 reactors. In the simulations the influent substrate flux is kept constant by decreasing the influent substrate concentration being injected into the substrate-receiving reactors to 2089/*n* mg/L, where *n* is the number of reactors into which the substrate is injected. Simulation results are presented for substrate concentration, active and non-active biomass, gas flow, dissolved TCE concentration, and the TCE degradation profiles in the column for each of the simulated cases in Figure 2.

Figure 2a presents changes in substrate concentration in the column. The figure shows that substrate degradation in the reactors is rapid leading to low concentrations in the effluent; however, injection of substrate increases the influent substrate concentration in the downgradient reactor. Figures 2b and 2c present results for active and non-active biomass attached to the soil in each of the reactors (expressed here as *equivalent liquid phase biomass* (g-protein/L-pore liquid). These can be converted to g-protein/kg-soil by dividing the g-protein/L-pore liquid values by $\rho_b/S_w\theta$, where ρ_b is the bulk density of soil, S_w is the water saturation and θ is the porosity of the porous media. In all columns the active and non-active biomass profiles closely follow the substrate concentrations profiles.

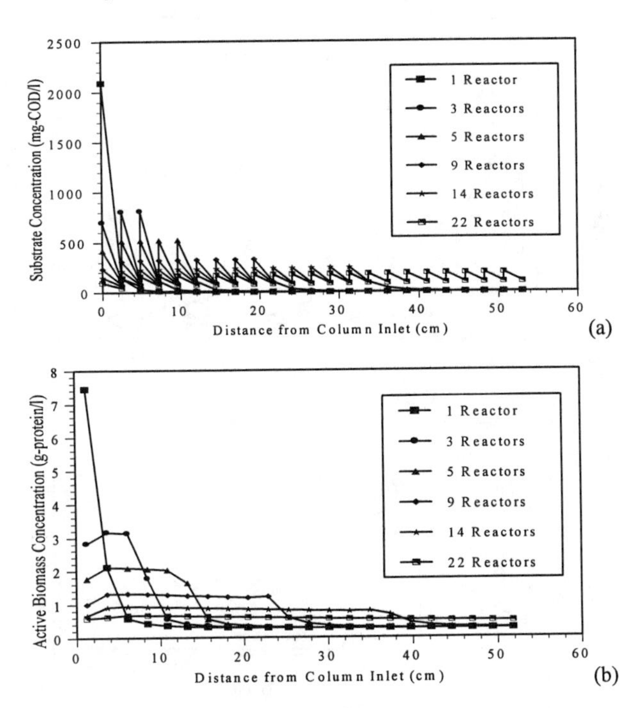

FIGURE 2. Effect of substrate injection pattern on TCE degradation. (a) Substrate profile. (b) Active biomass profile.

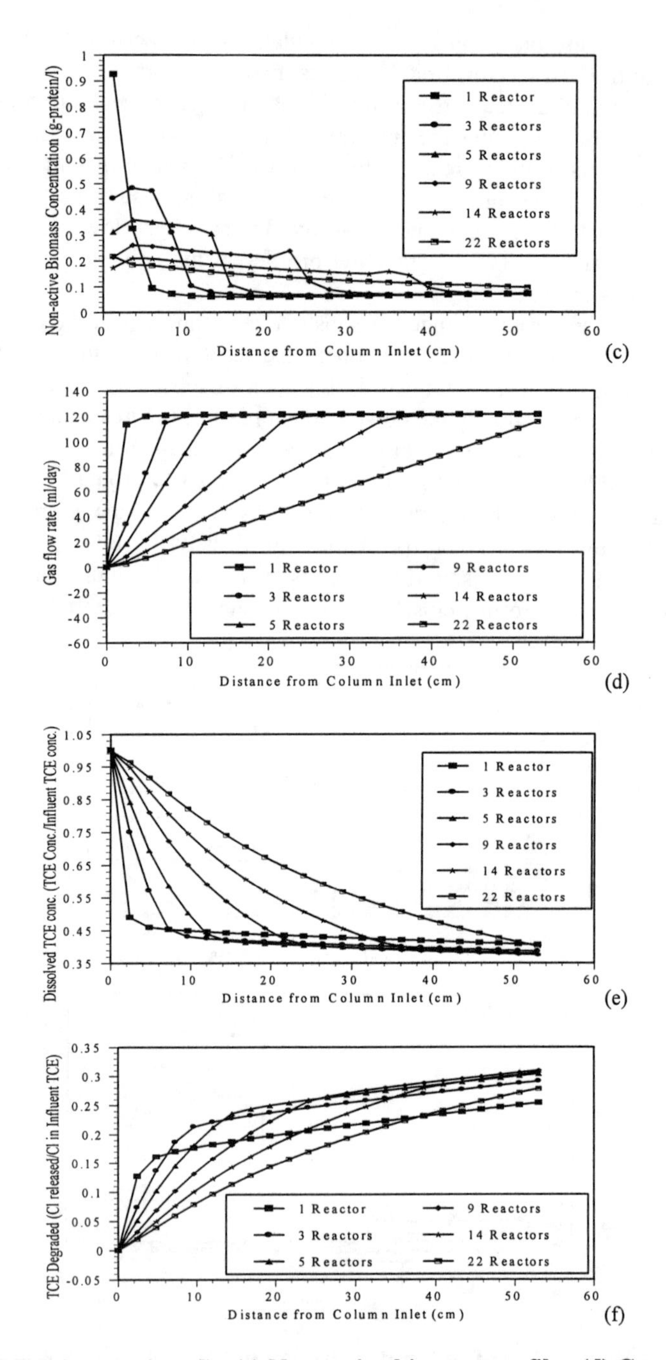

FIGURE 2. (continued). (c) Non-active biomass profile. (d) Gas flow profile. (e) Dissolved TCE profile. (f) TCE degradation profile.

The active biomass concentration in the downgradient reactors is approximately the same in all columns. The gas flow rates in different sections of the columns are presented in Figure 2d. Figures 2e and 2f show results for the dissolved TCE concentrations and the total TCE degraded in the columns. Unlike gas flow rates, these profiles are nonlinear and asymptotically approach a limiting value.

Effect of Gas Phase on TCE Degradation. The effect of including and ignoring the gas phase on TCE degradation was investigated using the proposed model. Results for two sets of simulations are presented. Simulation results are presented for Column 1 which is 53 cm long with an influent COD of 2089 mg/L, and Column 2 is double the length and receives twice the substrate concentration of Column 1.

Columns 1 and 2 were modeled as 22 and 44 reactors, respectively. Substrate were injected into multiple non-contiguous reactors so that every reactor with injected substrate was followed by two reactors in which no substrate was injected; seven of the 22 reactors in Column 1 and 14 of the 44 reactors in Column 2 were injected the substrate. Simulation results are presented in Figure 3. For Column 1 ignoring the effect of the gas phase results in TCE degradation of 47%, while including the effect of the gas phase decreases the predicted

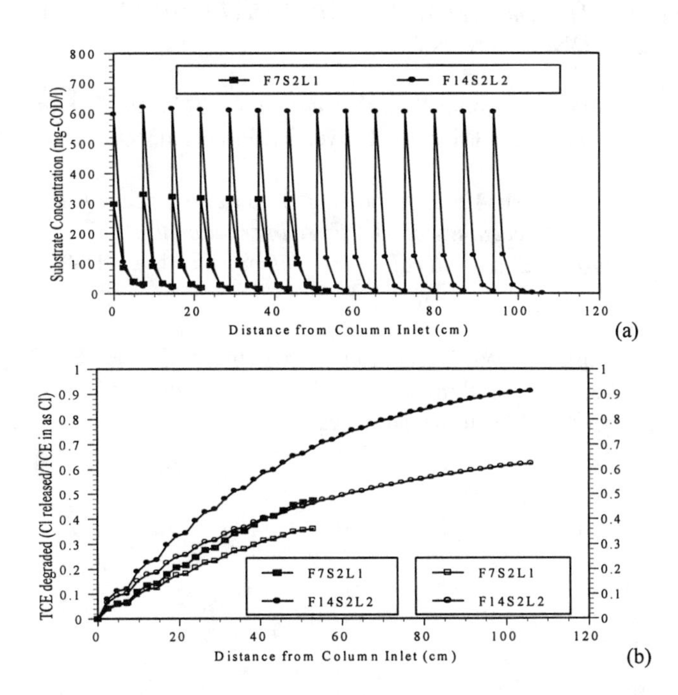

FIGURE 3. Importance of include gas phase transport for volatile chemicals (a) Substrate distribution. (b) TCE degradation profile.

TCE degradation to 35%. Thus, ignoring the gas phase will result in an inflated estimate for TCE degradation. For Column 2 the effect is even more pronounced; TCE degradation decreases from 92% to 62% upon inclusion of the effect of the gas phase.

CONCLUSION

Anaerobic degradation of TCE in soils is governed by the activity and concentration of the attached biomass, and the concentration of dissolved TCE remaining in the liquid phase. Numerical simulations presented in this paper show that achieving a more uniform active biomass distribution along the entire column by injecting the substrate over a larger region of the column can result in increasing the total amount of TCE degraded. It is also shown that volatilization is an important mechanism by which TCE may be removed from contaminated water and that excessive gas production can result in decreasing the total amount of TCE degraded in soils.

REFERENCES

Hermanowicz, S.W. and J.J. Ganczarczyk. 1985. "Mathematical Modeling of Biological Packed and Fluidized Bed Reactors." In S.E. Jørgensen and M.J. Gromiec (Eds.), *Mathematical Models in Biological Waste Water Treatment*, pp. 473-524. Elsevier, New York, NY.

Singhal, N. 1995. *Anaerobic Dehalogenation of TCE in Continuous Flow Systems*. PhD Dissertation, Princeton University, Princeton, NJ.

Singhal, N. and K. Odish. 1998. "Modeling Bacterial Adhesion And Shearing In Porous Media." in *Proceedings of The First International Conference on Remediation of Chlorinated and Recalcitrant Compounds*, May 18-21. Monterey, California.

Singhal, N., P. Jaffé, and W. Maier. 1997. "Modeling the Anaerobic Degradation of Volatile Chemicals in Porous Media." in *Proceedings of CHEMECA 97*, September, paper #EN4. Rotorua, New Zealand.

REDUCTIVE DECHLORINATION VERSUS ADSORPTION OF TETRACHLOROETHYLENE IN FLUIDIZED BED REACTORS

Sébastien Marcoux and Jim Nicell (McGill University, Montréal, Québec, Canada)
André Beaubien (Gama Innovation Inc., Brossard, Québec, Canada)
Serge Guiot (Biotechnology Research Institute, Montréal, Québec, Canada)

ABSTRACT: In this study, the PCE treatment efficiency of two anaerobic fluidized bed reactors (AFBRs) were investigated and compared under different hydraulic loads. Granular activated carbon (GAC) and Biolite™ were used as a support media in the first and second reactor, respectively. The results showed that the GAC AFBR achieved close to 100% PCE removal under all the loading conditions studied. Both adsorption and dechlorination accounted for this total PCE removal. With the exception of one hydraulic loading, adsorption was the main PCE removal mechanism for the duration of this investigation for the GAC reactor. The maximum PCE removal efficiency achieved by the Biolite™ AFBR was approximately 70%. Dechlorination was the only removal mechanism for this reactor. The highest specific chloride ion production rate for both reactors was achieved under the lowest loading condition and methanol activity.

INTRODUCTION

Tetrachloroethylene (also known as perchloroethylene, PCE) is one of the volatile organic compounds which is most frequently detected in groundwaters used as drinking water sources. In addition to being used as dry cleaning-solvent and as a degreaser, PCE also serves as a starting material for other man-made chemicals such as fluorocarbons. Agencies worldwide have classified PCE as being possibly carcinogenic.

It has been demonstrated that PCE can be reductively dehalogenated under methanogenic conditions (De Bruin *et al.*, 1992). In this process, PCE, the electron acceptor, is sequentially reduced to TCE, DCE isomers (*cis*-1,2-DCE, *trans*-1,2-DCE, or 1,1-DCE), VC, ethylene and ethane. An electron donor such as methanol, acetate, or hydrogen must be available to provide reducing equivalents for reductive dechlorination and for cell growth.

The anaerobic fluidized bed reactor (AFBR) has been shown to rapidly and efficiently degrade PCE (Carter and Jewell, 1993). Many parameters such as the hydraulic retention time (HRT), the organic load and the supporting media used for cell immobilization influence the PCE treatment efficiency and still need to be investigated in order to optimize treatment efficiency. These variables form the basis for this research.

The objective of this study is to investigate the ability of an anaerobic fluidized bed reactor to dechlorinate PCE using two different supporting media; granular activated carbon (GAC) and Biolite™ (particles of expanded clay). The performance of each reactor was assessed by evaluating overall PCE removal. For the GAC anaerobic fluidized bed reactor, the role played by adsorption in the total PCE removal was also examined. The details of this research are provided in Marcoux (1997).

MATERIALS AND METHODS

Parallel experiments were performed using two separate AFBRs, reactor 1 (R1) and reactor 2 (R2), operating simultaneously at 35°C. The AFBRs were glass columns with cone-shaped inlets and with working volumes of 15.8 L. A recycling line ensured the fluidization of the bioparticles. Granular activated carbon (GAC) (Sigma, St-Louis, MO) and Biolite™ (Degrémont, France) were used in R1 and R2, respectively. The specific gravity of GAC and Biolite™ equalled 1.64 g/g and 2.10 g/g, respectively. For the GAC AFBR and Biolite™ AFBR, the upflow velocities equalled 4.02 m/hr and 7.95 m/hr, respectively. These upflow velocities resulted in an initial static bed expansion of 30% and were kept constant throughout the entire experiment. A nutrient solution and PCE-methanol solution were fed to each reactor.

Anaerobic sludge used for inoculating the reactors was obtained from a food industry (Champlain Industries, Cornwall, Ontario). Methanol served as the electron donor and as the carbon source for the microorganisms. The methanol:PCE mass concentration ratio was kept constant at 10:1 for the entire experiment. The influent PCE and methanol concentrations were equal to 10 mg/L and 100 mg/L, respectively.

PCE, TCE, DCEs and VC in the effluent were analyzed using the GC headspace method. In the off gas, concentrations of PCE and its metabolites were determined using a gas chromatograph and FID. The inorganic chloride concentration in the effluent was measured using the Hach Mercuric Thiocyanite Method and a Hach DR/3000 spectrophotometer set at 455 nm. The methanol concentration in the reactors and during the activity test was determined using a gas chromatograph and FID.

This study was divided in four phases. The only variable parameter was the HRT. It varied from 24 hrs (phase 1), to 48 hrs (phase 2), to 12 hrs (phase 3), and to 18 hrs (phase 4). The duration of each phase was determined by the time taken by each reactor to reach steady-state in terms of inorganic chloride and aqueous PCE concentration. The entire experiment lasted 130 days.

RESULTS

During PCE addition, the methanol activity for both reactors was highest at a HRT of 0.75 day and lowest at a HRT of 2 days. The methanol activity was lower at a HRT of 0.5 day than at a HRT of 0.75 day for both reactors.

Figure 1 shows the variation of the percentage of total PCE removed with respect to the hydraulic retention time of each phase. The removal of PCE and its metabolites by escaping into the gas phase represented less than 1% for both reactors during all HRTs except for the Biolite™ AFBR (R2) at a 0.5 day HRT where it approximately equalled 2% of the total PCE removal. TCE was the only metabolite present in measurable quantities in the reactors' effluent with a maximum value of 0.99 mg/d in the Biolite™ reactor. For the GAC AFBR (R1), the total PCE percentage removal was greater than 99.3% for each HRT. For the Biolite™ AFBR, the total PCE percentage removal reached a maximum value of 69.5 ± 13.1% at a HRT of 2 days.

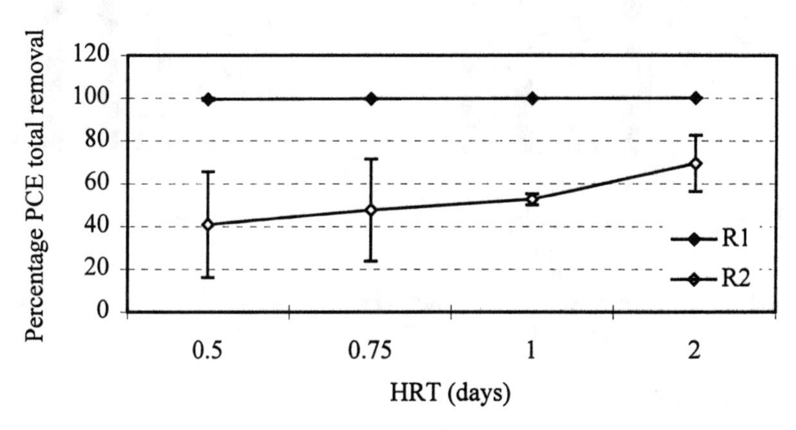

FIGURE 1. Percentage of total PCE removal versus HRT.

The maximum specific chloride ion production for GAC AFBR (R1) occurred at a 2 day HRT (phase 2). The minimum specific chloride ion production occurred at a HRT of 0.75 days (phase 4). These values, in PCE equivalents, equalled 9.82 ± 1.56 and 5.58 ± 0.89 mg PCE/g prot/d. For the Biolite™ AFBR (R2), the specific chloride ion production decreased constantly from phase 1 to phase 4 (from 7.46 ± 1.76 to 4.21± 1.63 mg PCE/g prot/d).

Figures 2(a) and 2(b) show the percentage of total PCE removal and the percentage of PCE removed by dechlorination. For the GAC AFBR, the percentage of PCE being adsorbed to the GAC was always higher than the percentage of PCE being dechlorinated (totally or partially) with the exception of the 2 day HRT where the dechlorination of PCE was higher than the total PCE removed. For the Biolite™ AFBR, the percentage of PCE removal due to dechlorination almost equalled the percentage of PCE total removal at HRTs of 1 and 2 days.

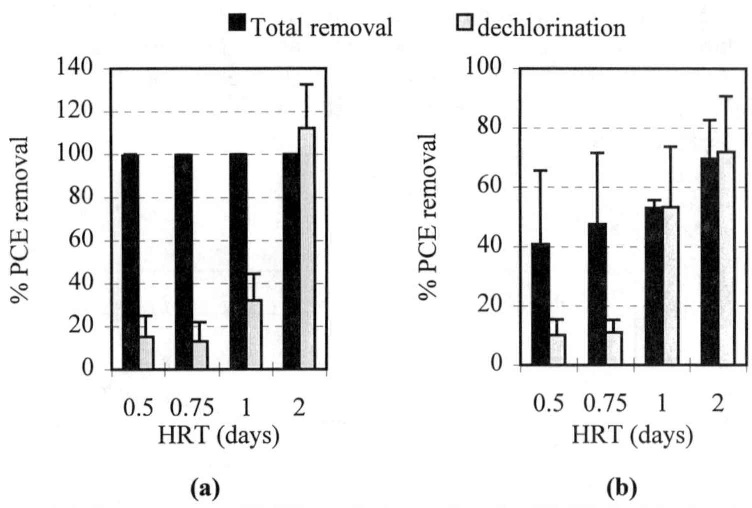

FIGURE 2. Percentage of PCE total removal and of PCE dechlorination for (a) GAC AFBR (R1) and (b) Biolite™ AFBR (R2).

DISCUSSION

PCE Removal. The GAC AFBR experienced more than 99.5 % PCE removal for the entire experiment. It is obvious that adsorption of PCE onto GAC particles played a major role in the removal efficiency of the system because of the imbalance between the chloride ion production and the total PCE removed (Figure 2(a)). The presence of PCE metabolites in the gas and liquid phase as well as chloride ions confirmed PCE biodegradation.

Chloride Ion Production. The specific chloride production rate was better at high HRTs where the methanol activity was lower than at low HRTs where the methanol activity was higher. From these results, it appears as if a low methanol activity favored the dechlorination of PCE. Two hypotheses may explain why low methanol activity might favor dechlorination. First, a low methanol activity might favor the acetogenic metabolism of methanol and increase the H_2 available for the dechlorinators. Second, the slow release of H_2 during low methanogenic activity might favor the dechlorination over the hydrogenotrophilic methanogens, as dechlorinators could use H_2 at lower concentration than the hydrogenotrophilic methanogens (Fennell *et al.*, 1997).

From the formulated observations, it is suggested that the methanol load to be provided for optimal PCE removal should result in a low methanol activity, but an activity that is high enough to provide sufficient H_2 for dechlorination.

Role of Adsorption and Dechlorination. As seen in Figure 2(a), at a 1 day HRT, most of the PCE was adsorbed on the GAC particles and the remainder was biodegraded. In this case, the rate of PCE adsorption was greater than the rate of

PCE biodegradation. This low rate of dechlorination might be explained by the high methanol activity of the reactor. This methanol activity value was also higher for the GAC AFBR than for the Biolite™ AFBR while the specific chloride production was lower for the GAC AFBR than for the Biolite™ AFBR.

At a HRT of 2 days, when the methanol activity was at its lowest, the specific chloride production in the GAC AFBR tends to be higher than its maximum possible chloride ion production. This is explained by a PCE biodegradation rate higher than a PCE adsorption rate. The higher dechlorination rate (which shifted the equilibrium) caused a desorption of the PCE that had been adsorbed in the previous phase so as to maintain a PCE equilibrium between the solid and liquid phase. As seen from Figure 2(b), for the GAC AFBR, the removal of PCE by dechlorination at a HRT of 1 and 2 days is equal to the total PCE removed. These results demonstrate that the PCE removal mechanism was through biodegradation. When the Biolite™ AFBR was operated at a HRT of 0.5 and 0.75 days, approximately 30% and 35%, respectively, of the PCE was removed by adsorption. One potential explanation would be that PCE adsorbed on the recently replaced inflow part of PCE inflow Viton tubes. This adsorption did not affect the results of the first two phases since the interior tubing wall had probably reached saturation. The higher standard deviations of the total PCE removal show that at low HRTs, where the average PCE percentage removal is lower, the system performance fluctuated.

For the Biolite™ AFBR, the chloride ion production at HRTs of 1 and 2 days can be considered as the total PCE removal since both values are equal. At HRTs of 0.5 and 0.75 days, if the total PCE removal was represented by the chloride ion production, it would be equal to approximately 10% of the influent PCE concentration (Figure 2(b)).

Applications of GAC versus Biolite™ in an AFBR. The specific chloride ion production was more or less the same for both reactors. Therefore, one can conclude that in terms of chloride ion production, both systems resulted in a very similar performance. Since the Biolite™ AFBR never achieved 100% PCE removal, it had reached its maximum dechlorination capacity under the highest specific chloride production measured. This corresponds to a value of 6.40 ± 1.51 mg Cl^-/g prot/d. However, since the GAC AFBR achieved a maximum chloride ion production of 8.46 ± 1.34 mg Cl^-/g prot/d which equalled to a PCE removal greater than 100%, it is possible that the system had not reached its optimum PCE removal capacity.

CONCLUSIONS

The following conclusions can be drawn from these experiments:

(1) The GAC reactor achieved a higher PCE percentage removal than the Biolite™ reactor. PCE metabolites were measured in very low concentrations in the effluent and in the gas phase of both reactors.

(2) The specific chloride production was similar for the GAC and Biolite™ AFBRs under the same loading conditions for both reactors. Both reactors achieved the highest percentage of PCE removal by dechlorination under the lowest methanol activity.
(3) For the GAC AFBR, with the exception of the 2 day HRT, adsorption was the main PCE removal mechanism.

ACKNOWLDEGMENTS

The expertise of Dr. J.A. Hawari and the technical support of the Analytical Chemistry Group of the Biotechnology Research Institute were greatly appreciated.

REFERENCES

Carter, S. R., and W. J. Jewell. 1993. "Biotransformation of Tetrachloroethylene by Anaerobic Attached-films at low Temperatures." *Water Research.* 27(4): 607-615.

De Bruin, W. P., J. J. Kotterman, M. A. Posthumus, G. Schraa, and A. J. B. Zehnder. 1992. "Complete Biological Reductive Transformation of Tetrachloro-ethene to Ethane." *Applied and Environmental Microbiology.* 58(6): 1996-2000.

Fennell, D. E., J. M. Gossett, and S. H. Zinder. 1997. "Comparison of Butyric Acid, Ethanol, Lactic Acid, and Propionic Acid as Hydrogen Donors for the Reductive Dechlorination of Tetrachloroethylene." *Environmental Science and Technology*. 31(3): 918-926.

Marcoux, S. 1997. "A Comparison of Two Anaerobic Fluidized Bed Reactors for the Treatment of Tetrachloroethylene." M. Eng. Thesis, McGill University, Montreal, Quebec, Canada.

ANAEROBIC DEGRADATION OF PCE AND TCE DNAPLS BY GROUNDWATER MICROORGANISMS

R. Brent Nielsen (University of California, Berkeley, CA U.S.A.)
J. D. Keasling (University of California, Berkeley, CA U.S.A.)

ABSTRACT: Groundwater microorganisms enriched from a site contaminated with trichloroethene (TCE) reductively dechlorinate tetrachloroethene (PCE) and TCE to ethene. Kinetic studies indicate that PCE dechlorination follows a first-order dependence at low PCE concentrations and a zero-order dependence at high PCE concentrations. TCE and vinyl chloride (VC) dechlorination studies indicate a first-order dependence at all TCE and VC concentrations. At subsaturating PCE and TCE concentrations, nearly stoichiometric amounts of the toxic intermediate, VC, accumulated prior to its dechlorination to ethene. However, at saturating chlorinated ethene concentrations, a rapid conversion to ethene occurred without significant accumulation of VC.

INTRODUCTION

Contamination of groundwater by chlorinated hydrocarbons is a major problem through out the United States (Dyksen and Hess, 1982; Vogel et al., 1987). The chlorinated hydrocarbons, PCE and TCE, are frequently detected as contaminants (Dyksen and Hess, 1982; Vogel et al., 1987; Lee et al., 1988). A 1980 study of drinking water systems throughout the United States reported that 31 municipal wells in California and wells in 15 Pennsylvania communities were closed due to TCE contamination (Lee et al., 1988). PCE and TCE have been used as degreasing agents, in the production of silicon wafers, in the dry cleaning industry, and in household cleaning products (Love and Eilers, 1982; Vogel et al., 1987). Their common occurrence as groundwater contaminants results from poor handling and disposal practices (Dyksen and Hess, 1982; Vogel et al., 1987). Furthermore, VC, an intermediate of anaerobic PCE and TCE dechlorination, is a known carcinogen. Since PCE and TCE are suspected carcinogens and due to VC toxicity, these compounds are regulated by the United States Environmental Protection Agency and the Safe Drinking Water Act of 1986.

An additional problem associated with groundwater contamination by PCE and TCE results from their formation of dense-nonaqueous-phase-liquids (DNAPLs) (Mackay and Cherry, 1989; Oolman et al., 1995). TCE and PCE DNAPLs will pool above non-permeable zones in groundwater aquifers and provide a source of contamination that can persist for decades.

The objective of this work is to examine the kinetics of PCE and TCE dechlorination over a wide range of concentrations and investigate PCE and TCE dechlorination under saturating conditions. The results obtained should help to more effectively evaluate the use of anaerobic bioremediation of PCE and TCE contaminated sites.

MATERIALS AND METHODS

Culture Cultivation. Groundwater samples were collected anaerobically from a TCE-contaminated site (Santa Clara County, CA). 50 mL of groundwater were transferred to 160-mL serum bottles containing 50 mL of a minimal salts medium (Bolesch et al., 1997). Using glucose as the electron donor and TCE as the electron acceptor, a mixed culture of microorganisms capable of anaerobically dechlorinating TCE completely to ethene was obtained (Bolesch et al., 1997). Culture enrichments were maintained by repeatedly adding glucose and either TCE or PCE to anaerobic culture-containing serum bottles. The headspace of each bottle was purged prior to the glucose and chlorinated ethene additions.

Analytical Techniques. Gas chromatography was used to measure PCE, TCE, VC, ethene, and methane headspace concentrations. An HP 6890 GC fitted with a capillary column (AT-624, Alltech, Deerfield, IL or HP-624, Hewlett Packard, San Fernando, CA for PCE and TCE; Poraplot-Q, Hewlett Packard, San Fernando, CA for VC, ethene, and methane) and a Varian 3700 GC fitted with a packed column (Porapack-Q, Alltech, Deerfield, IL for VC, ethene, and methane) were used to separate the components prior to detection. Both a flame ionization detector and an electron capture detector were used to detect the compounds. The peak areas obtained were converted to concentrations by preparing standard curves for each component of interest. Liquid concentrations were calculated using Henry's Law and published Henry's constants (Gossett, 1987). Protein levels were determined using a modified Bradford assay (Bradford, 1976). The absorbance was converted to protein concentrations using a standard curve prepared with bovine serum albumin. Glucose concentrations were determined using a Trinder assay (Sigma, St. Louis, MO) (Trinder, 1969).

Kinetic Studies. For each kinetic study, 6-mL aliquots of medium containing bacteria were added to 10-mL serum bottles under anaerobic conditions. Glucose was supplied as the electron donor. Varying amounts of PCE, TCE, and VC were added as electron acceptors. The bottles were incubated at room temperature on an orbital shaker operating at 150 rpm. Headspace samples were analyzed at frequent intervals. The initial degradation rates were obtained from the slope of the time-concentration curves.

Degradation Studies. Degradation rates under saturating and subsaturating conditions were determined using 160-mL serum bottles with 100-mL of medium containing bacteria. Subsaturating conditions were obtained by the addition of 0.21 µmol of PCE and 0.21 µmol of TCE to separate bottles. Saturating conditions were obtained by adding 0.5 mmol of PCE and 2.8 mmol of TCE to separate bottles. Glucose was used as the electron acceptor. The bottles were incubated at 30°C. Headspace samples were analyzed for PCE, TCE, VC, ethene, and methane.

RESULTS AND DISCUSSION

Complete conversion of TCE to ethene was achieved using a mixed culture enriched from a TCE contaminated groundwater site (Figure 1). TCE was dechlorinated to cis-dichloroethene which was subsequently dechlorinated to VC. Once TCE and cis-DCE were utilized, VC was dechlorinated completely to ethene (Bolesch et al.,1997).

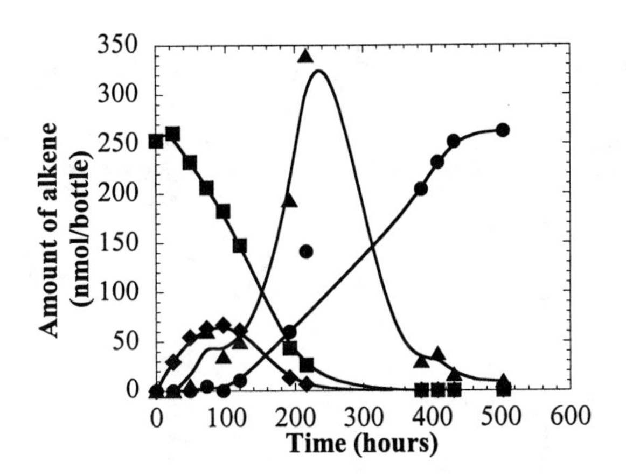

FIGURE 1. TCE degradation by enriched groundwater microorganisms. Cultures from a groundwater microcosm were enriched in medium containing TCE and glucose. TCE (■), cis-DCE (◆), VC (▲), Ethene (●). (Data taken from Bolesch et al. 1997. *Ann. New York Acad. Sci.* 829:97-102)

The kinetics of PCE and TCE dechlorination was examined by determining the initial rates as a function of initial substrate concentration. Since dechlorination of PCE followed a first-order dependence at low concentrations and a zero-order dependence at high concentrations (Figure 2), estimates for V_{max} and K_S can be obtained. These values were determined to be 1150 nmol/mg/h and 252 μM, respectively. TCE and VC dechlorination followed a first-order dependence on substrate concentration (Figure 2). As a result, only values for the ratio V_{max}/K_S can be obtained from the data. The values for this ratio were determined to be 2.24 L/g/h and 1.61 L/g/h, respectively. Sufficient amounts of glucose remained in each bottle at the conclusion of each study and indicates that glucose was not a limiting factor in the degradation of the chlorinated ethenes.

The results from these kinetic studies give insight into the chlorinated ethene dependence of PCE, TCE, and VC dechlorination. Furthermore no substrate inhibition occurs at high substrate concentrations. This implies that anaerobic biodegradation in regions of high chlorinated ethene concentrations is possible. Such regions would exist near PCE and TCE DNAPLs.

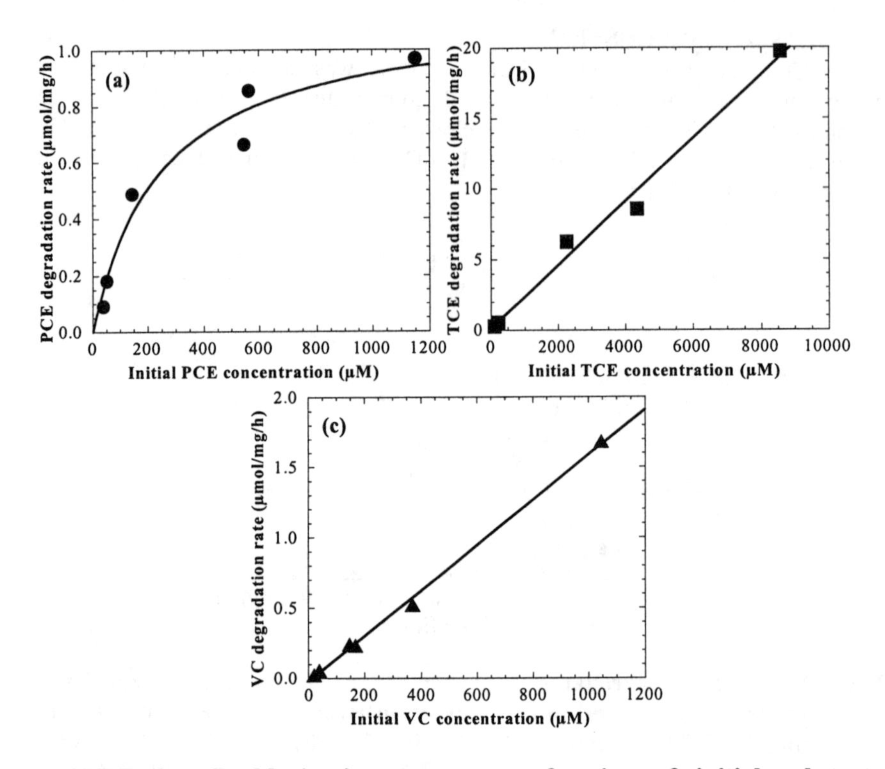

FIGURE 2. Dechlorination rates as a function of initial substrate concentration. (a) PCE degradation. (b) TCE degradation. (c) VC degradation. (Data taken from Nielsen and Keasling. 1998. *Biotechnol. Bioeng.* Submitted.)

The dechlorination behavior under saturating and subsaturating conditions was compared. The production of VC and ethene was used as an indicator of dechlorination since it is assumed that both PCE and TCE are rapidly converted to VC and/or ethene and since the aqueous concentration of PCE and TCE remain constant as long as a separate chlorinated ethene phase exists. Under subsaturating PCE and TCE conditions, VC accumulated and was then dechlorinated to ethene (Figure 3). However, under saturating PCE and TCE conditions, a rapid production of ethene resulted with little or no accumulation of VC (Figure 3). In the case of TCE dechlorination, the VC production rate was 23 times smaller under saturating conditions than under subsaturating conditions and the production of ethene was 120 times larger under saturating conditions than under subsaturating conditions. Likewise, the production of VC under subsaturating PCE conditions was approximately 50 times smaller than under subsaturating conditions and ethene production was approximately 30 times larger under saturating conditions than under subsaturating PCE concentrations. Methane production rates (data not shown) from this study are inhibited at high chlorinated ethene concentrations. Under subsaturating PCE and TCE concentrations, the methane production rate was 3.5 μmol/h whereas under saturating conditions it was essentially zero (Nielsen and Keasling, 1998).

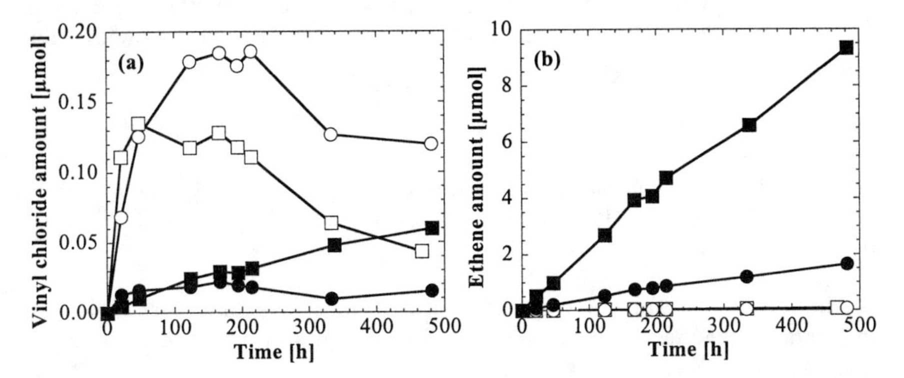

FIGURE 3. VC, ethene, and methane production during reductive dechlorination of PCE and TCE under saturating and subsaturating conditions. Saturating PCE conditions (●). Subsaturating PCE conditions (○). Saturating TCE conditions (■). Subsaturating TCE conditions (□). (a) VC production. (b) Ethene production. (Data taken from Nielsen and Keasling. 1998. *Biotechnol. Bioeng.* Submitted)

The rapid conversion of PCE and TCE to ethene with little accumulation of VC under saturating conditions is an interesting and unexpected result. One explanation is that of competition for electron between different organisms. For example, at high PCE and TCE concentrations methane production becomes inhibited and more electrons become available for PCE and TCE dechlorination by the ethene dechlorinating organisms. Separate studies showed that as the TCE concentration increased, the VC and methane production rates decreased and the ethene production rates increased (Nielsen and Keasling, 1998). These results tend to support the idea that more electrons become available to the ethene dechlorinating organisms as the methane producing organisms become inhibited.

CONCLUSIONS

Since these studies were performed with mixed cultures obtained from TCE contaminated groundwater, the results give insight into PCE and TCE degradation at such a site. One could expect PCE, TCE, and VC dechlorination to follow the chlorinated ethene dependencies indicated above when the chlorinated ethene is the limiting component. Furthermore, dechlorination in regions near PCE and TCE DNAPLs should result in little VC accumulation. Since VC is a toxic intermediate of PCE and TCE dechlorination under anaerobic conditions, the risk associated with anaerobically remediating PCE and TCE DNAPLs may be significantly less than would otherwise be expected.

ACKNOWLEDGEMENTS

We thank the University of California Water Resources Center (W-849), Geomatrix Consultants, Inc., San Francisco, CA, the Toxic Substance Research and Teaching Program, and the Lawrence Berkeley National Laboratory for support.

REFERENCES

Bolesch, D. G., R. B. Nielsen, et al. 1997. "Complete reductive dechlorination of trichloroethene by a groundwater microbial consortium." *Ann. New York Acad. Sci.* **829**: 97-102.

Bradford, M. M. 1976. "A rapid and sensitive method for the quantitation of microgram quantities of protein utilizing the principles of protein-dye binding." *Anal. Biochem.* **72**: 248-254.

Dyksen, J. E. and A. F. Hess. 1982. "Alternatives for controlling organics in groundwater supplies." *J. Am. Water Works Assoc.* **74**: 394-403.

Gossett, J. M. 1987. "Measurement of Henry's law constants for C_1 and C_2 chlorinated hydrocarbons." *Environ. Sci. Technol.* **21**(2): 202-208.

Lee, M. D., J. M. Thomas, et al. 1988. "Biorestoration of aquifers contaminated with organic compounds." *CRC Critical Review in Environmental Control* **18**(1): 29-89.

Love, O. T. and R. G. Eilers. 1982. "Treatment of drinking water containing trichloroethylene and related industrial solvents." *J. Am. Water Works Assoc.* **74**: 413-425.

Mackay, D. M. and J. A. Cherry. 1989. "Groundwater contamination: pump-and-treat remediation." *Environ. Sci. Technol.* **23**(6): 630-636.

Nielsen, R. B. and J. D. Keasling. 1998. "Reductive dechlorination of chlorinated ethene DNAPLs by a culture enriched from contaminated groundwater." *Biotechnol. Bioeng.* Submitted.

Oolman, T., S. T. Godard, et al. 1995. "DNAPL flow behavior in a contaminated aquifer: evaluation of field data." *Ground Water Monitoring and Remediation* **15**(4): 125-137.

Trinder, P. 1969. "Determination of glucose in blood using glucose oxidase with an alternative oxygen acceptor." *Ann Clin Biochem* **6**.

Vogel, T. M., C. S. Criddle, et al. 1987. "Transformation of halogenated aliphatic compounds." *Environmental Science & Technology* **21**: 722-736.

REMOVING RECALCITRANT VOLATILE ORGANIC COMPOUNDS USING DISACCHARIDE AND YEAST EXTRACT

Honniball, James H., Delfino, Thomas A., and Gallinatti, John D.,
Geomatrix Consultants, Inc., San Francisco, California, U.S.A.

Abstract: Volatile organic compounds (VOCs) that leaked from underground storage tanks affected a site in the San Francisco Bay Area. After the tanks were removed, the site was remediated in 1991 by excavating 3700 cubic yards of affected soil. Groundwater monitoring following remediation detected residual VOC concentrations in a single monitoring well, at 2.1 micrograms per liter (µg/L) of vinyl chloride (VC). The Maximum Contaminant Level (MCL) for VC in California is 0.5 µg/L. In order to avoid the need for long-term monitoring at the site, in situ reductive bioremediation was implemented to degrade the residual VOC. Background oxidation-reduction (redox) potential (Eh) was measured in site monitoring wells before a disaccharide and yeast extract solution was injected into the VOC-affected well. Eh ranged from 298 to 348 millivolts (mV), with the higher reading in the affected well. One month after injection, Eh measurements indicated that the site had become more reducing (161 to 261 mV), with the lower value in the affected well. Successive monthly redox measurements indicated that the groundwater at the affected well was becoming even more reducing (86 mV). VC concentrations decreased to 0.59 µg/L. The effect of lowering the redox potential in the injection well also resulted in lowering the redox potential in all of the adjacent monitoring wells, beyond the area that was in direct contact with the injection solution. The lowering of the redox potential in the adjacent wells lasted up to three months after injections. These results suggest that modification of the geochemical environment may have extended as much as 100 feet beyond the location of the injected solution. After a second injection and one year of monitoring, Eh in the affected well decreased to 84 mV, and VC was not detectable at the laboratory reporting limit of 0.5 µg/L. The site is now closed.

INTRODUCTION

Enhanced bioremediation of halogenated organic compounds has been a goal of researchers for nearly a decade. Recent years have seen development of such methods and they are now going to the field for trial. We conducted a field trial of a reductive dechlorination bioremediation method developed with the University of California, Berkeley (Bolesch, et al., 1995). During the field trial, redox potential exhibited behavior that could have relevance to full scale use of this and other similar enhanced bioremediation methods.

Two underground storage tanks were removed from the site of the field trial in August 1985. The site was remediated in 1991 by excavating 3700 cubic yards of affected soil from the source area.

The site consists predominately of silty clay and clayey silt, which exhibit relatively low hydraulic conductivities. Monitoring well A located near the vicinity of the tanks was constructed in the shallowest saturated sand layer, which occurs at a depth of 27 feet, is approximately 18 inches thick, and consists of silty sand and sandy silt.

Monitoring well A is the only monitoring well that has shown detections of VOCs since 1992 (Figure 1). The detected concentrations of VOCs in groundwater reflected very low residual concentrations in close proximity to the excavated area. To eliminate the need for continued long term monitoring, reductive in situ bioremediation was selected to degrade these residual VOCs.

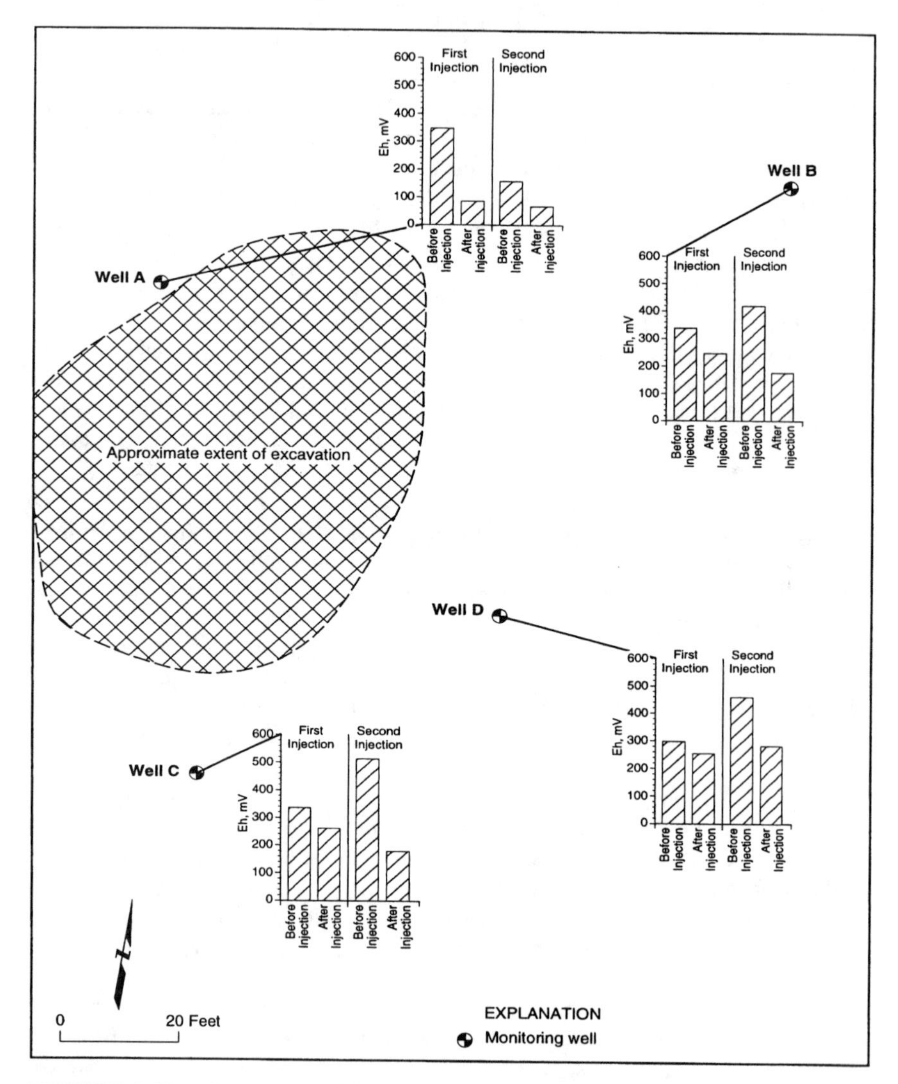

FIGURE 1. Site plan and Eh intensities before and after injections.

The presence of VC and moderately low redox suggested that redox conditions in the aquifer were not low enough. To lower the redox further, a disaccharide was added to the groundwater. Previous studies (Bolesh et al., 1995) indicated that the presence of glucose and various components can dechlorinate trichloroethene completely to ethene. The disaccharide acts as a reducing agent/secondary substrate (increasing reducing intensity by donating electrons), thereby cultivating an enriched endogenous microbial population capable of reducing the chlorinated organic compound in groundwater.

MATERIALS AND METHODS

Three-percent sucrose supplemented with 0.3 percent yeast extract was mixed with municipal tap water. The tap water was aerated for 24 hours to remove trihalomethanes before being used to dissolve the sucrose and yeast extract. Two injections were made from 11 through 13 July 1995 and 2 through 28 May 1996 of approximately 72 and 190 gallons, respectively. The solution was siphoned into monitoring well A through Teflon-coated vinyl tubing. A laboratory pinch clamp controlled the rate of siphoning.

Redox potential was measured in site monitoring wells (A, B, C and D) before and after the solution was injected. Redox potential was selected as an appropriate measure of the effectiveness of the carbohydrate injection. A decrease in the redox potential would correspond to an improved environment for reductive dehalogenation.

Because of site geology (silty clay and clayey silt) and well construction, purging with a bailer was selected as the best alternative to obtain groundwater from monitoring well A versus other low flow methods. This required a modification from our typical flow-through cell method of measuring redox potential. Groundwater was decanted from a polyethylene disposable bailer into a beaker with a stopcock. The stopcock was placed near the bottom of the 500-milliliter beaker and groundwater was allowed to overflow the beaker into a 5-gallon bucket to simulate a flow-through cell. Redox potential was measured in the beaker by using a silver/silver-chloride electrode in a saturated potassium chloride reference cell. Redox potential was corrected to an Eh by adding 199 mV (Stumm and Morgan, 1996), the voltage of the reference cell relative to the standard hydrogen electrode.

An independent commercial laboratory analyzed groundwater samples for purgeable chlorinated organic compounds by U. S. Environmental Protection Agency Method 8010. Samples collected prior to November 1995 were preserved with hydrochloric acid and refrigeration. However, after the second injection of sucrose and yeast extract, groundwater samples produced excessive frothing in the purge-and-trap step of Method 8010, which interfered with the performance of the analysis. In August 1996, it was discovered that preservation with hydrochloric acid worsened frothing. Samples collected in August 1996 and after were preserved with refrigeration only.

RESULTS

Measurements of Redox Potential. Figure 2 presents measurements of redox potential in site monitoring wells. Before injection of the carbohydrate yeast extract in July 1995, Eh measurements ranged from 298 to 348 mV. During the three months following injection, the Eh measurements indicated decreased redox potential in all monitoring wells. Although the Eh in well A continued to decrease through November 1995, the Eh in the distant monitoring wells began to rise. By February 1996, 7 months after the first injection, the Eh in all wells had risen to levels similar to or greater than prior to injection. A second injection event of disaccharide and yeast solution in May 1996 resulted in a renewed lowering of Eh throughout the site monitoring wells. Following the second event, the lowering of the Eh again persisted for approximately three months in the distant wells. However, a distinctively low Eh was maintained in well A for at least one year following injection.

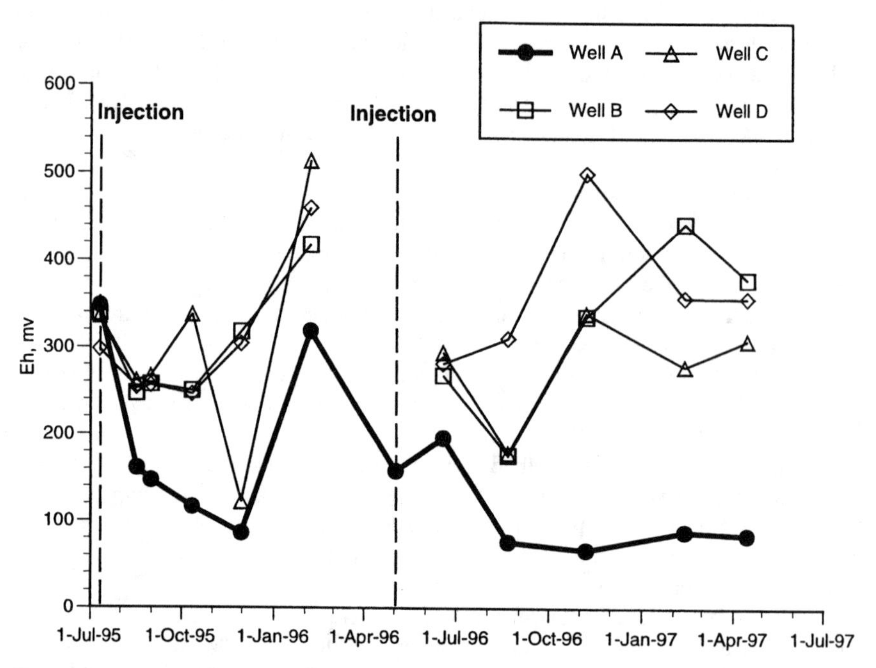

FIGURE 2. Monitoring well Eh intensities during in situ bioremediation.

Analytical Results. The results from the sampling events after injections are presented in Table 1. In December 1994 before injections, the only VOC of concern in monitoring well A was VC at 3.1 µg/L. After the first injection, VC decreased to 0.59 µg/L in November 1995. In February 1996, concentrations of VC increased to 2.9 µg/L. After the second injection (May 1996), VC was not detected above the laboratory reporting limit of 0.5 µg/L in August 1996 and November 1996. VC was detected at 0.5 µg/L in February 1997 sampling event. In April 1997, confirmatory samples for VC were taken from all monitoring wells. No VC was detected in any of the wells above the laboratory reporting limit of 0.5 µg/L.

TABLE 1. The residual VOC compounds in well A before and after in situ remediation with respect to redox potential.

Date	1,1-DCA (µg/L)	Vinyl Chloride (µg/L)	Chloro-ethane (µg/L)	Eh (mV)
12/6/94	4.6	3.1	<0.5	342
First injection event				
11/30/95	3.1	0.59	1.6	86
2/8/96	5.1	2.9	1.9	319
Second injection event				
8/21/96	0.7	<0.5	<0.5	76
11/7/96	<0.05	<0.5	<2	67
2/12/97	2.8	0.5	<2	88
4/15/97	NR	<0.5	NR	84

Notes:

µg/L = micrograms per liter.
mV = millivolts.
Eh is reference to a silver/silver chloride electrode and is calculated using the Nernst Equation: Eh = E(SSCE) + 199 mV.
NR = analyte not required by the regulatory agency.

DISCUSSION AND CONCLUSION

The criteria for closure were that groundwater meet state MCLs for VC. The MCL for VC is 0.5 µg/L in California.

The low concentration of VC in monitoring well A indicated incomplete degradation of a VOC. In identifying an appropriate technique for removing VC in the vicinity of monitoring well A, redox potential was of considerable importance. Background data indicated that prior to injections, the site was predominantly moderately reducing, with Eh ranging from 306 mV to 451mV, or within the lower range of manganese reduction and approaching iron reduction, an anaerobic environment. An anaerobic environment is identified when the redox potential (Eh) of the system is less than 740 millivolts (mV), referenced to a standard hydrogen electrode, at 25 degrees Celsius (Nyer and Duffin, 1997). However, complete reductive dehalogenation of VC may require an Eh less than –300mV. Nyer and Duffin (1997) have suggested that the optimal range for reductive dehalogenation is between –150 to –350 mV.

Because the site subsurface was primarily reducing, it was advantageous not to resist the Eh of the system by removing VC through aerobic means in this predominately clay soil. By using a strong reducing agent such as sugar and yeast extract, the Eh in the vicinity of monitoring well A was lowered to induce indigenous bacteria to mineralize VC. The results indicate successful lowering of redox potential, a requirement for reductive dechlorination (Vogel and McCarty, 1985; DiStefano et al., 1991; Freedman, 1989).

Figure 1 shows the relative difference in Eh after injections were made into monitoring well A. The injected solution is estimated to extend only to a radius of approximately 2.6 and 4.2 feet during the first and second injections, respectively. However, the results suggest that the injection resulted in a measurable decrease in Eh at distances of approximately 100 feet from the injection location. The decrease in Eh observed in distant monitoring wells over approximately 3 months, appears to have been a relatively short-term phenomenon (Figure 2).

After Eh was lowered, concentrations of VOCs decreased in monitoring well A, with VC concentrations (0.59 μg/L) approaching the MCL. However, a subsequent return of VC to a concentration of 2.9 μg/L required a second injection of disaccharide/yeast extract solution to complete the degradation process. After lowering the Eh considerably in monitoring well A, VC was no longer detected above the laboratory detection limit of 0.5 μg/L. Similar to the first injection event, Eh was lowered in all monitoring wells for a period of two to three months. The lower Eh condition was maintained in the injection well for a period of at least one-year following the second injection (Figure 2). More research is required to fully understand the processes involved in redox buffering of the system in the vicinity of the injection well.

With the addition of a reducing agent, the redox potential of the system was successfully lowered to enhance reductive in situ bioremediation in the subsurface, VC was removed and the cleanup goal obtained. The site has been closed by the regulating agency.

REFERENCES

Bolesch, D.G., Delfino, T.A., and Keasling, J.A. 1995. Complete Anaerobic Dechlorination of Trichloroethene by a Groundwater Enrichment Culture.

Department of Chemical Engineering, University of California, Berkeley, California.

DiStefano, T.D., Gossett, J.M. and Zinder, H. 1991. "Reductive Dechlorination of High Concentration of Tetrachloroethene to Ethene by an Anaerobic Enrichment Culture in the Absence of Methanogenesis." *Applied Environmental Microbiology. 57*: 1187-2292.

Freedman, D.L., and Gossett, J.M. 1989. "Biological Reductive Dehalogenation of Tetrachloroethylene and Trichloroethylene to Ethylene Under Methanogenic Conditions." *Applied Environmental Microbiology. 55*: 2144-2155.

Nyer, E.V., and Duffin, M.E. 1997. The State of the Art of Bioremediation. *Groundwater Monitoring and Remediation.* Spring. 17:2: 64-69.

Stumm, W., and Morgan, J.J. 1996. Aquatic Chemistry. Chemical Equilibria and Rates in Natural Waters. Third Edition. John Wiley and Sons, Inc. New York.

Vogel, T.M., and McCarty, P.L. 1985. "Biotransformation of Tetrachloroethylene to Trichloroethene, Dichloroethylene, Vinyl Chloride and Carbon Dioxide Under Methanogenic Conditions." *Applied Environmental Microbiology. 49*: 1080-1083.

A COMBINED ANAEROBIC AND AEROBIC MICROBIAL SYSTEM FOR COMPLETE DEGRADATION OF TETRACHLOROETHYLENE

Tae Ho Lee, Michihiko Ike, and ***Masanori Fujita***
(Osaka University, Suita, Osaka, Japan)

ABSTRACT: Anaerobic dechlorination of tetrachloroethylene (PCE) and aerobic co-oxidation of trichloroethylene (TCE) were studied for constructing an effective bioremediation system for complete PCE degradation. An anaerobic microbial enrichment culture was constructed for efficient PCE dechlorination. This enrichment culture could dechlorinate 150 mg/liter PCE to 1,2-*cis*-DCE (*cis*-DCE) within 2 days. From a series of batch tests, Monod kinetic coefficients for PCE dechlorination were determined; *k* (the maximum specific substrate removal rate) was 0.93 mg-PCE/mg-VSS/day and Ks (the Monod half velocity constant) was 109 mg-PCE/liter. On the other hand, phenol degrading bacteria isolated from a variety of environmental samples were examined to evaluate their activity of TCE co-oxidization. Among 14 strains, *Pseudomonas* putida BH, was selected as the most efficient strain and was adapted to higher TCE concentration. Finally, an efficient bacterial culture of strain BH which could co-oxidize 20 mg/liter of TCE within 2 days was established. Its Monod coefficients *k* and Ks for TCE degradation were also determined as 0.1 mg-TCE/mg-VSS/day and 10.3 mg-TCE/liter, respectively. A sequential anaerobic/aerobic reactor system using these PCE dechlorinating enrichment culture and TCE co-oxidizing phenol-degrader was constructed.

INTRODUCTION

Chlorinated ethenes, represented by trichloroethylene (TCE) and tetrachloroethylene (PCE), are widely used in various industries and have become common groundwater contaminants as a result of discharge of wastewaters, leachates from landfills and leakage from storage tanks. Because these compounds are toxic and are believed to be carcinogenic, techniques for eliminating them from contaminated sites are needed. Recently, cost-effective bioremediation techniques that can completely mineralize chlorinated compounds have been applied on site and in situ.

It is generally believed that biological degradation of PCE can occur only under anaerobic conditions. PCE can be reductively transformed into TCE, dichloroethylene (DCE) isomers, and then vinyl chloride (VC), and eventually ethene or ethane. However, such less-chlorinated catabolic products seem to persist longer under anaerobic condition than the parent compound PCE. Particularly, the highly toxic and carcinogenic metabolite, VC, is often accumulated. The dechlorination of VC to ethene seems to start after complete PCE depletion (Maymo-Gatell *et al.* 1997).

TCE and its reductive metabolites are susceptible to an oxidative process. A variety of oxygenases which catalyze oxidation of ammonia, aliphatics such as methane, propane, and propene, and aromatics such as phenol and toluene have been demonstrated to be able to co-oxidize TCE and less-chlorinated metabolites. Methane oxygenase was more effective for the biotransformation of 1,2-*trans*-dichloroethylene (*trans*-DCE) and VC than that of TCE and *cis*-DCE (Hopkins *et al.* 1993a). For TCE and *cis*-DCE degradation, phenol has been reported as an effective primary substrate in lab-scale and filed evaluations (Shurtliff *et al.* 1996, Hopkins *et al.* 1993b).

In order to completely and rapidly degrade PCE, sequential anaerobic /aerobic processes have been recently proposed. In this system, PCE is dechlorinated into TCE and/or DCEs under anaerobic conditions and then these metabolites are completely mineralizing into CO_2, HCl and cell materials under aerobic conditions. Hence, the combination of these anaerobic PCE-dechlorinating and aerobic TCE or *cis*-DCE-mineralizing systems can lead to an efficient and complete PCE degradation process.

The goal of this study was to construct an efficient sequential anaerobic /aerobic microbial system for complete PCE degradation. A soil enrichment and a phenol degrading strain, *Pseudomonas* putida BH, were selected for this purpose after a comparative evaluation of a variety of microbial samples. Based on the results of the batch tests of PCE dechlorination and TCE degradation, an anaerobic /aerobic reactor was designed and constructed.

MATRIALS AND METHOD

Soil and sludge sampling. To establish PCE dechlorinating enrichments, nine soil or ditch sludge samples from uncontaminated and contaminated sites (Chiba and Osaka, Japan) were used as inoculants. For isolating TCE co-oxidizing phenol degraders, six forest soil and two field soil samples were taken from Osaka, Japan. Every soil sample was taken from 5-10 cm depth for aerobic microbial samples and 20-40 cm depth for anaerobic microbial samples at each sampling site. Sludge samples were well mixed before sampling.

Batch tests for PCE dechlorination. For PCE dechlorination, 27-ml serum bottles containing each 10 ml of basal medium (Lee *et al.*1997) was used. Bottles were capped with Teflon-lined butyl rubber septums and fixed with aluminum crimp caps. The medium was purged with N_2 gas (>99.9%) before the PCE injection. PCE was injected into the bottles to achieve the initial concentrations of 10-150 mg/liter (these norninal concentrations were estimated by ignoring equilibrium partitioning between gas and liquid phase; the actual aqueous concentrations were 4-58 mg/liter considering Henry's law constant 0.933 at 30℃; Gossett 1987). Bottles were incubated at 30℃ on an orbital shaker water bath at 100 rpm.

Batch tests for TCE dechlorination. For TCE degradation, phenol-degrading bacteria were cultivated in 20 ml of a minimal salt medium (per liter; K_2HPO_4 1.0g, $(NH_4)_2SO_4$ 1.0g, $MgSO_4 7H_2O$ 0.2g, $FeCl_3$ 0.01g, $CaCl_2$ 0.05g, NaCl 0.05g) containing phenol (125-500 mg/l) as a sole carbon and energy source. Experiments were performed in closed batch systems (120-ml serum vials) with relatively large headspace to keep adequate oxygen amount for phenol and TCE oxidization. The bacterial cultures at late-log to stationary growth phase were harvested by centrifugation (10,000 rpm for 10 min. at 4℃), washed with phosphate-potassium buffer (pH 7.5), and resuspended in the same buffer at an approximately turbidity of 2.0 at 600 nm (resting-cell). The vials containing 20 ml of the resting cell received ca. 1-20 mg/liter TCE (nominal concentrations; the actual concentrations were 0.33-6.75 mg/liter considering Henry's law constant 0.392 at 25℃; Gossett 1987) and incubated at 25℃ with reciprocal shaking at 100 rpm.

Analytical methods. PCE and TCE were measured by head-space analysis with an electron capture detector (ECD) or a flame ionization detector (FID) equipped gas chromatography (GC-14A, Shimadzu, Tokyo, Japan). DCEs and VC were measured with a gas chromatograph mass spectrometer (GC/MS; GC-17A-QP-5000, Shimadzu). The chloride ion was detected with an ion chromatography (IC; DX-300 system, Dionex Co., California, USA). Phenol was analyzed with a high performance liquid chromatography (HPLC; CCPE, Tosoh, Tokyo, Japan). The details of the analysis were described in Lee *et al.* 1997.

RESULTS AND DISCUSSION

Efficient PCE dechlorination by a soil enrichment. Although most of all samples showed PCE dechlorinating activity, effective enrichments were not obtained from non-polluted samples, suggesting poor presence of effective PCE dechlorinating microbes in ordinary environments. Since a soil sample from a solvent-contaminated site showed the highest potential of PCE dechlorination, its culture was serially transferred into a fresh medium for enrichment.

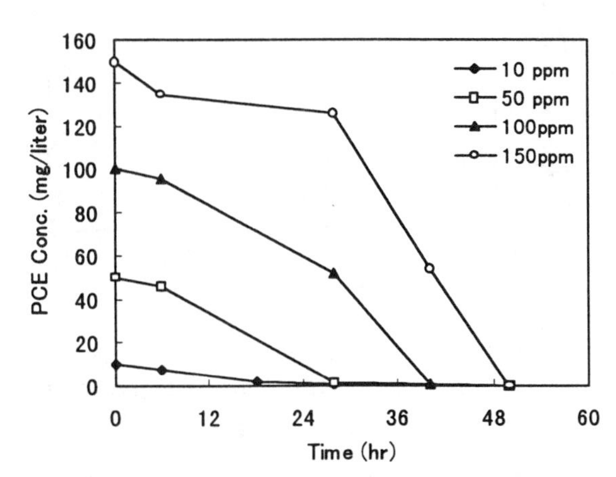

FIGURE 1. PCE dechlorination by the efficient enrichment culture under a wide range of initial concentrations.

Finally, we have constructed a stable enrichment which could stoichiometrically dechlorinate 150 mg/liter of PCE into *cis*-DCE within 2 days. During PCE dechlorination by this enrichment, although a transient accumulation of TCE was observed, VC was not accumulated at a detectable concentration.

The enrichment exhibited PCE dechlorinating activity under a relatively wide range of initial PCE concentration (10-150 mg/liter; Figure 1), temperature (15-40℃ at pH 7) and pH (6.0-9.0 at 25℃). PCE dechlorinating rate of the enrichment under an optimized condition (30℃, pH 7) was comparable with that of the previously-reported, most efficient bacterial cultures (Lee *et al.* 1997). Monod kinetic coefficients for the PCE dechlorination were determined; k (the maximum specific substrate removal rate) was 0.93 mg-PCE/mg-VSS/day and Ks (the Monod half velocity constant) was 109 mg-PCE/liter (the actual Ks may be lower because the dechlorination activity was inhibited at PCE concentration higher than 150 mg/liter). Since it may be very difficult to create alternate anaerobic/aerobic conditions in situ, the PCE bioremediation using the enrichment should be performed in a bioreactor on site.

TCE degradation by phenol degrading bacteria. TCE degrading properties of 14 distinct types of phenol-degrading bacteria were evaluated. Among them, 13 strains showed 20-100 % removal of 1 mg/liter of TCE when grown on phenol. However, when they were grown without phenol, no strain showed significant TCE biodegradation. It was suggested that most of phenol degrading bacteria may have ability to degrade TCE, if grown on or induced with phenol.

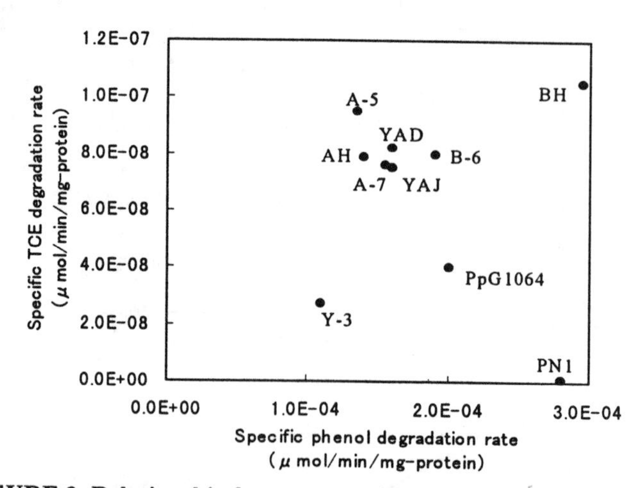

FIGURE 2. Relationship between specific phenol degradation rate and specific TCE degradation rate of phenol-degrading bacteria

Although the general trend was observed that bacterial strains with the higher phenol degrading activities showed the higher initial TCE degradation rates

(Figure 2), the extents (ultimate removal) of TCE degradation had no relationship with their phenol degrading activity (data not shown). This indicates that the stability of TCE degrading activity depends on the strain, probably due to the differences in the stability of the phenol catabolic enzyme(s) of each strain and TCE or its metabolite(s) toxicity.

Among tested phenol degrading bacteria, *P.* putida BH, which showed the highest TCE degrading activity, was selected and was adapted to higher TCE concentration. Finally, an efficient bacterial culture of BH which could co-oxidize TCE as high as 20 mg/liter within 2 days was established. The maximum specific TCE degradation rate *k* was 0.1 mg-TCE/mg-VSS/day and the Monod half velocity constant Ks was 10.3 mg-TCE/liter.

Although phenol has been reported as an effective primary substrate for co-oxidative TCE and *cis*-DCE degradation in lab-scale and filed evaluation, addition of phenol for biostimulation can lead to another contamination problem by phenol itself. Therefore, this efficient TCE degrading bacterial culture was applied to aerobic reactor system.

Construction of a sequential anaerobic/aerobic reactor system. Based on the results mentioned above, an anaerobic/aerobic reactor combining the anaerobic PCE dechlorinating enrichment and the TCE degrading BH culture was designed and constructed for complete degradation of PCE .

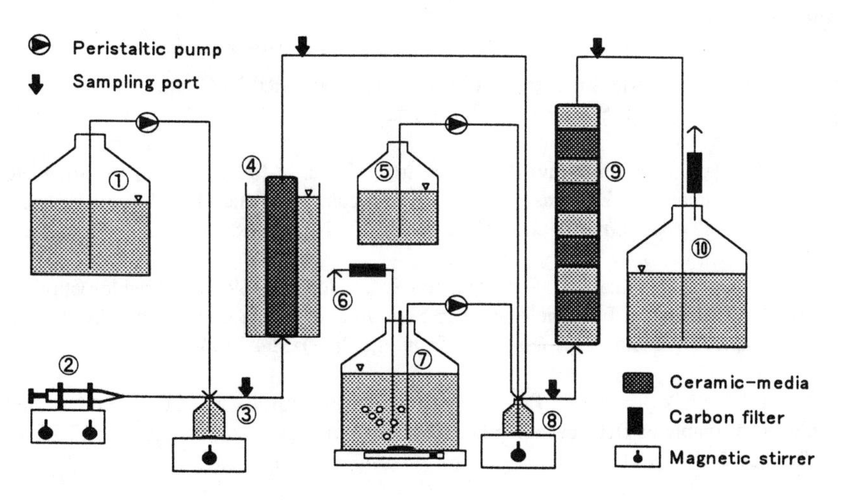

FIGURE 3. Schematic diagram of a sequential anaerobic/aerobic reactor system for complete degradation of PCE. ① Medium for anaerobic culture; ② Syringe pump for injection of PCE; ③ Influent mixing bottle; ④ Anaerobic column; ⑤ Phenol solution; ⑥ Air; ⑦ Medium for aerobic culture; ⑧ Influent mixing bottle; ⑨ Aerobic column; ⑩ Effluent tank

Figure 3 shows a schematic diagram of two up-flow glass column system. All connections were made of stainless-steel, glass, viton and Teflon tubings to minimize abiotic loss of chlorinated compounds. The volume of anaerobic and aerobic column were 707 ml (length 360 mm, inner diameter 50 mm) and 1881 ml (length 550 mm, inner diameter 66 mm), respectively. Both columns were filled with nearly cubical (10 mm × 10 mm × 10 mm) and porous ceramic-media and the occupied volume were about 500 ml for anaerobic and 800 ml for aerobic columns so that the actual volume of anaerobic and aerobic column were 400 ml and 1350 ml, respectively. The primary feed solution (basal medium) for anaerobic culture pumped with a peristaltic pump and a PCE solution injected with a syringe pump were mixed before fed upward through the anaerobic column. The effluent of the anaerobic column was mixed with the basal medium for aerobic culture and with separately pumped 100-200 mg/liter phenol solution for up-flow aerobic column. The basal medium was purged with air purified by activated carbon filter, before mixing. The sampling port to detect remaining and resulting products of chlorinated compound, phenol and carbon source were located at the influent and effluent lines of both columns. The system was operated at 30±2℃ for anaerobic and at 20±2℃ (room temperature) for aerobic columns and pH 7.0±0.2 for both. The reactor system is being successfully operated at present.

REFERENCE

Gossett, J. M. 1987. "Measurement of Henry's Law Constants for C_1 and C_2 Chlorinated Hydrocarbons." *Environ. Sci. Technol.* 21(2): 202-208.

Hopkins, G. D., L. Semprini, and P. L. McCarty, 1993a. "Microcosm and In Situ Field Studies of Enhanced Biotransfomation of Trichloroethylene by Phenol-Utilizing Microorganisms" *Appl. Environ. Microbiol.* 59(7): 2277-2285.

Hopkins, G. D., J. Munakata, L. Semprini, and P. L. McCarty, 1993b. "Trichloroethylene Concentration Effects on Pilot Field-Scale In-Situ Groundwater Bioremediation by Phenol-Oxidizing Microorganisms." *Environ. Sci. Technol.* 27(12): 2542-2547.

Lee, T. H., M. Ike, and M. Fujita, 1997. "Characterization of an Anaerobic Soil Enrichment Capable of Dechlorinating High Concentration of Tetrachloroethylene." *Water Sci. Technol.* 36(6-7): 117-124.

Maymo-Gatell, X., Y. Chien, J. M. Gossett, and S. H. Zinder. 1997. "Isolation of a Bacterium that Reductively Dechlorinates Tetrachloroethene to Ethene." *Science*. 276(6): 1568-1571.

Shurtliff, M. M., G. F. Parkin, L. J. Weathers, and D. T. Gibson. 1996. "Biotransformation of Trichloroethylene by a Phenol-Induced Mixed Culture." *J. Envir. Engrg.* ASCE. 122(7): 581-589

EFFECT OF TWEEN SURFACTANTS ON THE MICROBIAL REDUCTIVE DECHLORINATION OF HEXACHLOROBENZENE

Daniel H. Yeh, Kurt D. Pennell and ***Spyros G. Pavlostathis***
Georgia Institute of Technology, Atlanta, Georgia

ABSTRACT: The effect of Tween surfactants on the reductive dechlorination of hexachlorobenzene (HCB) was examined using three anaerobic HCB-dechlorinating, enriched cultures. The mixed cultures were developed at 22°C using contaminated estuarine sediments as inoculum. One culture received glucose and the other two lactate as the carbon source. One of the lactate-fed cultures exhibited slow dechlorination activity while the other actively dechlorinated HCB. Reductive dechlorination assays were performed at surfactant concentrations ranging from 10 to 1000 mg/L. In general, increasing surfactant concentrations inhibited reductive dechlorination. When dechlorination was not inhibited, HCB was sequentially dechlorinated to 1,3- and 1,4-dichlorobenzene. The extent of the surfactants' impact on HCB dechlorination varied. In the cases of the glucose-fed and the actively-dechlorinating lactate-fed cultures, Tween 61 and 65 were the least inhibitory and dechlorination proceeded at a surfactant concentration of 1000 mg/L. Unexpectedly, the dechlorination rate was enhanced in the slowly-dechlorinating lactate-fed culture in the presence of Tween 40 and 60 at 10 and 50 mg/L, as well as Tween 81 and 85 at concentrations up to 200 mg/L. These results demonstrate the great variability in the biological impact of nonionic surfactants, even among those from the same homologous series.

INTRODUCTION

Natural attenuation of organic contaminants in soils and sediments typically occurs slowly due to the hydrophobic and recalcitrant nature of these compounds. In recent years, the use of surfactants to enhance the bioavailability of sorbed organic contaminants has been examined (Rouse *et al.*, 1994). Although most of these studies have focused on the degradation of hydrocarbons under aerobic conditions, limited information exists on the effect of surfactants on anaerobic biotransformation processes, especially reductive dehalogenation (Liu *et al.*, 1996; Van Hoof and Jafvert, 1996). Typically, nonionic surfactants have been used in bioavailability studies because of their relatively low toxicity compared to ionic surfactants. The selection among nonionic surfactants, however, has often been based primarily on the physical/chemical attributes of the surfactants (e.g., extent of contaminant solubility enhancement), with less emphasis placed on the direct compatibility of the surfactant with the specific microbial system. As a result, it is often difficult to determine whether biological or physicochemical mechanisms are responsible for the observed enhancement or inhibition of biotransformation in soil/sediment systems amended with surfactants.

In a previous study it was found that food-grade, polyoxyethylene (POE) sorbitan esters, commercially available as the Tween series, were less inhibitory to mixed, methanogenic cultures than several linear alcohol ethoxylates or alkylphenol ethoxylates (Yeh *et al.*, 1998). The Tween series are generally considered to be non-toxic and therefore should be ideal for use in bioavailability-enhancement applications. However, even among the Tween surfactants, great differences in properties exist. The objective of the work presented here was to evaluate the effect of Tween surfactants on the microbial reductive dechlorination of hexachlorobenzene (HCB) using two anaerobic consortia derived from estuarine sediments contaminated with polychlorinated organic compounds. Since the investigation focused specifically on the biological effect of Tween surfactants, the experiments were conducted with mixed, enriched cultures free of sediment solids in order to eliminate complications from contaminant and surfactant sorption as well as contaminant bioavailability.

MATERIALS AND METHODS

Surfactants. The Tween series (20/21/40/60/61/65/80/81/85) consists of nine surfactants containing different types of fatty acids (lauric, palmitic, stearic, and oleic), ethylene oxide (EO) group numbers ranging from 4 to 20, and hydrophile-lipophile balance (HLB) numbers ranging from 9.6 to 16.7. The surfactants were obtained from ICI Americas (Wilmington, DE) and used without further purification.

Culture Development. Two HCB-dechlorinating, methanogenic, mixed cultures were developed using estuarine contaminated sediments (Prytula and Pavlostathis, 1996). The cultures were maintained at 22°C in 9-L (6-L liquid volume) sealed glass reactors with a 14-day fill-and-draw cycle resulting in an average hydraulic retention time of 84 days. After over six months of operation, the cultures were sediment-free. One culture was fed glucose while the other was fed lactate, leading to the development of two different consortia. Both cultures have been maintained for over two years. At the time of this experiment, the dechlorination activity of the lactate-fed culture was low. A new, HCB-dechlorinating, lactate-fed culture was redeveloped using a previously enriched and hexachloro-1,3-butadiene dechlorinating culture (Booker *et al.*, 1997) and the effect of surfactants on both lactate-fed cultures, in addition to the glucose-fed culture, was investigated.

Dechlorination Assays. Dechlorination assays were conducted in 26 mL glass serum tubes sealed with TFE-lined septa and aluminum crimps. The tubes contained 20 mL of culture, HCB dissolved in methanol, an electron donor (glucose or lactate), and surfactants at levels ranging from 10 to 1000 mg/L. The reference series contained no surfactant. Sacrificial sampling was done over time by injecting 2 mL of isooctane containing 1,3,5-tribromobenzene as an internal standard into each unopened tube, which was then shaken vigorously to extract and concentrate the chlorinated benzenes from both the gas- and liquid-phases. Excess gas production and methane content were measured using previously reported methods (Prytula and Pavlostathis, 1996). The tubes were then centrifuged to facilitate the separation of the isooctane layer which was analyzed using a gas chromatography unit equipped with an electron capture detector (ECD).

RESULTS AND DISCUSSION

The presence of the Tween surfactants did not negatively impact the extent of methanogenesis. Generally, an initial lag period was observed at the high surfactant concentration (1000 mg/L), with the more inhibitory surfactants (especially Tween 21) requiring a longer acclimation period. Table 1 lists the total gas production in the 1000 mg/L surfactant series for each culture at the end of the incubation period. The surfactants were degraded over time, as evidenced from the increased volume of gas produced at the end of the incubation period as compared to that of the reference.

TABLE 1. Total gas production (mL @ 22°C, 1 atm) by the three cultures at the end of the incubation period (Surfactant concentration 1000 mg/L).

Series	Glucose-fed [a]	Slow lactate-fed [b]	Active lactate-fed [c]
Reference	8.6	6.6	5.2
Tween 20	12.0	N.A.[d]	9.4
Tween 21	12.2	N.A.	1.3
Tween 40	11.7	10.9	9.4
Tween 60	10.2	10.9	10.3
Tween 61	12.5	N.A.	8.9
Tween 65	10.4	N.A.	7.6
Tween 80	8.3	N.A.	6.5
Tween 81	14.6	14.8	11.3
Tween 85	10.5	11.5	12.1

Incubation period (days): [a] 72, [b] 53, [c] 48; [d] N.A., not assayed.

Figure 1 shows the HCB dechlorination profiles in the glucose-fed (A-C) and slowly-dechlorinating, lactate-fed (D-F) cultures at different initial surfactant concentrations. In the case of the glucose-fed culture, the rate of HCB dechlorination was not affected in any of the surfactant-amended series at 10 mg/L. At an initial surfactant concentration of 200 mg/L, certain surfactants slightly decreased the rate of HCB dechlorination. The observed rate of HCB dechlorination was as follows: Reference > T61 > T65 > T60 > T81 > T85 > T40 > T80 > T20 > T21. In general, HCB dechlorination was inhibited with increasing initial surfactant concentrations. As the surfactant concentration increased, the formation of products was delayed and complete inhibition of dechlorination occurred at a surfactant concentration of 1000 mg/L, except for Tween 61 and 65, for which dechlorination still occurred, albeit at a lower rate (Fig. 1C).

For those cases where the presence of certain surfactants affected the rate of HCB dechlorination, the extent of dechlorination and product distribution were not impacted. Figure 2 shows the product distribution over the course of the incubation for the reference, Tween 61 at 10 and 1000 mg/L, and Tween 80 at 10 and 1000 mg/L in the glucose-fed culture. Contaminant mass balance -- indicated by the dotted lines -- confirms that the extraction procedure was efficient and contaminant losses did not occur. The presence of surfactants at 10 mg/L did not impact the dechlorinates HCB led to results very similar to that of the glucose-fed culture (i.e., ranging from no enhancement to some inhibition of HCB dechlorination).

The two least inhibitory surfactants for the glucose-fed and the actively dechlorinating lactate-fed cultures were the stearic acid-containing Tween 61 and 65, both with low HLB values (9.6 and 10.5, respectively) which indicate high lipophilicity. Further evaluation of these two surfactants shows that they sorb strongly to the biomass. Possibly, the favorable biological effects of these two surfactants are attributed to the strong association and compatibility of the surfactants with the cell membrane components.

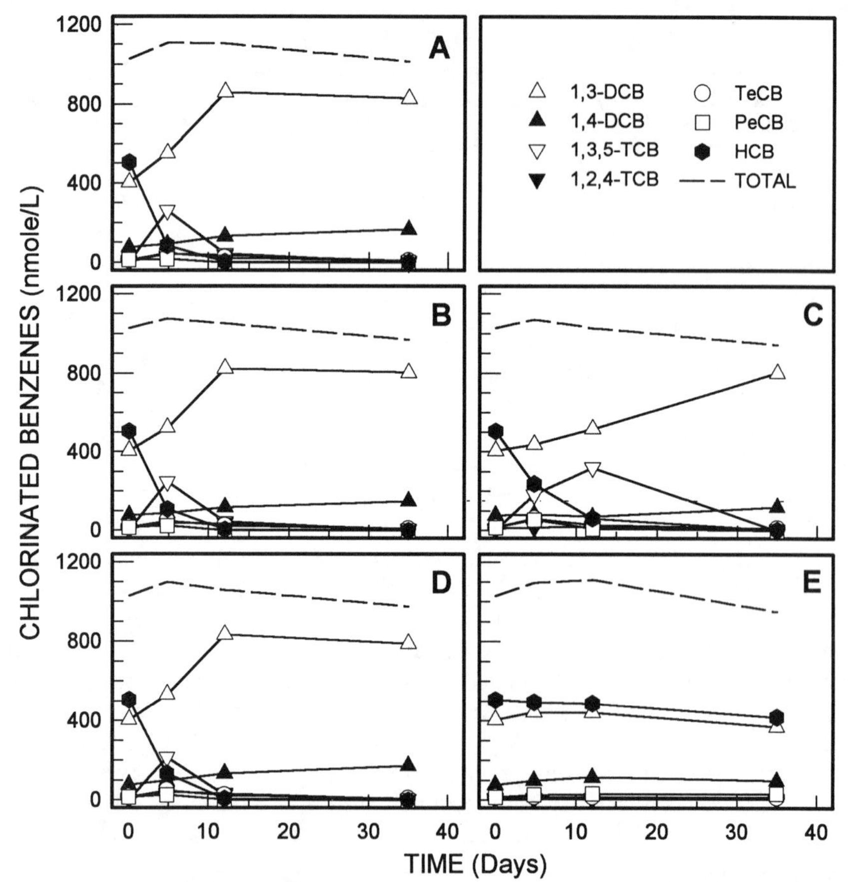

FIGURE 2. HCB dechlorination product distribution in the glucose-fed culture: Reference (A); Tween 61 at 10 (B) and 1000 (C) mg/L; Tween 80 at 10 (D) and 1000 (E) mg/L.

It has been suggested in the literature that contaminants partitioned into the surfactant micellar phase may not be available to the microorganisms. The higher gas production in the surfactant-amended series as compared to that of the reference series was attributed to surfactant biodegradation, which in turn indicates that the dechlorination activity of the culture. Although the HCB dechlorination rate in the 1000 mg/L Tween 61 series was slower, HCB dechlorination resulted in the production of 1,3- and 1,4-dichlorobenzene (DCB) similar to that of the reference (Fig. 2C and 2A, respectively). In contrast, HCB dechlorination was completely inhibited in the presence of Tween 80 at 1000 mg/L (Fig. 2E).

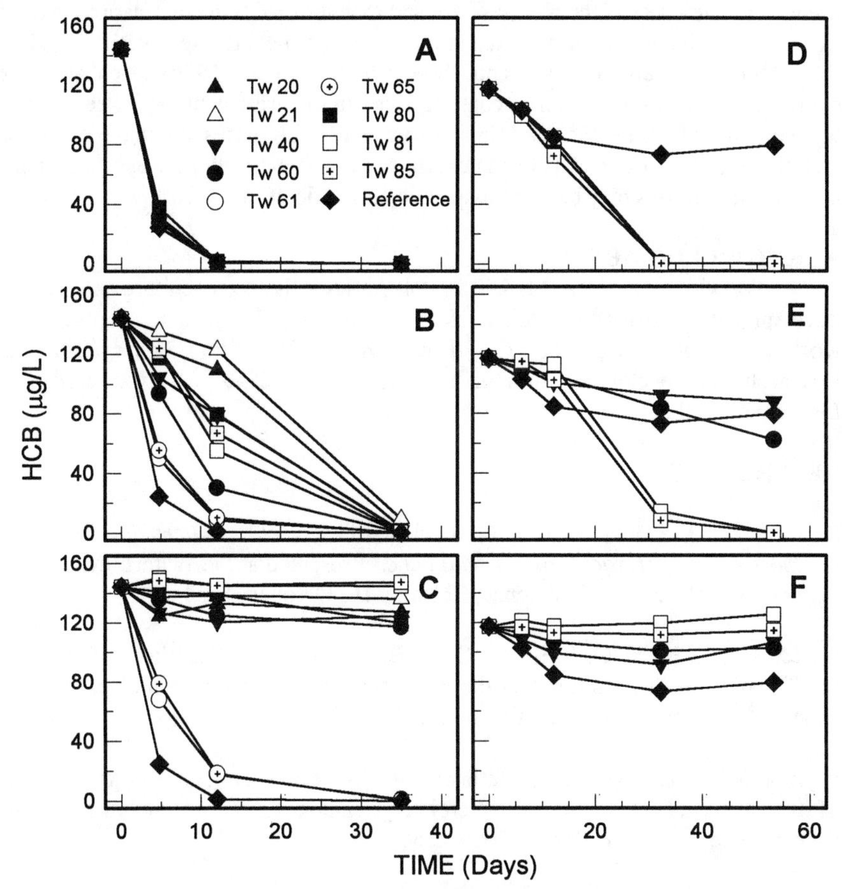

FIGURE 1. HCB dechlorination profiles in the glucose-fed (A-C) and the slowly-dechlorinating, lactate-fed (D-F) cultures in the presence of Tween surfactants at 10 (A, D), 200 (B, E), and 1000 (C, E) mg/L initial concentration.

The most surprising result of this study was that the HCB dechlorination activity of the slowly-dechlorinating, lactate-fed culture was enhanced in the presence of Tween 40 and 60 at 10 and 50 mg/L, as well as Tween 81 and 85 at concentrations up to 200 mg/L (Fig. 1D & 1E). As the exact cause for the slow HCB dechlorination of this culture was never established, it was not possible to determine why the Tween surfactants enhanced the dechlorination activity. Dechlorination assays in the presence of Tween surfactants with the second lactate-fed culture which actively surfactants were readily accessible to the microorganisms and that surfactant/cell contact was not a problem. Thus, the observed inhibition of HCB dechlorination at high surfactant concentrations (1000 mg/L) does not appear to be attributed to a low bioavailability of HCB in the surfactant micellar phase.

The addition of different Tween surfactants to mixed, dechlorinating cultures resulted in differences of the observed rate and extent of HCB dechlorination. Mixed results have also been reported in the literature on attempts to use Tween 80 and 85 (Van Hoof and Jafvert, 1996) and Tween 20 (Liu *et al.*, 1996) to enhance the biotransformation of sorbed-phase contaminants. In the present study, Tween 61 and 65 were found to be the least inhibitory. The differences among the Tween surfactants, as demonstrated in this study, show that compatibility between surfactants and microorganisms may be system-specific and difficult to forecast.

ACKNOWLEDGMENTS

We wish to thank Dr. Mark T. Prytula for his assistance in culture and method development, as well as ICI Americas for supplying the surfactant samples. This work was supported by the U. S. Environmental Protection Agency/Office of Research and Development through a contract no. R 825404-01-0 to the Georgia Institute of Technology.

REFERENCES

Booker, R., J. Gould, and S. G. Pavlostathis. 1997. *Microbial Reductive Dechlorination of Hexachloro-1,3-butadiene.* Presented at the Water Environment Federation 70th Annual Conference, Chicago, IL, October 18-22, 1997.

Liu, X., R. C. Sokol, O.-S. Kwon, C. M. Bethoney, and G.-Y. Rhee. 1996. "An Investigation of Factors Limiting the Reductive Dechlorination of Polychlorinated Biphenyls." *Environ. Toxicol. Chem.* 15:1738-1744.

Prytula, M. T., and S. G. Pavlostathis. 1996. "Effect of Contaminant and Organic Matter Bioavailability on the Microbial Dehalogenation of Sediment-Bound Chlorobenzenes." *Wat. Res.* 30:2669-2680.

Rouse, J. D., D. A. Sabatini, J. M. Suflita, and J. H. Harwell. 1994. "Influence of Surfactants on Microbial Degradation of Organic Compounds." *CRC Crit. Rev. Environ. Sci. Technol.* 24:325-370.

Van Hoof, P. L., and C. T. Jafvert. 1996. "Reductive Dechlorination of Chlorobenzenes in Surfactant-Amended Sediment Slurries." *Environ. Toxicol. Chem.* 15:1914-1924.

Yeh, D. H., K. D. Pennell, and S. G. Pavlostathis. 1998. *Toxicity and Biodegradability Screening of Nonionic Surfactants Using a Sediment-Derived Methanogenic Consortium.* Presented at the IAWQ 19th Biennial International Conference, Vancouver, Canada, June 21-26, 1998.

ENHANCED IN SITU REDUCTIVE DECHLORINATION

Erica S. K. Becvar (ARA, Tyndall AFB, Florida), Arthur Fisher (NAS Fallon, Fallon, Nevada), Guy Sewell (US EPA, Ada, Oklahoma), Victor Magar (Battelle, Columbus, Ohio), Jim Gossett (Cornell University, Ithaca, New York), and Catherine M. Vogel (U. S. Air Force Research Laboratory, Tyndall AFB, Florida)

ABSTRACT: Chloroethenes can be reductively dehalogenated. Hydrogen appears to be the direct electron donor. Studies with site core materials from a tetrachloroethene (PCE)-contaminated plume at Naval Air Station (NAS) Fallon, Nevada, have shown reductive dechlorination of chloroethenes to be stimulated by the addition of common fermentation products. Based on these results, a field treatability was initiated at NAS Fallon utilizing indigenous bacteria and added electron donors to promote in situ dechlorination of PCE. The field system includes injection of electron donors in various combinations in three treatment zones, isolated by barriers installed parallel to the groundwater flow path. Monitoring wells are sampled for parent compound dechlorination and dechlorination products; electron donor degradation, anaerobic fermentation products, and system stability. Transformation patterns and transport flow studies are being performed based on tracer studies, PCE removal, and the appearance of daughter products from the PCE dechlorination. After four months of operation, field monitoring of the system and laboratory analysis of collected field samples show evidence of an anaerobic environment with preliminary evidence of enhanced in situ reductive dechlorination. Continued operation of the system with nutrient injection is aimed at validating enhanced in situ reductive dechlorination as a remediation technology in a realistic field situation.

INTRODUCTION

Improper storage and disposal of chlorinated solvents have led to extensive soil and groundwater contamination. Chloroethenes can be reductively dechlorinated (Bario-lage et al. 86; Freedman and Gossett, 89; DiStefano et al. 91; Galli and McCarty, 89). A microbial culture capable of reductively dechlorinating PCE to ethene (ETH) with efficient use of electron donors has been isolated at Cornell University (Freedman and Gossett, 89; Maymó-Gatell et al. 97). Research at Cornell revealed that H_2 is the direct electron donor responsible for PCE dechlorination (DiStefano, et al. 91). Methanol (MeOH) and other reductants found to support dechlorination merely serve as H_2 precursors. Substrates such as butyrate, lactate, and ethanol-benzoate are not direct methanogenic substrates. They eliminate competition for the supplied donor itself and provide H_2 as a direct fermentation product. H_2 is produced slowly at low levels providing for complete mineralization of PCE, thus favoring dechlorination over competition for the substrates (DiStefano, et al. 92). These results suggest that strategies utilizing slow, steady H_2 delivery are best to stimulate and maintain reductive dechlorination.

In the field, studies with site materials and isolated test plots have shown reductive dechlorination of chlorinated solvents to be stimulated by the addition of electron donors (Gibson and Sewell, 92; Major and Cox, 92). Based on these results, this field effort utilizes indigenous bacteria and added electron donors to stimulate the degradation of PCE to ETH in the subsurface at NAS Fallon, Nevada (NASF). The field system consisting of five semi-enclosed treatment lanes is allowing researchers to investigate the addition of various electron donors to enhance reductive dechlorination, in addition to investigating dechlorination through natural attenuation and iron electrodes. A detailed understanding of the dechlorination process will lead to more efficient, cost-effective, and reliable strategies for the bioremediation of PCE and related compounds.

MATERIALS AND METHODS

Site Description. The Crash Crew Training Area (Site 1) at NASF consists of an unlined, earth-bermed burn fire-training pit, previously associated with two above-ground fuel storage tanks. The pit was used to burn an estimated 1.1 million gallons of flammable liquids (*i.e.*, fuel and lubricants). Sandy soils cover the site and extend to a depth of approximately 20 ft (1.2 m) below ground surface (bgs), with an intermittent 2-ft- (0.6 m) thick layer of clay-rich silts and sands at about 10 ft bgs. These layers form an unconfined aquifer. At the bottom of the unconfined aquifer is a sandy silt and clay layer that acts to impede contaminant movement from the surface aquifer to deeper aquifers. The clay layer is nearly 20 ft (6.1 m) thick across most of the site (ORNL, 94). The dissolved-phase plume at Site 1 contains both fuel and chloroethene related constituents (Table 1).

TABLE 1. Contaminant concentrations and general groundwater chemistry.

Contaminant	Concentration (μg/L)	General Water Chemistry	Concentration
PCE	2.6 – 2130	pH	7.60 - 9.11
TCE	9.9 – 675	Conductivity	3,750 - 48,900 μmhos
DCEs	1.0 – 2130	Total Alkalinity	569 - 1965 mg/L
VC	1.1 – 3.8	O-P	0.74 - 3.08 mg/L
Toluene	1.3 – 56.4	Cl^-	661 - 15,100 mg/L
Benzene	1.2 – 242	SO_4^{-2}	386 - 8,650 mg/L
Ethylbenzene	2.0 – 152	NO_2^- (N)	< 0.20 mg/L
Xylenes	1.2 – 450	NO_3^- (N)	< 2.68 mg/L

Field System Setup. The field site consists of five parallel, 25-ft- (7.6 m) long biotreatment lanes (Lanes A through E), separated by 20-ft- (6.1 m) deep, high-density polyethylene (HDPE) barriers. The barriers are installed approximately 4 ft (1.2 m) into the 20-ft-deep clay layer. The layout of the treatment lanes and corresponding injection, extraction, and groundwater monitoring wells is shown in Figure 1. The treatment lanes are oriented in the direction of the groundwater flow. Groundwater flow through the five lanes is hydraulically controlled using a single downgradient extraction well for all five lanes and five injection wells located at the upstream end of each lane. The downgradient extraction well pump rate is approximately 200 gallons per day (gpd) (756 liters per day [Lpd]), and 10 gpd (37.3 Lpd) is injected into each of the groundwater injection wells.

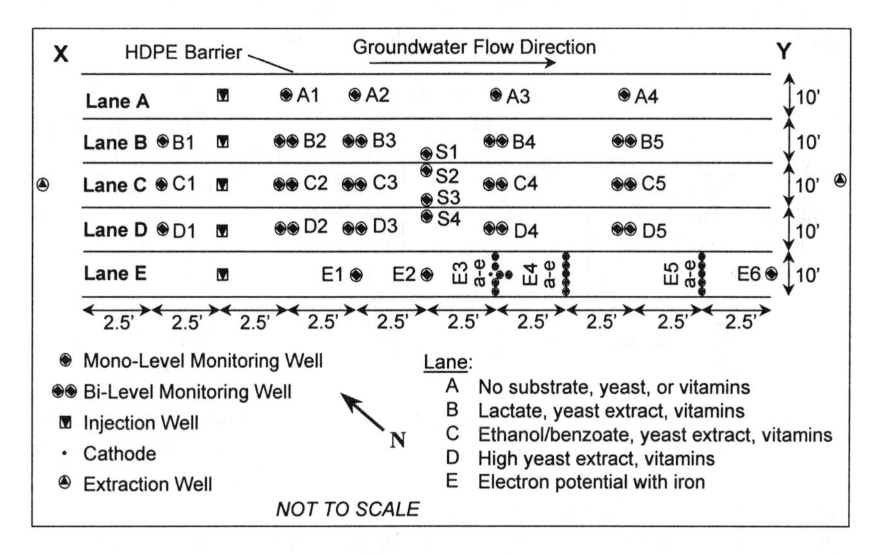

FIGURE 1. Site 1 Treatment Lane Configuration and Injection Well Extraction Well, and Monitoring Well Layout.

Lane A is used as a control lane and has four mono-level monitoring wells downgradient of the Lane A injection well. The control lane is operated without adding electron donors or nutrients. Lanes B, C, and D are fed alternative electron donors; nutrient feed solutions are blended with influent water from the upgradient extraction well during injection. Each of these lanes has four bi-level monitoring wells located downgradient of their respective injection wells. All five lanes have mono-level wells, located 5 ft (1.5 m) upgradient of the injection wells. Lanes B, C, and D have mono-level side wells (S wells), located on either side of the HDPE barriers that separate these lanes. Mono-level wells are screened from 9 to 10 ft (2.7 m to 3 m) below ground surface (bgs), and bi-level wells are screened at 9 to 10 ft (2.7 m to 3 m) and 11 to 12 ft (3.4 m to 3.7 m) bgs. All wells are 1-inch-diameter, stainless steel, direct push wells. An iron electrode was installed in Lane E where the iron acts as an anode, giving off electrons which go toward the reduction of hydrogen ions to dissolved H_2 gas. Hydrogen is expected to contribute to the reductive dechlorination of PCE. Lane E is being used by the US Environmental Protection Agency (EPA) in conjunction with the US Air Force and the US Navy and its discussion is not included in this paper.

Feed Schedule. Initial electron donor concentrations are 540 mg/L for lactate, and 140 mg/L and 170 mg/L for ethanol (EtOH) and benzoate, respectively. Concentrations are modified as needed and are based on NASF soil microcosm studies. Over 16 g/L lactate, 8 g/L benzoate, or 8 g/L EtOH would be required to satisfy the total $SO_4^{=}$ burden in each lane. Because cost and the potential for clogging the aquifer render such high electron donor concentrations prohibitive, the added electron donors are not expected to satisfy the electron donor-demand for $SO_4^{=}$ reduction. Vitamin and yeast extract concentrations are shown in Table 2. The high yeast extract concentration is applied to Lane D.

TABLE 2. Influent and yeast extract concentrations.

Vitamin/ Yeast Extract	Concentration (mg/L)	Vitamin/ Yeast Extract	Concentration (mg/L)
pyridoxine hydrochloride	0.05	d-biotin	0.01
thiamin hydrochloride	0.025	folic acid	0.01
DL-calcium pantothenate	0.025	riboflavin	0.025
p-aminobenzoic acid	0.025	nicotinic acid	0.025
high yeast extract	200	lipoic acid	0.025
yeast extract amendment	20	vitamin B_{12}	0.025

Tracer Test. Two tracer tests were conducted in series. In both tests, a freshwater tracer was injected at 10 gpd (37.8 Lpd) for a one- to two-week period into Lane C. Freshwater was expected to result in reduced total dissolved solids concentrations in Lane C, including chloride and other anions and cations. Field monitoring parameters included conductivity, temperature, pH, dissolved oxygen (DO), and oxidation-reduction potential (ORP). Additional samples were sent to the US EPA (Kerr Research Laboratory, Ada, Oklahoma). These were analyzed for anions (sulfate, total nitrates [nitrate + nitrite], and chloride), dissolved organic carbon, alkalinity, pH, and conductivity.

Fresh water tracer test results were inconclusive regarding groundwater transport in Lane C at Site 1. Currently, bromide is being investigated as a tracer. Modifications have been made to detect bromide above background chloride levels. A groundwater model describing the treatment lanes will be used to simulate groundwater transport and the tracer results at the site. Laboratory results will be compared with field sampling to assess groundwater flowrates at the site.

Sampling and Analysis. On-site field system monitoring analysis consists of conductivity, pH, temperature, ORP, and DO. The US EPA (Kerr Research Laboratory, Ada, Oklahoma) is performing laboratory analyses of field samples.

Laboratory analysis for organics include PCE, dechlorination by-products (TCE, DCE, and VC), and electron donor concentrations. The headspace gas chromatography/mass spectroscopy (GC/MS) of chloroethenes uses US EPA, Robert S. Kerr Environmental Research Center (RSKERC), standard analytical method RSKSOP-148 for the analysis. The HPLC analysis of acetic acid uses a Dionex ICE-ASI IonPac column and an AMMS-ICE MicroMembrane Suppressor in the analysis. The Suppressor reagent used is 5 mM tetrabutylammonium hydroxide and the eluent is 1.0 mM heptafluorobutyric acid. The flowrate is 0.8 mL/min for the eluent and 1.0 mL/min for the Suppressor reagent.

Inorganic laboratory analyses include $SO_4^{=}$, NO_3^{-}, iron, DO, pH, alkalinity, and conductivity. The methods used for the inorganic analyses are EPA

Method 353.1 for NO_3^- and NO_2^-; EPA Methods 120.1, 310.1, and 150.1 for pH; and Waters capillary electrophoresis Method N-601 for chloride and $SO_4^=$. DOC analysis uses the US EPA RSKERC standard analytical method RSKSOP-102.

Total fuel carbon samples are analyzed by purge and trap/GC-PID:FID using the US EPA RSKERC standard analytical method RSKSOP-133 as reference. The GC/MS analysis for phenols and aliphatic/aromatic acids uses US EPA RSKERC standard analytical method RSKSOP-177 for the extraction and derivatization. The dissolved gas analysis uses US EPA RSKERC standard analytical method RSKSOP-175 and US EPA RSKERC standard analytical method RSKSOP-194 for reference.

RESULTS

System startup at Site 1 began July 1997 with the freshwater tracer test. Nutrient injection in the treatment lanes began October 1998 and is scheduled for completion in August 1998. For the purposes of this paper, only data pertinent to the enhanced in situ reductive dechlorination will be discussed.

Field Monitoring. On-site field system monitoring consists of conductivity, pH, temperature, ORP, and DO. Of these parameters, only temperature, conductivity, and ORP showed definite trends. Temperature generally decreased across all lanes during the first four months of operation; this may be attributed to the onset of winter. Conductivity generally decreased across all lanes. This may be due to rainwater infiltration during winter months. ORP levels in the shallow (10-ft deep) wells increased slowly, from approximately –220 mV to –160 mV, and decreased in the deeper (12-ft deep) wells from –30 to –110 mV. These changes could be due to vertical mixing due to the increased groundwater flow rates at the site. However, the increase in the shallow in ORP values was not sufficient to indicate the loss of the anaerobic environment.

Laboratory Analysis. Inorganic laboratory analysis of field samples consists of $NO_2^- + NO_3^-$; bromide and chloride ion concentrations; NH_3, O-P, alkalinity, conductivity, pH, SO_4^{-2}, and DOC. No direct correlations can be drawn at this time from the $NO_2^- + NO_3^-$, bromide ion concentration, NH_3, O-P, and pH. However, alkalinity and DOC generally decreased across all lanes during the first four months of operation. However, considering the lower alkalinity and DOC levels of the injected water for each lane, the system may be experiencing a dilution phenomenon which, with continued system operation, may reach equilibrium. Conductivity, chloride ion concentration, and SO_4^{-2} generally experienced an increase during the first four months of operation. These increases may be attributed to the greater concentration of these parameters in the injected water for each lane, or to vertical mixing. Increases in conductivity can also be tied to the increase in chloride concentration and is in direct agreement with the field monitoring of the system.

Analysis for chlorinated solvents in the laboratory includes PCE, TCE, 1,1-DCE, c- and t-DCE, and VC. In general, there appear to be slow decreases in PCE, TCE, and *c*-DCE, without a corresponding increases in VC (Figure 2). This maybe attributed to dilution (injection chloroethene concentrations are lower than their original concentrations in each lane) or vertical mixing (chloroethene concentrations were vertically stratified at the onset of the study). There is no indication at this time of enhanced dechlorination at the site, based on chloroethene intermediate metabolite concentrations. This is surprising considering the promising evidence from laboratory analysis which shows the enhancement of the in situ anaerobic environment.

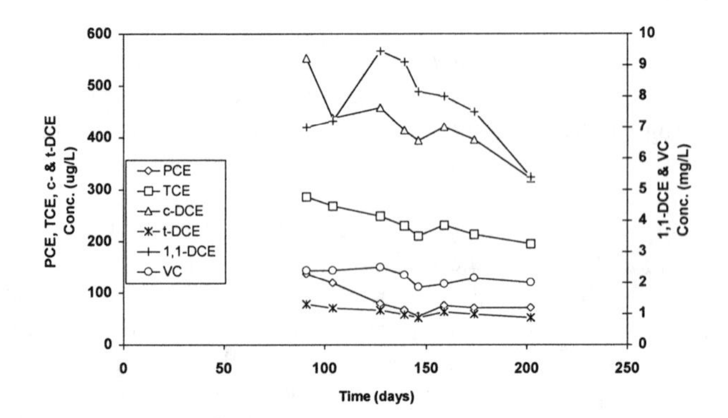

FIGURE 2. Dechlorination in Lane B after Four Months of Operation.

DISCUSSION

After four months of operation, field monitoring of the system and laboratory analysis of collected field samples show evidence of an anaerobic environment with preliminary evidence suggesting enhanced in situ reductive dechlorination. Although no direct positive correlation can be made at this time between decreases in parent compound chloroethenes and the increases in daughter or byproducts, other parameters show indication of an increasingly anaerobic environment. These promising indicators lead us to believe that continued operation of the system with nutrient injection will lead to enhanced in situ reductive dechlorination of chloroethenes.

ACKNOWLEDGEMENTS

This work is supported in part by the US Navy Facilities Engineering Services Center (NFESC); NAS Fallon, Nevada; and the US EPA National Risk Management Research Laboratory, Ada, Oklahoma. The authors wish to thank Roger Johnson (Manpower, Inc.), Raj Krishnamoorthy (NAS Fallon), and the Nevada Division of Environmental Protection (DEP) for their continuing invaluable assistance in this effort.

REFERENCES

Bario-Lage, G., F. Z. Parsons, R. S. Nassar, and P. A. Lorenzo. 1986. "Sequential Dehalogenation of Chlorinated Ethenes." *Environ. Sci. Technol.* 20(1): 96-99.

DiStefano, T. D. , J. M. Gossett, and S. H. Zinder. 1991. "Reductive Dechlorination of High Concentrations of Tetrachloroethene to Ethene by an Anaerobic Enrichment Culture in the Absence of Methanogenesis." *Appl. Environ. Microbiol.* 57: 2287-2292.

DiStefano, T. D., J. M. Gossett, and S. H. Zinder. 1992. "Hydrogen as an Electron Donor for the Dechlorination of Tetrachloroethene by an Anaerobic Mixed Culture." *Appl. Environl. Microbiol.* 58: 3622-3629.

Freedman, D. L., and J. M. Gossett. 1989. "Biological Reductive Dechlorination of Tetrachloroethylene and Trichloroethylene to Ethylene under Methanogenic Conditions." *Appl. Environ. Microbiol.* 55: 2144-2151.

Galli, R., and P. L. McCarty. 1989. "Biotransformation of 1,1,1-Trichloroethane, Trichloromethane, and Tetrachloromethane by a *Clostridium sp.*" *Appl. Environ. Microbiol.* 55: 837-844.

Gibson, S. A., and G. W. Sewell. 1992. "Stimulation of Reductive Dechlorination of Tetrachloroethene in Anaerobic Aquifer Microcosms by Addition of Short-chain Organic Acids or Alcohols." *Appl. Environ. Microbiol.* 58: 1392-1393.

Major, D. W., and E. E. Cox. 1992. "Field and Laboratory Evidence of In Situ Biotransformation of Chlorinated Ethenes at Two Distinct Sites: Implications for Bioremediation." *In situ Bioremediation Symposium, Niagara-on-the-Lake, Canada*, pp. 48-56.

Maymó-Gatell, X., Y. Chien, J. M. Gossett, and S. H. Zinder. 1997. "Isolation of a Bacterium That Reductively Dechlorinates Tetrachloroethene to Ethene." *Science* 276:1568-1571.

Oak Ridge National Laboratory. 1994. *Remedial Investigation Report Site 1 Section.*

Microbial Reductive Dechlorination of PCE in a Quasi Two-Dimensional Sandbox Model

Olaf A. Cirpka (Institut für Wasserbau, Universität Stuttgart, Germany)
Gerhard Bisch (Institut für Wasserbau, Universität Stuttgart, Germany)

ABSTRACT Microbial reductive dechlorination belongs to the naturally occurring processes tetrachloroethene (PCE) undergoes in aquifers. It contributes significantly to natural attenuation of chlorinated solvents and has been discussed to be stimulated for enhanced bioremediation. Experiments for the stimulation of reductive dechlorination were carried out in a large-scale quasi two-dimensional sandbox model inocculated with *dehalospirillum multivorans* transforming PCE to *cis*-1,2,-dichloroethene (DCE) and a mixed culture transforming DCE to ethene. Ethanol was used as electron donor. The stimulation of *dehalospirillum multivorans* was successfull, whereas stable dechlorination of DCE could not be achieved.

INTRODUCTION

In the past years, microbial transformation of chlorinated ethenes has become an important issue in the context of natural attenuation. It has been known for a long time that less chlorinated compounds such as dichloroethene (DCE) and chloroethene (VC) occur at sites contaminated by PCE or TCE. More detailed investigations showed that at certain sites complete dechlorination to ethene is achieved and that the dechlorination rates are high enough to balance the mass flux of PCE or TCE released into the groundwater (Lee et al., 1995).

Complete dechlorination of PCE by succesive microbial reduction has first been proofed for methanogenic systems (Vogel and McCarty, 1985). However, it has been shown that complete reductive dechlorination is not dependent on methanogenic conditions (DiStefano et al., 1991). Highly enriched mixed cultures and pure cultures have been isolated which couple the first two steps of reductive dechlorination of PCE to growth (Holliger et al., 1993; Neumann et al., 1994; Scholz-Muramatsu et al., 1995). For further dechlorination a mixed culture containing only a few organisms could be enriched (Granzow et al., 1996). None of these cultures have been methanogenic.

As suitable electron donors for reductive dechlorination of PCE, several compounds have been reported: glucose, acetate, formate, methanol, sucrose, proprionate and ethanol among others. DiStefano et al. (1992) explained this variety with an electron transfer from the primary electron donors to molecular hydrogen in a first step which then is utilized for the dechlorination of the chlorinated ethenes. The first step may be performed by organisms different from the dechlorinaters. That is, if e.g. ethanol is injected into an aquifer, some accompanying organisms may transform it to molecular hydrogen and acetate, which are used as electron donor and carbon source, respectively, by the dechlorinating organisms. In contrast to other electron donors such as glucose, ethanol is a rather poor substrate for most competitive anaerobic organisms, escpecially methanogens. This makes ethanol rather attractive for the stimulation of the dechlorination process. Ethanol is also easy to handle and cheap.

In the present study, microbial reductive dechlorination of tetrachloroethene using ethanol as electron donor was investigated under conditions comparable to aquifers. The experiments were conducted in a large-scale quasi two-dimensional sandbox model. In contrast to field studies, the structure of the porous medium was well known and all mass fluxes could be measured accurately. In contrast to studies in mixed reactors or soil columns, hydraulic control was comparable to *in-situ* conditions with the possibility of different compounds bypassing each other. The goal of the study was to identify the impact of hydraulics on microbial activity in the context of reductive dechlorination. This may be of interest both for intrinsic and enhanced bioremediation of chlorinated ethenes.

EXPERIMENTAL SET-UP

Construction of the Artificial Aquifer The experiments were conducted at the research facility for subsurface remediation *VEGAS* at the Institut für Wasserbau, Universität Stuttgart, Germany in a large-scale container made of stainless steel. The dimensions (LxHxW) of the container were 10.10m x 0.70m x 0.20m. The domain could be cooled to typical groundwater temperatures of $\approx$ 10 °C.

The artificial aquifer was filled with a fine Quartz sand ($K = 1.6 \cdot 10^{-3} m/s$) and a medium-grain Quartz sand ($K = 3.0 \cdot 10^{-4} m/s$) in a definite block structure shown in Fig. 1. The porosity was 0.40. Tracer studies were carried out in a second sandbox homogeneously packed with the coarser sand, in order to estimate the dispersivities of the material. In these studies values of 0.5 *mm* and 0.1 *mm* were identified for the longitudinal and transverse dispersivity, respectively.

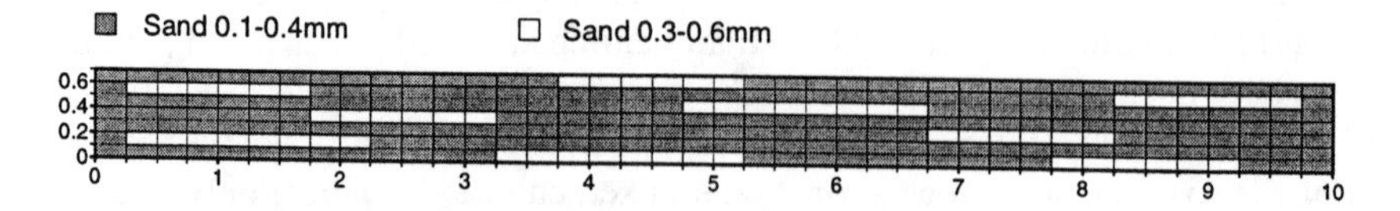

FIGURE 1: Distribution of the soil materials in the artificial aquifer.

The artificial aquifer was operated under confined conditions. The piezometric head was controlled at the outflow boundary by a constant head tank. The volumetric flux was controlled at the inflow boundary by a gear pump. The tap water used for the experiments was degasified by a vacuum stripper in which the total gas pressure was regulated to 50 *hPa*. The principle set-up of the experimental plant during the stimulation experiments is shown in Fig. 2.

Perforated pipes made of stainless steel penetrating the entire width of the domain were used as monitoring wells. The horizontal distance of the monitoring wells was 0.50 *m* and the vertical 0.10 *m*, yielding 140 monitoring points within the domain. Additional monitoring points were placed in the inlet and the outlet of the system. Water samples could be taken through Viton^{TM} septa at one end of the monitoring wells.

Stimulation Experiment The artificial aquifer was inocculated with two cultures transforming chlorinated ethenes. The first one was a pure culture of *dehalospirillum multivorans*, an organism dechlorinating PCE via TCE to DCE and coupling this process to growth (Scholz-Muramatsu et al., 1995). The second culture contained several organisms. This culture is capable to dechlorinate DCE via VC to ethene (Granzow et al., 1996). The inocculation was done by injecting 50 *ml* of both suspensions into each monitoring well. The inocculation was on November 19, 1996. The mixed culture was re-inocculated on December 29, 1997.

The water recharge into the aquifer inlet was regulated to 70 ml/min. A mixture of ethanol and water in the volumetric ratio of 1:1 was injected into this water with a flow rate of 0.2 ml/min leading to an ethanol concentration of 25 $mmol/l$. Water loaded with PCE at saturation was continuously injected into the monitoring well at $x = 0.75\ m$, $z = 0.35\ m$ with a flow rate of 10 ml/min. For this purpose degasified water was pumped through a mixing reactor containing PCE as non-aqueous phase liquid.

Initially no attempts were made to control the *pH* in the system. As a consequence of acetogenic reactions, the *pH* decreased dramatically. Since the dechlorination rates had not been satisfactory in the first months of operation and *dehalospirillum multivorans* was known to dechlorinate best at neutral *pH* conditions (Neumann et. al., 1994), various possibilities of injecting buffering compounds were tested. Finally sodium sulfide was chosen, a rather untypical buffer but not disturbing reductive dechlorination. The final set-up of the artificial aquifer during the stimulation experiment is shown in Fig. 2.

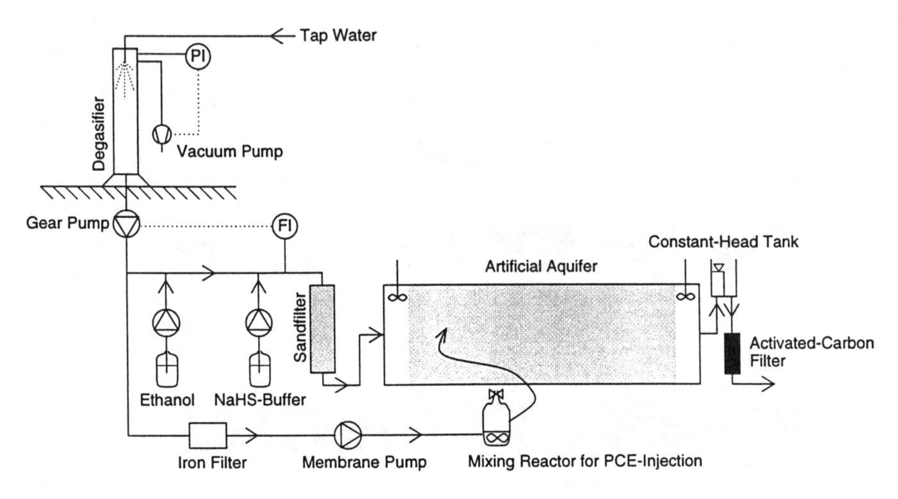

FIGURE 2: Principal set-up of the experimental plant during the stimulation experiment.

RESULTS AND DISCUSSION

Before inocculation hardly any anaerobic organisms were present in the artificial aquifer. With the inocculation the domain became microbially active. However, not all microbes introduced into the domain survived, and the growth rates of anaerobic organisms were rather low. As a consequence, it took about a month to achieve microbially active, anaerobic conditions and several months to achieve complete transformation of PCE to DCE.

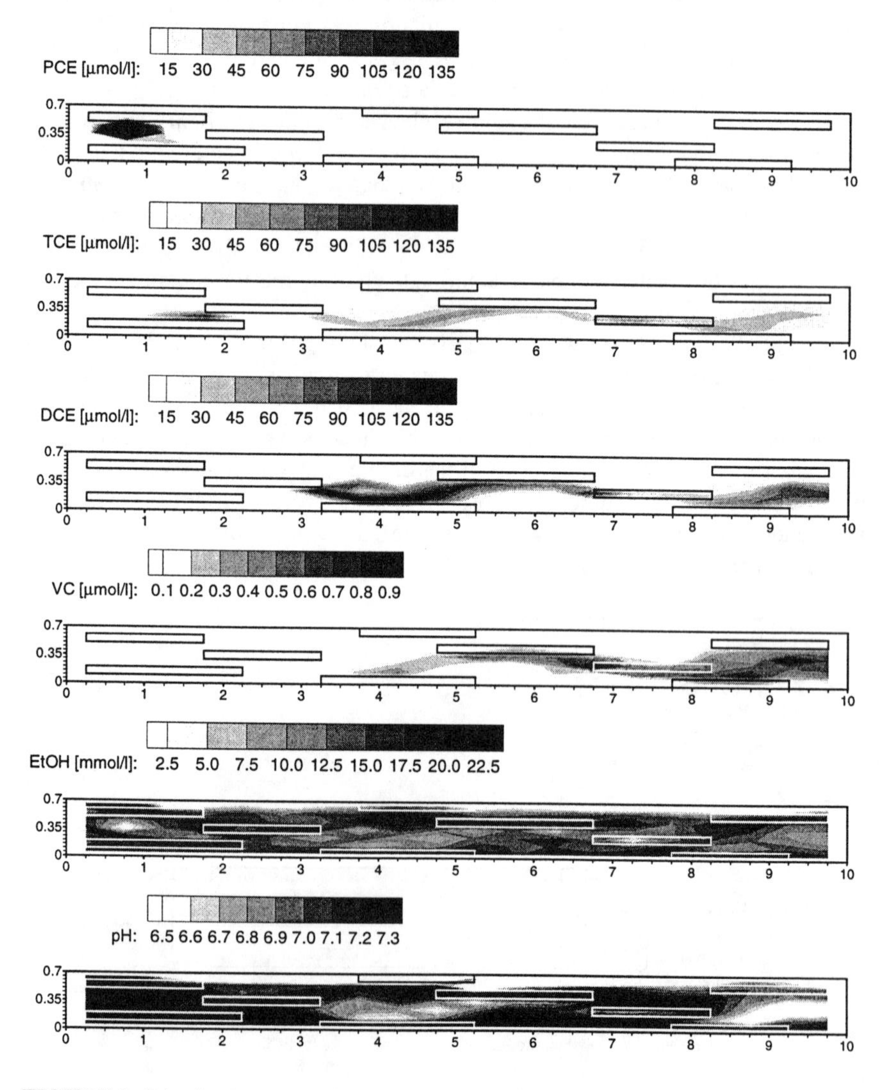

FIGURE 3: Distribution of chlorinated ethenes, ethanol and *pH* after 424 days of operation on January 19, 1998.

Fig. 3 shows the distribution of chlorinated ethenes after 424 days of operation on January 19, 1998. At this time the activity of *dehalospirillum multivorans* was satisfying, whereas the DCE-dechlorinating culture had not yet grown to larger densities. Chemical analysis of the aquifer outflow showed that 6.1 % of the PCE injected remained unchanged over the passage through the domain, 20.5 % was transformed to TCE, 72.7 % to DCE and 0.7 % to VC. Ethene was not measured.

The highest activity of *dehalospirillum multivorans* was in the region from the injection point to the end of the first high-permeability block. Since the plume passed by the monitoring wells within the high-permeability block, it's not clear where exactly the main transformation from PCE to DCE took place. The distribution of the VC concentration in Fig. 3 indicates that the highest activity of the DCE-dechlorinating organism was somewhere downstream of that of *dehalospirillum multivorans*. Note that in the weeks past the measurement shown in Fig. 3, the VC concentrations decreased indicating decay of the DCE-dechlorinating culture.

Fig. 3 includes also the distribution of the *pH*. Due to the injection of NaHS as buffer, the *pH* was in the widest part of the domain between 6.5 and 7.3. At the top of the domain the *pH* decreased to values below 5.0. However, this was a special region in which gas was accumulated. A gaseous phase was formed from H_2S produced from buffering. Eventually the gas saturation was sufficient for gas flow so that the lower density of gas led to its accumulation in the top region of the domain. In this region the environmental conditions were different to the rest of the domain, maybe influenced by lower flow velocities of water due to gas-clogging.

CONCLUSIONS

In the present study, microbial reductive dechlorination of tetrachloroethene was investigated in a large-scale sandbox model. The experimental plant was operated continuously over more than 15 months. Ethanol was used as electron donor introduced into the aquifer inlet, whereas PCE in aqueous solution was injected into a well so that the interaction of these compounds was dependent on mixing. However, due to the presence of high-permeability lenses focussing the flow field and thus enhancing transverse dilution of the plume, mixing of the compounds did not limit the microbial activity. This was confirmed by sufficiently high concentrations of ethanol in the center of the plume.

Both reductive dechlorination and competitive consumption of ethanol led to an acidification of the artificial aquifer. Since no natural buffer was present in the system, the *pH* decreased to values below 5.5 before additional buffers were added. In order to stabilize the *pH* in the neutral range, NaHS was successfully introduced into the aquifer inlet.

Once anaerobic conditions had been established, *dehalospirillum multivorans* dechlorinating PCE via TCE to DCE could be stimulated sufficiently well. Apparently the transformation was completed within a travel distance of 2 *m* which corresponded to a travel time of approximately one day. This reaction rate is in the range of what has been measured in mixed reactors containing the same organism (Granzow et al., 1996).

In contrast to this, the DCE-dechlorinating culture could not be established. As a consequence, reductive dechlorination was incomplete leading to the accumulation of *cis*-1,2-dichloroethene. This is a very typical end product of natural attenuation.

Note that transforming PCE to DCE at field sites would be no improvement. DCE is cancerous and even harder to be cleaned up in on-site treatment plants for extracted groundwater than PCE. Therefore, at the current state the authors cannot recommend the stimulation of complete microbial reductive dechlorination as an approach for *in-situ* remediation at field sites contaminated by PCE.

ACKNOWLEDGEMENTS

The present study was conducted in the framework of the interdisciplinary project "Stimulierung der vollständigen reduktiven Dehalogenierung in einem Modellaquifer - Artificial AnAerobic Aquifer A^4" funded by the Deutsche Forschungsgemeinschaft under the grant Ko-528/14.

REFERENCES

T.D. DiStefano, G.M. Gossett, and S.H. Zinder. 1991. "Reductive dechlorination of high concentrations of tetrachloroethene to ethene by an anaerobic enrichment culture in the absence of methanogenesis." *Appl. Environ. Microbiol.* *57*(8):2287–2292.

T.D. DiStefano, G.M. Gossett, and S.H. Zinder. 1992 "Hydrogen as an electron donor for dechlorination of tetrachloroethene by an anaerobic mixed culture." *Appl. Env. Microbiol.* *58*(11):3622–3629.

S. Granzow, M. Eisenbeis, C. Windfuhr, P. Bauer-Kreisel, and H. Scholz-Muramatsu. 1996. "Entwicklung eines Biofilmreaktors zur vollständigen Dechlorierung von Tetrachlorethen mit *Dehalospirillum multivorans* und einer Dichlorethen-dechlorierenden Mischkutur als Starterkulturen für die vollständige Ddechlorierung von Tetrachlorethen." In *Proceedings zur Dechema-Jahrestagung '96, 21.-23.4.1996, Wiesbaden, Band 1*, page 446.

C. Holliger, G. Schraa, A.J.M. Stams, and A.J.B. Zehnder. 1993. "A highly purified enrichment culture that couples the reductive dechlorination of tetrachloroethene to growth." *Appl. Environ. Microbiol.* *59*:2991–2997.

M.D. Lee, P.F. Mazierski, R.J. Buchanan, D.E. Ellis, and L.S. Sehayek. 1995. "Intrinsic in situ anaerobic biodegradation of chlorinated solvents at an industrial landfill." In R.E. Hinchee, F.T. Wilson, and D.C. Downey, editors, *Intrinsic Bioremediation.*, pages 205–222.

A. Neumann, H. Scholz-Muramatsu, and G. Diekert. 1994. "Tetrachloroethene metabolism of *dehalospirillum multivorans.*" *Arch. Microbiol.* *162*:295–301.

H. Scholz-Muramatsu, A. Neumann, M. Messmer, E. Moore, and G. Diekert. 1995. "Isolation and characterization of dehalospirilum multivorans." *Arch. Microbiol.* *163*(1):48–56.

T.M. Vogel and P.L. McCarty. 1985. "Biotransformation of tetrachloroethylene to trichloroethylene, dichloroethylene, vinyl chlorid, and carbon dioxide under methanogenic conditions." *Appl. Environ. Microbiol.* *49*(5):1080–1083.

CONTINUOUS ON-LINE MONITORING OF ORGANOCHLORINE COMPOUNDS IN A FIXED BED BIOREACTOR BY AN OPTICAL INFRARED SENSOR

Victor Acha*, Marc Meurens#, Henry Naveau*, Spiros N. Agathos*
*Unit of Bioengineering, Catholic University of Louvain, Place Croix du Sud 2/19, 1348 Louvain-la-Neuve, Belgium. E-mail:acha@gebi.ucl.ac.be
Unit of Nutritional Biochemistry, Catholic University of Louvain, Place Croix du Sud 2/8, 1348 LLN, Belgium.

ABSTRACT: An attenuated total reflection-Fourier transform infrared (ATR-FTIR) optical sensor was developed to perform continuous on line measurements of trichloroethylene (TCE), tetrachloroethylene (PCE), carbon tetrachloride (CT) and hexachlorobutadiene (HCB) in a fixed-bed dechlorinating bioreactor. A carrier-supported methanogenic microbial community was acclimated to a mixture of these chlorinated hydrocarbons in the bioreactor. The optical sensor was based on an ATR internal reflection element (IRE) coated with an extracting polymer which continuously enriched the toxic compounds. During three days of continuous operation the bioreactor was monitored by means of the infrared sensor, which was coupled permanently to the system. The sensor tracked the progression of absorbance spectra over time without perturbing the process in any way. The concentration of toxic compounds in the bioreactor were predicted from the absorbance spectra using a PLS calibration model which was developed for all components simultaneously and evaluated by their correlation coefficient (R^2), standard error of prediction (SEP) and relative standard error of prediction (RSEP). To examine the accuracy and reproducibility of this ATR-FTIR sensor, the concentration values of the toxic compounds in the effluents from the dechlorinating bioreactor were checked against gas chromatography (GC) measurements. This sensor is a promising tool to perform routine continuous on-line monitoring of the dechlorinating bioreactor.

Keywords: Organochlorine compounds; chlorinated aliphatic hydrocarbons; aqueous effluents; ATR-FTIR; optical sensor; polymer coating; anaerobic dechlorinating bioreactor

INTRODUCTION

The sector of industrial monitoring requires at the present time sensors providing real-time measurements of fundamental parameters in order to analyse, model and control optimally the time-course of biological processes taking place, for instance, in bioreactors. The main purpose of sensors is to provide information concerning the process under study and/or the quality of the final products. The environment field uses already physical and chemical sensors (for temperature, pressure, turbidity, conductivity, pH, pO_2, pNH_3, etc) others are under development or ready for application (biosensors for BOD, pesticides, phenols, etc) (Locher et al., 1992; Victor et al., 1997; Urban and Jobst, 1998).

In our recent work we have built an optoelectronic infrared (IR) sensor to measure off-line traces of organochlorine compounds in the effluent from an anaerobic dechlorinating bioreactor (Acha et al., 1998). Based on our previous results, the optical sensor measurement system was subsequently connected directly to this bioreactor. This kind of configuration allowed continuous on line measurements in the bioreactor without any sample preparation. This work presents the results obtained from monitoring the bioreactor over three days of continuous operation using this novel sensor.

MATERIALS AND METHODS

The ATR-FTIR sensor. The optical sensor (Fig. 1) consisted of a trapezoidal zinc selenide (ZnSe) ATR Internal Reflection Element (IRE). This IRE was 49 mm long, 9.5 mm wide, and 3 mm thick, the angle of incidence was 45° for a refractive index of 2.4 (Graseby-Specac Inc., Orpington, Kent, England). The crystal was coated with a hydrophobic polyisobutylene (PIB) polymer of 5.8 μm thickness, with a refractive index of 1.5. The waterproof cover of the crystal was specially fabricated from stainless steel material equipped with an O-ring. All measurements were performed with a MIDAC interferometer (MIDAC Corporation, Costa Mesa, CA, USA) coupled to a personal computer working with Spectra Calc software (Galactic Industries Corporation, Salem, NH, USA). All spectra were obtained at a nominal 4 cm^{-1} resolution by coadding 125 scans. The background spectra were obtained using as a reference the empty ATR crystal covered with the polymer. All spectra were acquired at room temperature. The spectral processing software used for spectrometer calibration was PLS from Spectra Calc (Galactic Industries) with PLS modules for quantitative analysis.

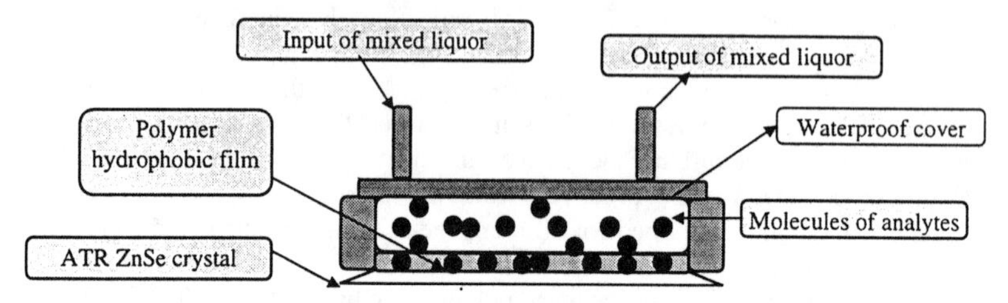

Fig. 1 Composition of the ATR-FTIR sensor

The sensor measurement system: the ATR-FTIR sensor and the bioreactor. The sensor measurement system (Fig. 2) was composed of the dechlorinating bioreactor coupled directly to the optical sensor inside the FTIR compartment by an accessory tubing made of a fluoroelastometer, Versinic™ (Vel, Louvain, Belgium). The material is resistant to aromatic

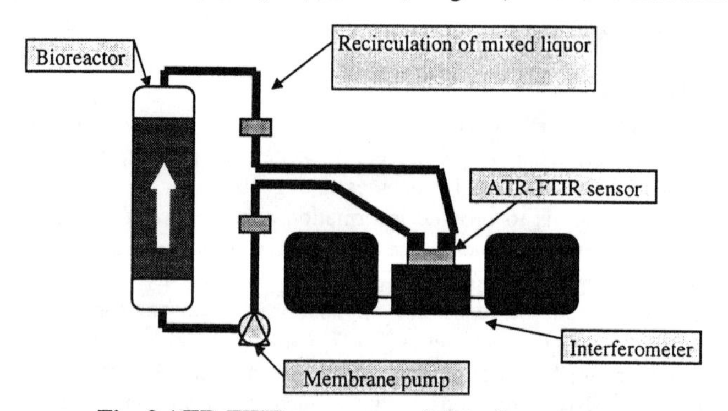

Fig. 2 ATR-FTIR sensor coupled to the measurement system

and chlorinated solvents. A membrane pump (Prominent, Heidelberg, Germany) was used to recirculate the mixed liquor in the system.

The bioreactor consisted of an elongated cylindrical glass tube with a diameter to height ratio of 1:15. and a total liquid volume of 4.4 L. The fixed bed contained a total of 1625.3 g of dry carrier (polyurethane foam entrapping magnetite). A methanogenic microbial consortium previously adapted to perform extensive dechorination on chloro-aliphatic compounds (Boucquey et al., 1995) was entrapped inside the carrier material. The bioreactor was operated in upflow mode, at 35°C. The mean hydraulic retention time was 30 days, and the recirculation rate was 15 d^{-1} based on total liquid volume of the bioreactor. The mixed liquor contained trichloroethylene (TCE), tetrachloroethylene (PCE), hexachloro-1,3-butadiene (HCB) and carbon tetrachloride (CT) (Janssen Chimica, Beerse, Belgium). These toxics were fed with a total space loading rate of 0.052 g/L.d. based on a mean hydraulic retention time of 30 days. A nontoxic cosubstrate (citric acid, ammonium and other salts plus vitamins) was introduced at 0.5 g COD/L.d. (Boucquey et al., 1995).

Standard Solutions for Spectrometer Calibration. A total of 13 mixtures of TCE, PCE, and CT were devised in the closed symmetrical experimental design obtained from a 3^2 full factorial design (Deming et al., 1993). The concentration range (0 - 60 mg/L) for the three analytes was selected to match typical values found during normal bioreactor operation. This pattern avoided any correlation between the concentrations of the individual matrix components. This was necessary for a valid estimation of the ability to measure each component in an independent manner.

Calibration. The treatment of the data was done using the multivariate method of partial least squares (PLS), a full-spectrum method based on a prior spectral decomposition through factor analysis. Thus this method uses the whole set of absorbance data. The objective of PLS is to obtain the spectrum of the sample mixture from a number of variable spectra (factors), and the different contributions of each of them (scores) that must be added to reconstruct the original spectrum. The absorbance and the concentration information data are used in the spectral decomposition (Martens, 1989). In order to select the optimum number of factors in the PLS regressions, the predictive residual sum of squares, PRESS (Eq. 1), was calculated with the calibration set (cross-validation with leave-one-out technique):

$$PRESS = \sum_{i=1}^{n} \left(y_p^i - y_m^i \right)^2 \qquad (1)$$

where n is the number of samples in the set, while y_p^i and y_m^i are the calculated and the measured concentrations, respectively. The optimal number of factors corresponds to the minimum value of PRESS (Martens, 1989).

RESULTS AND DISCUSSION

Bioreactor runs. The progressive acclimation of the carrier-supported microbial consortium to a mixture of chlorinated hydrocarbons TCE, PCE, CT and HCB in a fixed-bed bioreactor to arrive at continuous dechlorination was reported previously (Boucquey et al., 1995, Acha et al., 1998). This toxic mixture together with the cosubstrate were fed to the vessel three times a week. The concentration of toxic compounds (mg/L), volatile acids (g COD/L units) and mineral chlorides (mg/L) were checked regularly, the first two by GC and the latter by conductimetry.

ATR-FTIR Spectra. The ATR-FTIR sensor connected directly to the fixed bed bioreactor made possible the simultaneous and continuous on-line measurements of organochlorine compounds in the lower mg/L range. The molecules of organochlorine compounds diffused as expected into the polymer film while the influence of water absorption was negligible.

The bioreactor was monitored over a period of more than three days with the infrared sensor permanently coupled to the system. The coupling of the IR sensor to the bioreactor was done 2 h before introducing the cosubstrate plus organochlorine mixture. The bioreactor was fed with a toxic mixture of 0.052 g/L.d, CT 0.025 g/L.d, TCE, PCE and HCB 0.0091 g/L.d each).

Fig. 3 illustrates the absorbances of TCE (931, 841 and 762 cm^{-1}), PCE (910 and 777 cm^{-1}) as a function of polymer enrichment time from 1 to 125 min after coupling the optical sensor to the bioreactor. The absorbance of these two toxic compounds increased progressively from 1 to 125 min remaining constant when the enrichment equilibrium in the polymer film was attained. As observed in Fig. 3, no CT (789 cm^{-1}) nor HCB (852 and 979 cm^{-1}) absorption bands were detected in this period of time. As already noted in earlier experiments, HCB diffused extremely slowly and consequently it had no time enough to enrich the coating polymer. On the other hand CT is readily biodegraded and in this period of time it was probably already exhausted. After 125 min CT and HCB began to absorb.

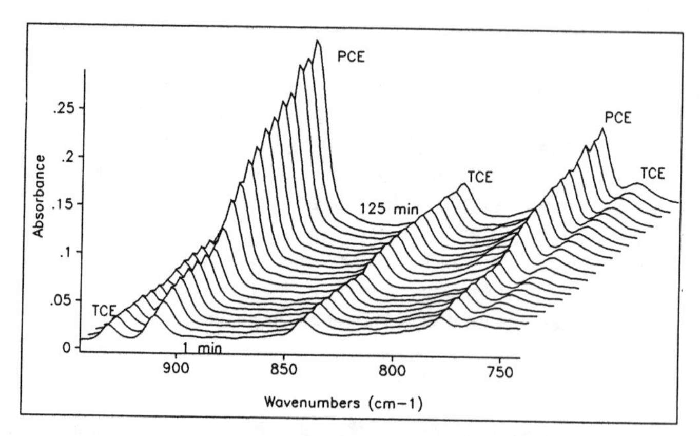

Fig. 3 Absorbances of TCE (931, 841 and 762 cm^{-1}) and PCE (910 and 777 cm^{-1}) as a function of polymer enrichment time after coupling the sensor to the bioreactor

Fig. 4 shows a typical spectrum of the bioreactor effluent after 23 h of operation. This figure shows the absorption bands of four toxic compounds from the bioreactor: TCE (931 and 841 cm^{-1}), PCE (910 and 777 cm^{-1}), CT (789 cm^{-1}) and HCB (852 and 979 cm^{-1}). It can be seen that HCB peaks overlapped with the TCE absorption bands, and also CT bands overlapped with the PCE bands, consequently the calibration multivariate method of partial least squares (PLS) needed to be used to resolve these overlapping absorption bands. On the other hand the apparent differences in signal sensitivity between the four toxic compounds were observed. The reason for this effect is the distinct difference in the enrichment behavior of these substances as reflected in the distribution coefficients ($f_{p/w}$) between aqueous solution and polymer coating. The $f_{p/w}$ is a function of the solubility in water and the boiling point of each compound (Shrader, 1994). Values of $f_{p/w}$ for TCE, PCE, CT and HCB were calculated to be 231, 1270, 297, and 9023 respectively. This explains clearly the behavior of HCB in our measurement system.

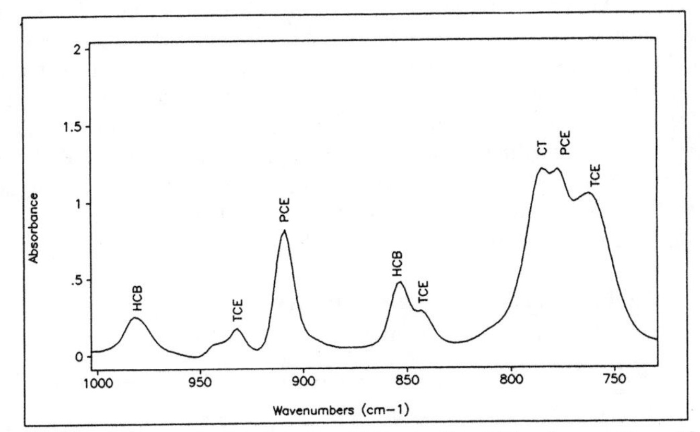

Fig. 4 Typical spectrum of the bioreactor effluent after 23 h of operation. Absorption bands of TCE (931, 841 and 762 cm^{-1}), PCE (910 and 777 cm^{-1}), CT (789 cm^{-1}) and HCB (852 and 979 cm^{-1})

After 125 min of bioreactor operation, the toxic mixture of TCE, PCE, CT and HCB was applied to the vessel at twice the initial mass rate in order to create a toxic shock. This can be seen in Figure 5. This figure shows the bioreactor performance in terms of analyte absorbances during more than 3 days of operation (from 1 to 89 h).

Following a transient increase in the concentration of the organochlorine compounds, a rapid consumption of these molecules was observed after introducing the toxic mixture to the vessel (Fig. 5 from 1 to 3h). Their concentrations decreased continuously over the next 17h (Fig. 5 from 3 to 20h). In the next period a new toxic mixture at the initial normal loading rate was applied again to the bioreactor (Fig. 5 after 20h). The toxic concentration increased modestly and then decreased progressively during the next 68 h until 89 h of total monitoring time (Fig. 5).

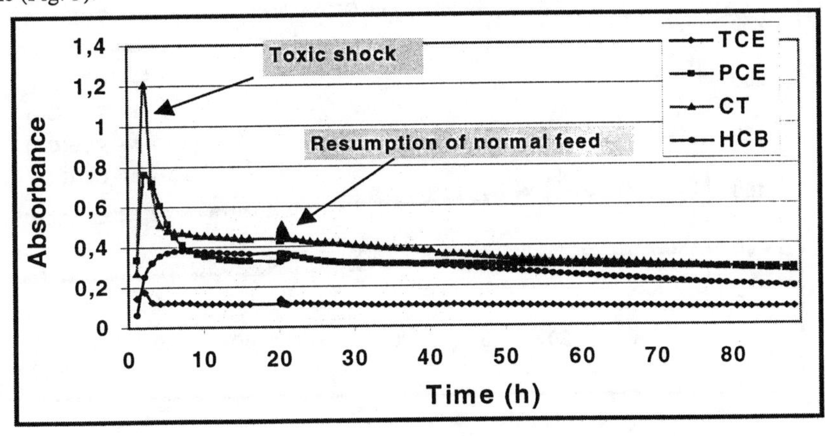

Fig. 5 Profiles of bioreactor effluent in terms of toxic compound absorbances as a function of time from 1 to 89h.

Multivariate regression. Thirteen absorbance spectra corresponding to the standard mixtures of the experimental design matrix were used for calibration purposes. For a valid assessment of the ability to measure each component in an independent manner, it is imperative that no correlation exists between the concentrations of the individual matrix components to avoid large systematic errors. No inter-analyte correlation was present within the data matrix used in this study. Table 1 shows the results obtained for each analyte with the PLS calibration model. The results in terms of R^2, standars error of prediction (SEP) and relative standard error of prediction (RSEP) appeared to be similar to the calibrations performed earlier during off-line toxic compound measurements (Acha et al. 1998).

TABLE 1. Calibration results using PLS model

Analyte	Correlation coefficient, (R^2)	Factors	SEP (mg/L)	RSEP (%)
TCE	0.997	4	0.65	3.5
PCE	0.996	6	0.76	4.0
CT	0.996	5	0.94	5.2

In order to verify the results obtained with the optical sensor, sample effluents taken at different dates from the dechlorinating bioreactor were checked for their TCE, PCE and TC concentration by gas chromatography (GC), which is the most commonly applied method for analyzing organochlorine compounds in aqueous solution. PLS calibration was also used to predict the concentration of the same effluents. Figure 6 shows the PLS results in predicting concentrations of TCE, PCE and CT compared to the GC report. The ATR-FTIR sensor measurements of toxic compounds followed the GC data adequately.

As expected the continuous on-line IR spectra measurements showed close correspondence to every variation of organochlorine concentration inside the bioreactor.

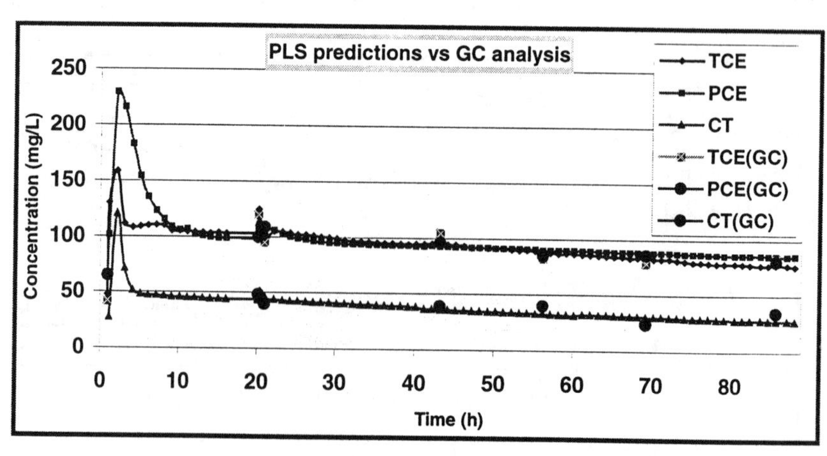

Fig. 6 Profiles of bioreactor effluent in terms of toxic compound concentrations as a function of time. PLS predictions vs. GC analysis

CONCLUSION

The IR sensor allowed the determination of the chemical composition simply by looking at the liquid effluent without separating out the individual components. The ATR-FTIR sensor served to obtain a valuable evolving fingerprint for the biodegradation of toxic compounds by tracking the progression of absorbance spectra over time without perturbing the process in any way.

The results obtained here show clearly that the ATR-FTIR optical sensor is effectively sensitive enough to the TCE, PCE, CT and HCB toxic compounds and to the fluctuations of their concentration inside the fixed bed bioreactor. The PLS predictions of the concentrations of the mixed chloro-organics were in good agreement with the GC measurements. Therefore this sensor can be used easily to perform routine continuous on-line monitoring of the dechlorinating bioreactor. Interfacing this ATR-FTIR sensor with a computer, the dechlorinating process could be controlled in real time, on the basis of data obtained on-line, thus saving costs and time.

REFERENCES

Acha, V., Meurens, M., Naveau, H., Agathos, S. N. 1998. "ATR-FTIR sensor development for measurements of chlorinated aliphatic hydrocarbons in aqueous effluents of a fixed bed bioreactor" (in preparation).

Boucquey, J. -B., Renard, P., Amerlynck, P., Modesto Filho, P., Agathos, S. N., Naveau, H., Nyns, E.-J. 1995. "High-rate continuous biodegradation of concentrated chlorinated aliphatics by a durable enrichment of methanogenic origin under carrier-dependent conditions". *Biotechnol. Bioeng.* **47**: 298-307.

Deming, S.N., Morgan, S. L. 1993. Experimental design: A chemometric approach, Elsevier, Amsterdam.

Locher, G., Sonnleitner, B., Fiechter A. 1992. "On-line measurement in biotechnology: Techniques", *J. Biotechol.* **25**: 23-53.

Martens, H.E., Naes, T. 1989. *Multivariate Calibration*, Wiley, New York.

Schrader, B. (Ed.) 1994. *Infrared and Raman spectroscopy. Methods and Applications*, VCH, 603-617.

Urban, G.A., Jobst G. 1998. "Sensor systems", *Microsystem Technology in Chemistry and Life Science*, Springer, **194**: 189-213.

Victor, S., Lin, Y., Motesharei, K, Dancil, K.P.S., Sailor, M.J., Ghadiri M.R. 1997. "A porous silicon-based optical interferometric biosensor", *Science*, **278**: 840-843.

BIOREMEDIATION OF PENTACHLOROPHENOL: A PILOT-SCALE STUDY

Harry R. Compton, U.S. EPA, Edison, New Jersey, USA
Gail Scogin, formerly of U.S. EPA, Atlanta, Georgia, USA
Waynon Johnson, U.S. EPA, Atlanta, Georgia, USA
T. Ferrell Miller, Ph.D., Michael F. Mohn, and Daniel G. Crouse, P.E.,
Roy F. Weston, Inc., Edison, New Jersey, USA

ABSTRACT: A pilot-scale solid-phase bioremediation study was implemented at a former woodtreating facility in Alabama to determine the optimum mixture of amendments required to reduce soil pentachlorophenol (PCP) levels to below 50 milligrams per kilogram. Studies were conducted to determine whether the soil contained PCP-degrading microbes. Six 5.5-metric ton pilot-scale test plots were constructed, of which four were amended with sawdust. The sawdust-amended plots showed the highest rate of PCP removal, with about 90% removed after 30 days. Two plots were bioaugmented with native PCP-degrading microbes. Soluble chloride in the soil was measured to monitor PCP degradation. A full-scale operation was successfully implemented based on the pilot-scale study results.

INTRODUCTION

The 30-acre (12-hectare) Fullco Lumber Co. site is an abandoned wood treatment facility located in northwest Alabama. Plant operations contributed several sources of soil contamination. These included process wastewater pumped to an unlined onsite pond that sometimes overflowed into a drainage ditch, frequent spills from the process tanks into a poorly constructed containment area, and drip areas for the treated lumber.

More than 6,880 cubic meters of soil contaminated with PCP, arsenic, and chromium were excavated and disposed of offsite at a cost exceeding $1 million. Based on the U.S. EPA cleanup goals of 50 mg/kg PCP and 100 mg/kg arsenic, the remaining volume of PCP contaminated soil was estimated to be at least 7,650 cubic meters.

EPA used bioremediation as the treatment option for the remaining contaminated soil based on the successful results from the pilot-scale studies conducted at this site. Preliminary cost estimates for bioremediation indicated that it was considerably less expensive than offsite disposal. Other factors also suggested that bioremediation would be a viable treatment method for PCP-contaminated soils. Typically, Alabama has a long growing season with warm weather occurring seven to eight months of the year. In addition, the site was large enough, with sufficient distance from residential areas, to build a 2-acre (0.84 hectare) biotreatment cell for full-scale operations.

METHODOLOGY AND RESULTS

Bench-Scale Studies. In late 1995, sixteen site soils were collected and screened for PCP-degrading activity. Laboratory screening protocols were developed from published reports (Saber and Crawford, 1985). Soil enrichment cultures were developed by the inoculation of 1 gram of test soil in Erlenmeyer flasks containing 100 mL of screening medium. The medium consisted of mineral salts supplemented with PCP as the sole carbon and energy source. Flasks were placed on a gyratory shaker and monitored for PCP removal and an accompanying decrease in culture pH. Enrichment cultures showing complete removal of PCP and a decrease in culture pH by day 7 were considered to have strong PCP-degrading activity (3 samples). Enrichment cultures requiring longer growth periods for PCP removal and a decrease in culture pH were considered to have weak PCP-degrading activity (3 samples). Enrichment cultures showing no reduction in PCP levels after 28 days were considered to be devoid of PCP-degrading activity (10 samples). Active cultures were subcultured at least four times to confirm PCP-degrading activity.

The three site locations exhibiting strong PCP-degrading activity were considered as possible sources of inoculum soil for pilot-scale landfarm studies. Soil from two of these locations (DR-Low-A3 and DIG 7) were used as inoculum soil in the treatability studies. Soil from another location (SB-12) was chosen as the test soil for the pilot-scale landfarm studies because of its desirable PCP contamination level and the lack of PCP-degrading activity.

Design and Setup of the Pilot-Scale Plots. Approximately 5.5 metric tons of a soil blend composed of 90% SB-12 soil and 10% DIG 7 soil (a source of PCP-degrading microbes) was added to each plot, and each was further amended with 55 kg of soil from location DR-Low-A3 as additional inoculum soil.

The contents of each of the six pilot-scale plots are shown in Table 1. An equivalent amount of calcium carbonate was added to each plot for pH control. Inorganic nitrogen and phosphorus sources consisted of ammonium nitrate and trisodium phosphate neutralized with sulfuric acid, respectively. Red oak sawdust was added to soil at rates ranging from 2 to 6% by weight.

TABLE 1. Pilot-Scale Plot Contents

Plot No.	Amendments
1	Calcium carbonate (18 kg)
2	Calcium carbonate (18 kg), ammonium nitrate (11 kg), trisodium phosphate (2 kg)
3	Sawdust (227 kg), calcium carbonate (18 kg)
4	Sawdust (318 kg), calcium carbonate (18 kg)
5	Sawdust (340 kg), calcium carbonate (18 kg), ammonium nitrate (11 kg), trisodium phosphate (2 kg)
6	Sawdust (136 kg), calcium carbonate (18 kg), ammonium nitrate (11 kg), trisodium phosphate (2 kg)

Moisture levels were maintained in sawdust-amended plots and non-sawdust amended plots at levels ranging from 12 to 20% and 10 to 12%, respectively. Each plot was covered with a removable plastic tarp. At day 0 and periodically throughout the study, nine composite samples from each plot were collected and analyzed for PCP and chloride content.

PCP-Degrading Activity of Pilot-Scale Plot Soils. Day 0 plot samples were screened in the laboratory for PCP-degrading activity using similar procedures developed for screening site soils. Broth from the final subculture was further evaluated for production of soluble chloride to determine the extent of PCP dechlorination.

PCP-degrading activity was not detected in Plots No. 1 and 2. This suggests that these landfarm plots did not contain PCP-degrading microbes, or that PCP-degrading microbes were present at such low levels that a 1-gram soil sample randomly collected may not contain PCP-degraders. Since reduction in PCP levels and increases in soil chloride were shown in Plot No. 1, it is apparent that PCP degraders were present, but at very low population levels. In Plot No. 2 soil, PCP reduction was initially demonstrated, but significant chloride production was not demonstrated within the first 90 days. Chloride production was demonstrated only after the test plot was bioaugmented with a purified PCP degrader.

Strong PCP-degrading activity was demonstrated by enrichment cultures developed from soil samples collected from Plots No. 3, 4, and 5. Results showed that primary cultures and all subcultures completely removed available PCP with an accompanying reduction in culture pH. Moreover, actual chloride production approached the theoretical level expected from 100 μg/mL of PCP. The strong PCP-degrading activity that was exhibited in shake flask culture was also reflected in each of the respective plots.

Studies showed that the enrichment culture developed from Plot 6 soil exhibited weak PCP-degrading activity. This weak activity was also reflected in the low level of PCP removal in Plot 6 soil in the first 29 days. PCP was rapidly degraded in Plot No. 6 soil between day 29 and day 57. These data suggest that a more aggressive PCP-degrading population may have replaced the original population. The evaluation of plot soil for PCP-degrading activity may provide valuable insights into the performance of a solid-phase process.

Bioaugmentation. Bioaugmentation experiments were initiated to enhance PCP degradation after it was determined that significant levels of PCP remained in Plot No. 2 soil after 57 days of incubation. A purified PCP-degrading isolate was bulk produced and introduced into Plot No. 2 soil at day 90. The effects of its introduction on the rate and extent of PCP removal and chloride production was determined.

The strain, designated 43-31, was one of several that was isolated from the active enrichment culture developed with soil from the SB-9A site location. The strain produced a characteristic yellow pigment. The culture was produced in shake flask culture using a mineral salts-glutamate-yeast extract medium supplemented with 50 μg/mL PCP. The PCP levels were reduced by approximately 90% within 96 hours.

The culture was concentrated by centrifugation, suspended in 50% volume/volume (v/v) aqueous glycerol, and stored in sterile plastic 50 mL conical centrifuge tubes. Initial viable cell counts were found to be 1.83 x 10^{10} colony forming units (CFU)/mL.

On the date of inoculation (day 90), the cell concentrate was suspended in approximately 10 L of tap water in a watering can, sprinkled uniformly throughout the plot, and rototilled into the test soil. Soil samples were collected before and after the addition of the inoculum and analyzed for PCP and soluble chloride.

PCP levels in Plot No. 2 soil had stabilized at approximately 140 mg/kg by day 90, with no significant changes in PCP concentration for 33 days. There were no significant changes in chloride concentration for the entire 90 day period. Chloride levels measured before and after inoculation were virtually identical. When PCP and chloride levels were measured in samples collected 36 days after inoculation, results showed that PCP levels had decreased and chloride levels increased. This trend persisted for an additional 36 days. By day 162, PCP levels had decreased to below the action level with an average value of 42 mg/kg. Thus, it appears that the addition of the bioaugmenting culture resulted in the successful reinitiation of PCP removal in a plot which had lost the ability to remove PCP. Moreover, it was only after the addition of the bioaugmenting culture that chloride levels began to increase in test soil. Bioaugmentation was thus beneficial in enhancing PCP removal in Plot No. 2 soil. However, the use of bioaugmentation was necessary only because the original recipe was not optimal for rapid growth of PCP-degraders.

PCP Results. Table 2 shows the PCP and chloride results of the pilot-scale study. Sawdust was the key ingredient in stimulating PCP-degradation in plot soils. PCP levels were reduced in Plots No. 3, 4, 5, and 6 by 81% to 96% by day 57. PCP levels in Plots No. 3 and 5 soil were reduced by 88% and 91%, respectively, as early as day 29. The major difference noted was the increased rate of PCP removal when test soil was amended with sawdust. Addition of inorganic nutrients to sawdust-amended soil did not significantly improve the rate of PCP removal.

TABLE 2. Pilot-Scale PCP and Chloride Concentrations (mg/kg)

	PCP							Chloride	
Plot No.	Day 0	14	29	57	90	126	162	Day 0	126
1	360	470	250	139	47	23	15	83	224
2	330	450	270	149	138*	78	42	81	144
3	300	100	35	34	19	21	SC	82	175
4	250	220	69	47	21*	8	25	84	220
5	240	63	21	10	8	5	SC	44	238
6	320	430	270	24	23	14	SC	70	228

* Microbes were introduced into these plots

SC The study was completed for these plots

Chloride was measured in test soil as a qualitative indicator of PCP degradation. Table 2 shows that degradation of PCP resulted in increases in soil chloride. Further studies are required to complete a chloride mass balance to determine if chloride production can be used as a quantitative indicator of PCP degradation.

The rate of PCP removal was considerably slower in the unamended Plot No. 1 and Plot No. 2 amended with inorganic nitrogen and phosphorus supplements. PCP levels were reduced by 61% and 55%, respectively, by day 57. PCP levels were reduced by 87% in Plot No. 1 soil by day 90 and by 96% by day 162. After bioaugmenting Plot No. 2 soil, PCP levels were reduced by 76% at day 126 and 87% at day 162.

Full-Scale Landfarm Operations. Full-scale operations were implemented to treat the contaminated soil based on the results obtained from the pilot-scale studies. Starting in August 1996, approximately 1,820 metric tons of contaminated soil was spread over 2 acres (0.84 hectare) to an average depth of 0.30 meters. Sawdust at 50 percent by volume (765 cubic meters) and 41 metric tons of limestone were added to the soil, and 33 metric tons of soil from the pilot-scale plots and 650 metric tons of soil containing active PCP-degrading microbes were added as inoculum soil. These amendments and soils were watered and mixed throughout the study period with a tractor-mounted rototiller. The landfarm plot was separated into four subplots (C1, C2, C3, and C4), and composite samples were collected from each. The initial and final PCP concentrations in the four subplots from Lift No. 1 are shown in Table 3. The average initial PCP concentration was 213 mg/kg, and the data indicate that three subplots met the site cleanup goal. The soil from subplots C2, C3, and C4 were excavated from the landfarm plot and placed at a separate area.

TABLE 3. Full-Scale Landfarm Plot Data for Lift No. 1

	PCP Concentrations (mg/kg)			
Subplot	**Day**	**Initial**	**Day**	**Final**
C1	0	214	348	51
C2	0	342	314	30
C3	0	147	227	8.8
C4	0	147	227	35

Lift No. 2 was started in September 1997. Approximately 4,130 metric tons of contaminated soil was spread over the area at a thickness of 0.53 meters. Sawdust was added at 75% v/v, and limestone was added at the identical ratio as that used for Lift No. 1. Soil (733 metric tons) from Subplot C1 was used as the inoculum for Lift No. 2. The tractor-mounted rototiller was used to homogenize the mixture. Table 4 indicates that after 57 days, the soil in the four subplots reached the cleanup level for the site. The rate of PCP degradation is much faster in Lift No. 2, which is probably because of the increased moisture content in the soil, the large volume of inoculum soil (20%) added, and a somewhat lower initial concentration of PCP.

TABLE 4. Full-Scale Landfarm Plot Data for Lift No. 2

	PCP Concentrations (mg/kg)			
Subplot	**Day**	**Initial**	**Day**	**Final**
C1	0	123	57	36
C2	0	163	57	41
C3	0	160	57	37
C4	0	168	57	24

CONCLUSIONS

Solid-phase bioremediation at pilot-scale was successful in attaining PCP removal goals. After 126 days, five of the six test plots showed greater than 93% PCP removal. Plot No. 2 showed 76% PCP removal. Initial PCP levels, ranging from 240 to 360 mg/kg, decreased to residual levels ranging from 5 to 23 mg/kg.

Sawdust, high soil moisture content, pH control, and inoculation with soil having PCP-degraders were the keys to successful landfarm treatment of PCP-contaminated soil. Other factors, including a long, wet growing season, the warm Alabama climate, soils with adequate nutrient value, and native soil requiring minimal pre-treatment, enhanced the success of the project.

Measurement of soil chloride was a good qualitative screening method to confirm and monitor PCP degradation in the pilot-scale study.

Bioaugmentation assisted in achieving treatment goals at this site. The bioaugmentation of pilot-scale Plot No.2 promoted further degradation of PCP. Bioaugmentation provided a viable contingency if the non-bioaugmented soil does not meet the treatment goal.

In the Lift No. 1 full-scale system, the PCP treatment goals of 50 mg/kg were met within 150 to 348 days. Lift No. 2 PCP treatment goals of 50 mg/kg were met in 57 days. Higher soil moisture content and seeding the soil in Lift No. 2 with native active PCP-degraders from Lift No. 1 affected the treatment times favorably by reducing the microbe growth curve.

Bioremediation expenses were approximately 50% of the costs associated with off-site disposal of soil. With economies of scale taken into consideration, by the time this project is finished, bioremediation is expected to cost approximately $76/cubic meter. Most of the costs were incurred with sample analysis and materials handling (moving and screening soil, building treatment cells, and maintaining the soil in the biotreatment cells).

REFERENCE

D. Saber and Crawford, R. L., 1985. "Isolation and Characterization of Flavobacterium Strains that Degrade Pentachlorophenol." *Appl. Environ. Microbiol. 50(6)*: 1512-1518.

MICROCOSM STUDIES OF BIOAUGMENTATION OF BUTANE AND PROPANE-UTILIZERS FOR IN-SITU COMETABOLISM OF 1,1,1-TRICHLOROETHANE

Pardi Jitnuyanont (Oregon State University, Corvallis, Oregon)
Lewis Semprini (Oregon State University, Corvallis, Oregon)
Luis Sayavedra-Soto (Oregon State University, Corvallis, Oregon)

Abstract: The cometabolism of 1,1,1-trichloroethane by augmented and non-augmented (indigenous) butane and propane-utilizers was evaluated in batch-fed microcosms containing groundwater and aquifer solids from the Moffett Field test site. *Indigenous* butane and propane-utlizers were stimulated through butane and propane addition. Butane and propane-utilizing mixed cultures obtained from the Hanford DOE site were used as inocula in a bioaugmentation study. The *indigenous* utilizers took 80-85 days to be stimulated, augmented strains took 5-23 days. Initially the augmented butane-utilizers were the most effective toward 1,1,1-TCA transformation, but eventually *indigenous* butane and propane-utilizers improved. The propane-utilizers had the highest transformation yields and were the most stable after 440 days of repeated stimulation. Subsequent bioaugmentation of butane-utilizers showed difficulty in growing inoculate strains in the groundwater microcosms. We suspected that the enrichments used for bioaugmentation adapted to nutrient rich conditions of the laboratory growth medium. This dependence was confirmed in a nutrient study. Microcosms with the highest media content (50%) and groundwater (50%) most effectively consumed butane and transformed 1,1,1-TCA. Microcosms with no media addition (100% groundwater) and 5% media (95% groundwater) lost 1,1,1-TCA transformation ability with repeated substrate and 1,1,1-TCA additions. Similar results were observed with augmented propane-utilizers. DNA fingerprinting of butane-utilizers using PCR methods showed variations in populations in microcosms for the different nutrient conditions. Successful bioaugmentation was achieved by enriching butane and propane-utlizers from Moffett microcosms. These strains effectively utilized butane and propane, transformed 1,1,1-TCA in 100% groundwater conditions (no added nutrients). The results help demonstrate the importance of using enrichments that perform well under the subsurface nutrient conditions.

INTRODUCTION

The Moffett Field Airfield (formerly known as Moffett Naval Air Station) is contaminated with various chlorinated aliphatic hydrocarbons (CAHs) including 1,1,1-TCA. The transformation of 1,1,1-TCA can be achieved by cometabolism. Previous studies demonstrated that methane (Broholm et al. 1990; Strand et al. 1990; Chang and Alvarez-Cohen 1996), propane (Keenan et al. 1993; Kim 1996; Tovanobootr 1997) and butane (Kim et al. 1997a; Kim et al. 1997b) are the possible substrates that can induce 1,1,1-TCA transformation. However, when methane was used as a growth substrate in the Moffett Field study, it faild to induce 1,1,1-TCA transformation (Roberts et al. 1990; Semprini et al. 1990). The Moffett Field groundwater conditions likely selected for the particulate methane monooxygenase enzyme (pMMO) instead of soluble methane monooxygenase enzyme (sMMO), which yields much higher 1,1,1-TCA transformation rates (Semprini 1997).

Our study investigated the ability of augmented and *indigenous* butane and propane-utilizers to transform 1,1,1-TCA in Moffett Field microcosms. Butane and propane-utilizers enriched from the Hanford DOE site were shown capable of transforming 1,1,1-TCA (Kim et al. 1997a). These strains were inoculated into Moffett Field microcosms to study and compare transformations in augmented and *indigenous* microcosms. Our study also evaluated nutrient requirements for effective bioaugmentation of selected enrichments.

MATERIALS AND METHODS

Soil Microcosms Setup. Batch-fed soil microcosms were used to mimic subsurface conditions. The microcosms (125 ml amber serum bottles) were composed of 15 ml of aquifer material and 50 ml of groundwater from the Moffett site, as described in by Kim et al. (1997).

Each microcosm was fed with gaseous butane or propane (4 to 5 mg) and 1,1,1-TCA (as saturated aqueous solution) to the desired concentration. After the transformation of substrates and 1,1,1-TCA in microcosms stopped, 60% of groundwater was exchanged with fresh groundwater to prevent accumulation of transformation products and to resupply nutrients. The procedure was repeated with 1,1,1-TCA mass gradually increased in each stimulation to the point where maximum transformation yields were observed.

Microcosms B1 and B2 were fed with butane and 1,1,1-TCA. Microcosm B2 was inoculated on day 57with butane-utilizing mixed culture enriched from Hanford DOE site. Microcosms P1 and P2 were fed with propane and 1,1,-TCA. Microcosm P2 was inoculated with propane-utilizers obtained from the Hanford DOE site on day 57, similarly to B2.

Nutrient Study. Batch-fed microcosms were set up with different ratios of groundwater to nutrient rich media. Microcosm B3 and BT3 contained 100% groundwater, B4 and BT4 contained 5% mineral salt media and 95% groundwater, B5 and BT5 contained 50% mineral salt media and 50% groundwater. 1,1,1-TCA was added to microcosms BT3, BT4 and BT5.

Analysis. 1,1,1-TCA was measured by injecting head space air from the microcosms into a GC with an electron captured detector (ECD) (Kim et al. 1997). The butane and propane analysis were similar, except that a flame ionization detector (FID) was used (Kim et al. 1997a).

DNA Fingerprinting. Comparison of the different culture was done by DNA fingerprinting by polymerase chain reaction (PCR). The mixed cultures were analyzed two primers, HP7 and GCA 12(Caetano-Anolles and Bassam 1993). PCR products were analyzed in agrose gel.

RESULTS AND DISCUSSION

No significant decrease of substrates (butane and propane) or 1,1,1-TCA was observed by day 57, indicating that no substrate utilization had started. On day 57 microcosms B2 and P2 were inoculated with butane and propane-utilzing mixed culture from Hanford DOE site, respectively. On day 60 butane utilization and 1,1,1-TCA transformation were observed in microcosm B2. Effective 1,1,1-TCA transformation observed in microcosms B2 almost instantly after inoculation indicated successful bioaugmentation. Other microcosms started to degrade substrates on day 80-85. The lag time of the *indigenous* microcosms was very long. *Indigenous* butane-utilizers (B1) had a lag time of 80 days. Bioaugmented propane-utlizers had a lag time of 23 days. With thses long lag periods it is difficult to conclude that the *indigenous* microorganisms were actually stimulated, since there is a possibility of laboratory contamination.

Microcosm B1 started to degrade butane on day 80, but did not show 1,1,1-TCA transformation. However, with successive feeding of butane and 1,1,1-TCA, microcosm B1 eventually transformed 1,1,1-TCA by day 130. The microcosms were operated for 440 days, while being exposed to increasing 1,1,1-TCA concentrations while the mass of butane fed was held constant. The transformation yields (mass of 1,1,1-TCA transformed/ mass of substrates utilized) of all microcosms is shown in Table 1 for different transformation periods. Microcosm B1, *indigenous* butane-utilizers, displayed low transformation yields initially, but yileds increased with successive feeding. The augmented butane-utilizers, microcosm B2, showed the highest transformation yields, but yields eventually decreased after 340 days. Microcosm P1 and P2 showed similar transformation yields from the beginning, which increased after repeated stimulations with increasing concentrations.

TABLE 1. Transformation yields of augmented and non-augmented (*indigenous*) butane and propane-utilizers over different time periods

	Transformation Yield (mg 1,1,1-TCA/ mg substrates)		
Microcosm	**Day 0-190**	**Day 190 to 340**	**Day 340 to 440**
B1	0.004	0.030	0.042
B2	0.070	0.077	0.041
P1	0.009	0.073	0.080
P2	0.008	0.071	0.080

Microcosm B1 and B2 had different lag times and activities, indicating that the strains stimulated in augmented and non-augmented microcosms were different. However, after 340 days, the transformation yields in both microcosms were similar. DNA fingerprints obtained from microcosms B1 and B2 on day 440 were similar in both microcosms. This may indicate that *indigenous* microorganisms had outcompeted the augmented microorganisms. Unfortunately, microbial samples for DNA fingerprinting were not available from the early stages of the study to help determine if different mixed populations initially existed. Microcosms P1 and P2 has similar lag times and transformation yields. Therefore, it was possible that the same strains were stimulated in both microcosms. Similar DNA fingerprints were obtained from both microcosms on day 440. It is possible that *indigenous* strains were stimulated in both microcosms from the beginning, or the *indigenous* strains eventually dominated in the augmented microcosm.

The presence of substrates (butane or propane) appeared to inhibit 1,1,1-TCA transformation. Figure 1 shows butane utilization and 1,1,1-TCA transformation in microcosms B1 and B2 at similar concentration conditions. 1,1,1-TCA transformation required a significantly longer time than butane utilization. The transformation was slow in the beginning (when butane concentrations were high) and accelerated when butane was almost completely utilized. Similar results were observed with propane-utlizers.

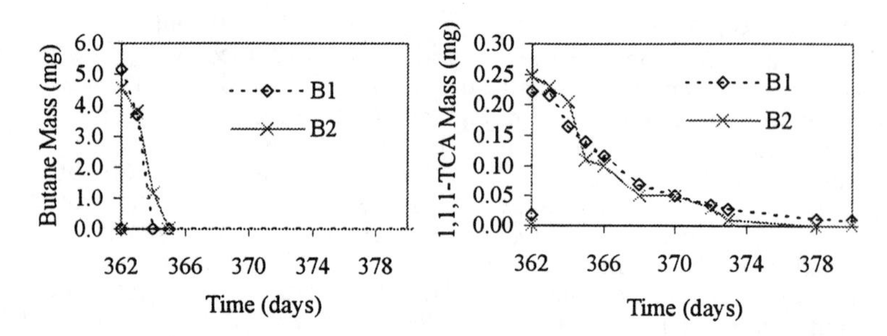

FIGURE 1. Butane utilization and 1,1,1-TCA transformation in microcosms B1 and B2 at similar concentration conditions.

The ability to reproduce the bioaugmentation results proved to be unsuccessful. The enrichments used for the second study were obtained from the same growth reactors 150 days after the first inoculation. These enrichments had been grown on mineral salt media. Thus, it was suspected that the enrichment selected for microorganisms that grew well in nutrient rich conditions. A study was performed to evaluate the effects of nutrient addition on bioaugmentation with these enrichments.

Microcosms B3 and BT3, which did not contain mineral salt nutrient, showed the slowest butane consumption rates, and the rates slowed with time (Figure 2). Microcosms B5 and BT5, which had the highest percentage of mineral salt nutrient (50%) showed the fastest and the most consistent butane consumption rates. The transformation of 1,1,1-TCA in microcosm BT3 (no nutrient) was the slowest and eventually stopped on the 4th stimulation. The 1,1,1-TCA transformation efficiency of microcosm BT4, which had 5% mineral salt nutrient, also decreased and eventually stopped on the 5th stimulation. Therefore, 5% mineral salt nutrients in groundwater was inadequate for these butane-utilizers to maintain their activities. Microcosm BT5, which had 50% mineral salt nutrient, was effective toward 1,1,1-TCA transformation throughout the study. Similar results were obtained with propane-utilizers.

FIGURE 2. Time required for butane and 1,1,1-TCA transformation

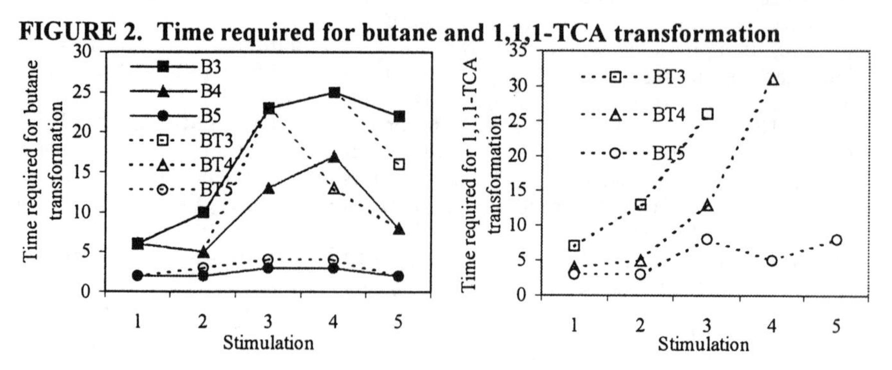

Microbial samples collected from each microcosms after 4th and 5th stimulations were analyzed by DNA fingerprinting. The DNA fingerprints showed the differences in the microcosms with different amounts of nutrients added. The microcosms with the same amount of nutrients, with and without

1,1,1-TCA exposure had similar DNA fingerprints. This suggested that the exposure to 1,1,1-TCA did not greatly affect the mixed cultures, but the mixed cultures that developed were affected by the amounts of nutrients present. Both the augmented butane and propane-utilizers required relatively high percentage of mineral salt nutrients to maintain their activities, which is impractical for bioremediation in the field.

Another bioaugmentation study was performed using butane and propane-utilizing strains, which were enriched from the Moffett groundwater microcosms, and were effective toward 1,1,1-TCA transformation (microcosms B1, B2, P1 and P2). The enrichments were raised in mineral salts growth media for 30 days to yield a high cell mass. The enrichments were then inoculated into fresh soil and groundwater microcosms without mineral salts addition, and fed with butane or propane and 1,1,1-TCA. In this case, the inoculated strains were effective and had no difficulties degrading butane or propane and transforming 1,1,1-TCA in the absence of mineral salt supplements.

CONCLUSIONS

Butane and propane-utilizers that effectively transformed 1,1,1-TCA could be stimulated in Moffett Field groundwater and aquifer materials. However, without bioaugmentation, the lag times were very long (80 days). Bioaugmentation significantly reduced the lag time and resulted immediately in effective 1,1,1-TCA transformation. Inoculating soil microcosms with laboratory grown enrichments can be problematic with respect to nutrient requirements. When the enrichments are raised on media, strains may be selected that grow well under these nutrient conditions. These strains may not perform well in groundwater nutrient poor conditions. This study indicates that it is important to use strains for bioaugmentation that perform well under the nutrient conditions of the site. Propane-utlizers enriched from microcosms P1 and P2 appeared to be the most efficient strains and do not require nutrient addition. Therefore, we feel these strains would be good candidates for bioaugmentation at Moffett Field to achieve 1,1,1-TCA transformation.

ACKNOWLEDGMENTS

This research was funded through the U.S. Air Force Armstrong Laboratory.

REFERENCES

Broholm, K., B. K. Jensen, T. H. Christensen, et al. 1990. "Toxicity of 1,1,1-Thrichloroethane and Trichlorethene on a Mixed Culture o f Methane-Oxidizing Bacteria". *Applied and Environmental Microbiology*. 56(8): 2488-2493.

Caetano-Anolles, G. and B. J. Bassam 1993. "DNA Amplification Fingerprinting Using Arbitrary Oligonucleotide Primers". *Applied Biochemistry and Biotechnology*. 42: 189-195.

Chang, H.-L. and L. Alvarez-Cohen 1996. "Biodegradation of Individual and Multiple Chlorinated Aliphatic Hydrocarbons by Methane-Oxidizing Cultures". *Applied and Environmental Microbiology* 1996. 62(9): 3371-3377.

Keenan, J. E., S. E. Strand and H. D. Stensel, Eds. 1993. "Degradation Kinetics of Chlorinated Solvents by a Propane-Oxidizing Enrichment Culture". *Bioremediation of Chlorinated and Polycyclic Aromatic Hydrocarbon Compounds*, pp. 1-13. Lewis Publisher.

Kim, Y. 1996. Personal communication.

Kim, Y., L. Semprini and D. J. Arp 1997a. "Aerobic Cometabolism of Chloroform and 1,1,1-trichloroethane by Butane-Grown Microorganisms". *Journal of Bioremediation*. 1(2): 135-148.

Kim, Y., L. Semprini and D. J. Arp 1997b. "Aerobic Cometabolism of Chloroform, 1,1,1-Trichloroethane, 1,1-Dichloroethylene, And The Other Chlorinated Aliphatic Hydrocargons by Butane Utilizing Microorganisms." Fourth International In Situ and On-Site Bioremediation Symposium, New Orleans, pp. 107-113. Battelle Press.

Roberts, P. V., L. Semprini, G. D. Hopkins, et al. 1990. "A Field Evalution of In-Situ Biodegradation of Chlorinated Ethenes: Part 1, Methodology an dField Site Characterization". *Ground Water*. 28(4): 591-604.

Semprini, L. 1997. "Strategies for the Aerobic Cometabolism of Chlorinated Solvents". *Current Opinions in Biotechnology*. 8(3):296-308

Semprini, L., P. V. Roberts, G. D. Hopkins, et al. 1990. "A Field Evaluation of In-Situ biodegradation of Chlorinated Ethenes: Part 2, Results of Biostimulation and Biotransformation Experiments". *Ground Water*. 28(5): 715-727.

Strand, S. E., M. D. Bjelland and H. D. Stensel 1990. "Kinetics of chlorinated hydrocarbon degradation by suspended cultures of methan-oxidizing bacteria". *Reserach Journal WPCF*. (62): 124-129.

Tovanobootr, A. 1997." Aerobic Cometabolism of Chlorinated Aliphatc Hydrocargons by Subsurface Microbes Grown on Methane, Propane and Butane from the McClellan Air Force Base.". M.S. Thesis, Oregon State University.

REDUCTIVE DECHLORINATION OF *CIS*-1,2-DICHLOROETHENE WITH AN ENRICHED MIXED CULTURE

Claudia Windfuhr , Silke Granzow and Heidrun Scholz-Muramatsu (Institute for Sanitary Engineering, University of Stuttgart, Germany)
Gabriele Diekert (Institute for Microbiology, University of Stuttgart, Germany)

ABSTRACT: Reductive dechlorination of *cis*-1,2-dichloroethene (DCE) is a crucial step for successful anaerobic bioremediation of sites contaminated with chlorinated ethenes. Whereas the reduction of tetrachloroethene (PCE) to DCE has been thoroughly investigated in mixed and pure cultures, little is known about the organisms which further degrade DCE to ethene. Experiments have been carried out to characterize an enriched mixed culture capable of dechlorinating DCE and to find out more about the involved microorganisms, their properties, requirements and limitations. The culture consists of five to six microscopically distinguishable morphotypes. It is capable of dechlorinating DCE, 1,1-dichloroethene, trichloroethene (TCE) and vinylchloride (VC), but not PCE, in the presence of e.g. formate, acetate and yeast extract. The DCE dechlorination activity is inhibited by several chlorinated methanes and ethanes. There is evidence, that the dechlorination of TCE, DCE and VC is carried out by only one organism in the mixed culture, but growth seems to be dependent on at least some of the accompanying organisms in the mixed culture.

INTRODUCTION

Chlorinated ethenes are common groundwater pollutants in most industrialized countries and remediation of contaminated sites has become increasingly important in the past years, because of the hazards to human health and the environment. Tetrachloroethene (PCE) is highly persistent under aerobic conditions and microbial attack can only occur under anaerobic conditions. Often a mixture of tetrachloroethene (PCE) and its anaerobic degradation products trichloroethene (TCE), *cis*-1,2-dichloroethene (DCE) and vinyl chloride (VC) is present at polluted sites, due to the natural microbial activity. Therefore microbial on-site or in situ remediation techniques seem to be a promising approach for the clean-up of sites contaminated with chlorinated ethenes.

The microbial reduction of PCE to DCE has been thoroughly investigated over the last years and several degrading organisms have been isolated in pure culture (Sharma and McCarty, 1996; Krumholz et al., 1996; Gerritse et al., 1996), some of which are able to couple the dechlorination of PCE to growth (Scholz-Muramatsu et al., 1995; Holliger and Schumacher, 1995; Wild et al., 1997). The enzymes involved in PCE and TCE dechlorination have been characterized, isolated (Neumann et al., 1995; Schumacher et al., 1997; Miller et al., 1997; Miller et al. 1998) and cloned (Neumann et al., 1997). Therefore much information is available to optimize the conditions in field applications according to the needs of the microbes involved in the dechlorination of PCE. Some organisms have even been shown to be well able to survive unfavourable conditions and the presence of other microbes, which might compete for growth substrates in natural environments or technical scale bioremediation reactors (Eisenbeis et al., 1997; Fennell et al., 1997; Cirpka and Bisch, 1998).

Some mixed cultures (Freedman and Gossett, 1989; DeBruin et al., 1992; Wild et al., 1995; Gerritse et al., 1997) and one pure culture (Maymo-Gatell et al., 1997) have to date been reported to be able to completely dechlorinate PCE to ethene, but so far only little is known about the mechanisms involved in the reductive dechlorination process from DCE to ethene. Since this conversion is a crucial step for successful anaerobic bioremediation of chlorinated ethenes, and the organisms involved appear to be less robust than those involved in PCE reduction (Cirpka and Bisch, 1998), more knowledge about the DCE- and VC-degrading organisms and their properties, requirements and limitations is necessary.

MATERIALS AND METHODS

The mixed culture was originally enriched from a contaminated site at a landfill in Eppelheim, Germany, with PCE-containing medium by Karin Böckle (DVGW, Karlsruhe). By cultivation under high PCE and ethene concentrations, methanogenic bacteria could be eliminated. Transfers over anaerobic solid media resulted in a pure culture, which could transform PCE via TCE to DCE, and a mixed culture, which had lost the ability to dechlorinate PCE, but could reduce DCE to ethene (Granzow, 1998).

Batch Experiments. Experiments were carried out in duplicates in 120 ml serum bottles sealed with teflon-lined butyl rubber septa under a gas atmosphere of N_2/CO_2 (10:1). The mixed culture was routinely cultivated in anoxic mineral medium (Scholz-Muramatsu et al., 1995) with formate (40 mM), acetate (20 mM), yeast extract (0.1%) and *cis*-1,2-dichloroethene (100 μM) and transfered into fresh medium about once a week. For the assays the same mineral medium was used with addition of the various substrates and chlorinated aliphatics. All cultures were incubated at 30°C on a rotary shaker at 140 rpm.

Analytical Procedures. The chlorinated ethenes, ethanes and methanes were analysed with a Perkin Elmer Autosystem XL gas chromatograph on a bonded phase fused silica capillary column by a flame ionization detector and an electron capture detector, using the headspace technique with a Headspace Sampler HS 40. All samples were taken from the liquid phase of the assays. Ethene concentration was determined from the gasphase of the assays with a Carlo Erba Fractovap Series 2350 gas chromatograph on a molecular sieve (5Å, 60-80 mesh) packed column by a flame ionization detector.

Bacterial growth was determined through the optical density at 578 nm with a Pharmacia Ultrospec III Spectrophotometer.

RESULTS AND DISCUSSION

Cultivation of the mixed culture under the conditions described above lead to a stable community of five to six microscopically distiguishable morphotypes, a thin and long rod-shaped non-motile bacterium, two kinds of spirillums of different width, two types of non-motile cocci of different diameter and sometimes a non-motile curved rod.

The mixed culture showed a stable ability to dechlorinate DCE when transfered into DCE-containing media (Figure 1) and lost the ability to dechlorinate after two passages over media without DCE. Various combinations of substrates could be utilized for growth (pyruvate, ethanol, glucose, formate/acetate/yeast extract), but only few allowed fast complete dechlorination to ethene combined with growth (formate/acetate/yeast extract, ethanol). Dechlorination was fastest with formate/acetate or H_2/acetate, but with those substrate combinations growth could only be observed in the presence of yeast extract. The optimum temperature for DCE dechlorination was 30°C.

None of the organisms in the mixed culture was able to grow on aerobic complex solid media, but some of them survived exposure to oxygen for five days and continued to grow after transfer into anaerobic medium. However, dechlorination was lost after even short exposure to oxygen.

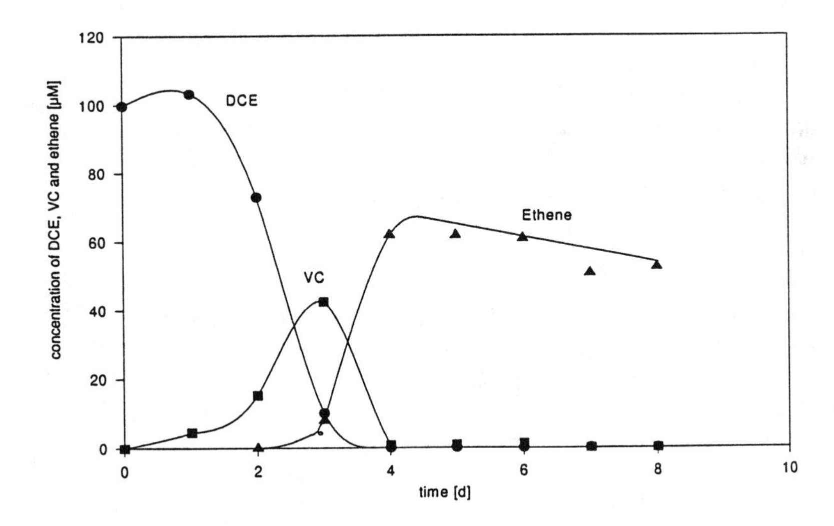

FIGURE 1. DCE dechlorination with the enriched mixed culture (in mineral medium with 40 mM formate, 20 mM acetate, 0.1% yeast extract)

The mixed culture could also dechlorinate TCE, 1,1-dichloroethene, 1,2-dichloroethane and VC to ethene. Dechlorination of DCE was inhibited completely in the presence of 150 μM tetrachloromethane, trichloromethane and 1,1,1-trichloroethane. Dichloromethane and 1,1-dichloroethane did not inhibit DCE dechlorination.

Various efforts to isolate the dechlorinating organism through dilution methods or anaerobic solid media failed so far. Also several attempts to enrich the dechlorinating organism (alternative electron acceptors, temperature variations, pasteurization etc.) were not successful, because the dechlorination activity was always the first to be lost.

So far it seems, the dechlorinating organism is a lot more sensitive to varying culture conditions, than the so far isolated PCE-degrading strains. It was possible to achieve a stable PCE dechlorination to ethene over several months in a laboratory-scale continuous-flow biofilm reactor inoculated with *Dehalospirillum multivorans* (Scholz-Muramatsu et al., 1995) and the DCE dechlorinating mixed culture. So the organism is able to survive and dechlorinate in a mixed population biofilm. But complete dechlorination could so far not be established for a longer period of time in a large-scale sandbox model for in-situ bioremediation (Cirpka and Bisch, 1998). Therefore, to be able to stimulate DCE dechlorination in various systems, more information about the properties of the involved organisms is neccessary.

ACKNOWLEDGEMENTS

The present study was conducted in the framework of the interdisciplinary project "Stimulierung der vollständigen reduktiven Dehalogenierung in einem Modellaquifer - Artificial AnAerobic Aquifer A^4" funded by the Deutsche Forschungsgemeinschaft under the grant Scho-537/1.

REFERENCES

Cirpka O. A., G. Bisch. 1998. "Microbial Reductive Dechlorination of PCE in a Quasi Two-Dimensional Sandbox Model." In *Proceedings of the Conference on Remediation of Chlorinated and Recalcitrant Compounds, May 18-21, 1998, Monterey*

De Bruin W. P., M. J. J. Kotterman, M. A. Posthumus, G. Schraa, A. J. B. Zehnder. 1992. "Complete Biological Reductive Transformation of Tetrachloroethene to Ethane" *Appl. Environ. Microbiol. 58*(6): 1996-2000

Eisenbeis M., P. Bauer-Kreisel, H. Scholz-Muramatsu. 1997. "Studies on the dechlorination of tetrachloroethene to *cis*-1,2-dichloroethene by *Dehalospirillum multivorans* in biofilms" *Wat. Sci. Tech. 36*(1): 191-198

Fennell D. E., J. M. Gossett, S. H. Zinder. 1997. "Comparison of butyric acid, ethanol, lactic acid, and propionic acid as hydrogen donors for the reductive dechlorination of tetrachloroethene" *Environ. Sci. Techn. 31*(3): 918-926

Freedman D. L., J. M. Gossett. 1989. "Biological Reductive Dechlorination of Tetrachloroethylene and Trichloroethylene to Ethylene under Methanogenic Conditions" *Appl. Environ. Microbiol. 55*(9): 2144-2151

Gerritse, J., V. Renard, T. M. Pedro Gomes, P. A. Lawson, M. D. Collins, J. C. Gottschal. 1996. "*Desulfitobacterium* sp. strain PCE1, an anaerobic bacterium that can grow by reductive dechlorination of tetrachloroethene or ortho-chlorinated phenols" *Arch. Microbiol. 165:* 132-140

Gerritse, J., G. Kloetstra, A. Borger, G. Dalstra, A. Alphenaar, J. C. Gottschal. 1997. "Complete degradation of tetrachloroethene in coupled anoxic and oxic chemostats" *Appl. Microbiol. Biotechnol. 48*(4): 553-562

Granzow S. 1998. "Isolierung und Charakterisierung eines neuen, strikt anaeroben, Tetrachlorethen-dechlorierenden Bakteriums." Ph.D. thesis, Universität Stuttgart, Stuttgart, Germany.

Holliger C., W. Schumacher. 1995. "Reductive Dehalogenation as a Respiratory Process" *Antonie van Leeuwenhoek 66*: 239-246

Krumholz L. R., R. Sharp, S. S. Fishbain. 1996. "A Freshwater Anaerobe Coupling Acetate Oxidation to Tetrachloroetylene Dehalogenation" *Appl. Environ. Microbiol. 62*(11): 4108-4113

Maymo-Gatell X., Y. Chien, J. M. Gossett, S. H. Zinder. 1997. "Isolation of a bacterium that reductively dechlorinates tetrachloroethene to ethene" *Science 276*(5318):1568-1571

Miller E., G. Wohlfarth, G. Diekert. 1997. "Comparative studies on tetrachloroethene reductive dechlorination mediated by *Desulfitobacterium* sp. strain PCE-S" *Arch. Microbiol. 168*(6):513-519

Miller E., G. Wohlfarth, G. Diekert. 1998. "Purification and characterization of the tetrachloroethene reductive dehalogenase of strain PCE-S" *Arch. Microbiol.*, in press

Neumann A., G. Wohlfarth, G. Diekert. 1995. "Properties of tetrachloroethene and trichloroethene dehalogenase of *Dehalospirillum multivorans*" *Arch. Microbiol. 163*: 276-281

Neumann A., G. Wohlfarth, G. Diekert. 1998. "Cloning and Sequencing of PCE Reductive Dehalogenase Gene from *Dehalospirillum multivorans*" *J. Bact.*, submitted

Scholz-Muramatsu H., A. Neumann, M. Meßmer, E. Moore, G. Diekert. 1995. "Isolation and characterization of *Dehalospirillum multivorans* gen. nov., sp. nov., a tetrachloroethene-utilizing, strictly anaerobic bacterium" *Arch. Microbiol. 163*: 48-56

Schumacher W., C. Holliger, A. J. Zehnder, W. R. Hagen. 1997. "Redox chemistry of cobalamin and iron-sulfur cofactors in the tetrachloroethene reductase of *Dehalobacter restrictus*" *FEBS Lett. 409*(3):421-425

Sharma P. K., P. L. McCarty. 1996. "Isolation and Characterization of a Facultatively Aerobic Bacterium That Reductively Dehalogenates Tetrachloroethene to *cis*-1,2-Dichloroethene" *Appl. Environ. Microbiol. 62*(3): 761-765

Wild A., W. Winkelbauer, T. Leisinger. 1995. "Anaerobic dechlorination of trichloroethene, tetrachloroethene and 1,2-dichloroethene by an acetogenic mixed culture in a fixed-bed reactor" *Biodegradation 6*: 309-318

Wild A., R. Hermann, T. Leisinger. 1997. "Isolation of an anaerobic bacterium which reductively dechlorinates tetrachloroethene and trichloroethene" *Biodegradation* 7(6): 507-511

PENTACHLOROPHENOL DEGRADATION USING LIGNIN PEROXIDASE PRODUCED FROM THE *PHANEROCHAETE CHRYSOSPORIUM* IMMOBILIZED IN POLYURETHANE FOAM.

Sue Hyung Choi, Eun Song, Ken W. Lee, Seung-Hyun Moon, and Man Bock Gu.
(Kwangju Institute of Science and Technology, Kwangju, Korea)

ABSTRACT: Lignin peroxidase(LiP) production and its usage for biodegradation of pentachlorophenol(PCP) were studied. Experiments for lignin peroxidase production have been conducted by aerobic fermentation of *P. chrysosporium* in order to find the optimum conditions for a previously mentioned parameter, shear stress. It has been found that immobilization on polyurethane foam is a good means for overcoming the high shear sensitivity. Scanning Electron Microscopy(SEM) pictures showed that this strain was very well attached and growing on the polyurethane foams. The production of lignin peroxidase by free and immobilized cultures of *P. chrysosporium* was studied. Free pellets of this strain produced a little of lignin peroxidase(about 30 unit/liter), but the lignin peroxidase production was increased by immobilizing the cells on polyurethane foam (about 250 unit/L). Cell-free culture fluids harvested from nitrogen limited cultures were used to degrade pentachlorophenol (PCP). The culture fluid containing lignin peroxidase underwent the pentachlorophenol degradation and produced intermediates. Biodegradation of PCP(30ppm) by cell-free culture fluid was detected by HPLC analysis. Analysis of two reaction mixtures (PCP 30 ppm + Culture fluid 40 unit/L + [H_2O_2 0.5 mM] or not) during 6 hours showed disappearance of about 70 % pentachlorophenol(PCP). Unidentified intermediates as byproducts were detected by HPLC assay in the pentachlorophenol(PCP) treatment using cell-free culture fluids.

INTRODUCTION

Lignin peroxidase has strong potential for environmental application such as degradation of aromatic compounds.(Cripps et al., 1990; Kennedy et al., 1990). Lignin peroxidase is so able to attack many different recalcitrant chemicals including PCP as well as lignin, one of the main components of the woods and grass(Knapp, 1985; Chung et al., 1995). In spite of the usefulness and important features of the enzymes mass production and industrial and environmental application of lignin peroxidase have been quite limited and delayed long time. High shear sensitivity was found to be one reason for lower production of lignin peroxidase(Kirk et al., 1978). In order to overcome shear stress effects, cell immobilization on polyurethane foam has been adopted in this study.

Pentachlorophenol has been used extensively as a fungicide, insecticide or herbicide in wood treatment plants. The disposal of pentachlorophenol has resulted in the contamination of ecosystem due to its stability in the environment. There are many studies on degradation of pentachlorophenol using pure enzymes or whole cells(Knapp, 1985; Chung et al., 1995).

Objective. Obtaining the culture conditions which allow *P. chrysosporium* to overproduce lignin peroxidase and show reasonable application of cell-free culture fluids for PCP treatment.

MATERIALS AND METHODS

Organism and culture condition. *P. chrysosporium* ATCC 24725 was maintained at 39℃ on the malt-agar plate. Spores were harvested and filtered through glass wool. Then the spore concentration was determined by measuring absorbance at 650 nm(an absorbance of 1.0 cm^{-1} is approximately 5×10^{6} spores/mL). Growth medium was same as Tien and Kirk used(Kirk et al., 1988), except that veratryl alcohol and Tween 80 were added again after two days passed. Nitrogen concentration ratio was modified as described in results and discussions.

Immobilization. A supporting matrix for immobilization was polyurethane foam and 4 grams of it was placed in each 250 mL Erlenmeyer flask and autoclaved at 120℃ for 20 min. After inoculation with 5×10^{7} spores in 80 mL media, the cultures were flushed with pure oxygen for 2 min every day.

Scanning Electron Microscopy. Mycelia on the polyurethane foam was fixed for 24 hours in 24 % chloroform solution, and then washed with Phosphate-buffered saline solution. Finally mycelia were soaked in 20, 40, 60, 80 and 100 % ethanol and dried. Before taking SEM pictures, the polyurethane foam was gold-filmed thinly.

PCP degradation. Culture fluids harvested from the nitrogen limited cultures containing lignin peroxidase were used to degrade the PCP. Two samples(0.85 mL each) were used to detect the degree of the oxidation of PCP and formation of intermediates. One of them consisted of culture fluid(Lip 40 unit/L), PCP 30 ppm, and H_2O_2 0.5 mM, and the other one included culture fluid(Lip 40 unit/L) and PCP 30 ppm, but H_2O_2. The amount of PCP remaining in reaction mixtures was detected by HPLC.

Enzyme activity. Lignin peroxidase activity was measured using the veratryl alcohol by the method described previously (Kirk et al., 1988).

RESULTS AND DISCUSSION

Lignin peroxidase secretion by pellets and immobilized cells of *P.*

chrysosporium. In the presence of polyurethane foam as solid matrix, lignin peroxidase production by *P. chrysosporium* were markedly enhanced and cultivation times for maximum activity were shorten. When immobilized on polyurethane foam, *P. chrysosporium* produced lignin peroxidase (activity = 300 unit/L) after 7 days, as compared to about 30 unit/L for free pellets after 11 days. The productivity increased about 10-fold(Figure. 1). It seems

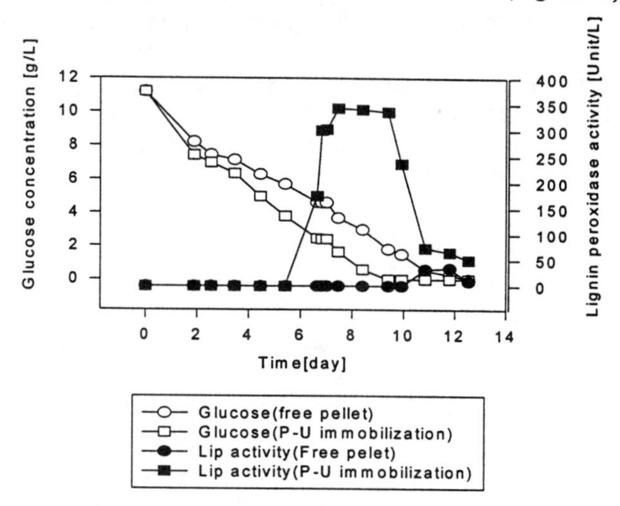

FIGURE 1. Enzyme activity of lignin peroxidase in the nitrogen limited culture using free pellet and immobilized cell

to us that immobilization on the polyurethane foam reduces the shear stress effect and so results in higher cell mass and enzyme production. At the end of culture, mycelia immobilized on polyurethane foam was photographed by scanning electron microscopy(SEM). Figure 2 showed that

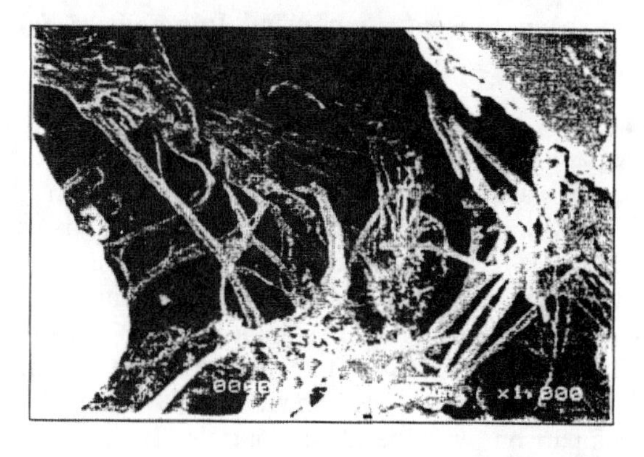

FIGURE. 2 A picture of *P. chrysosporium* immobilized in polyurethane foam(x1000) by Scanning Electron Microscope(SEM)

the fungi was very well attached to the solid matrix and was growing on the polyurethane foam.

Lignin peroxidase production and nutrient consumption by immobilized cultures in different nitrogen concentrations. In the limiting nitrogen condition(1.2 mM ammonium tartrate), lignin peroxidase production started when nitrogen was starved even the glucose in the media was not fully consumed. On the other hand, the lignin peroxidase in sufficient nitrogen condition(24 mM ammonium tartrate) was produced after glucose was fully consumed. In this case it is not clear whether nitrogen consumption was depleted or not. The activity of lignin peroxidase produced in both cases was about 250 unit/L(Figure 3).

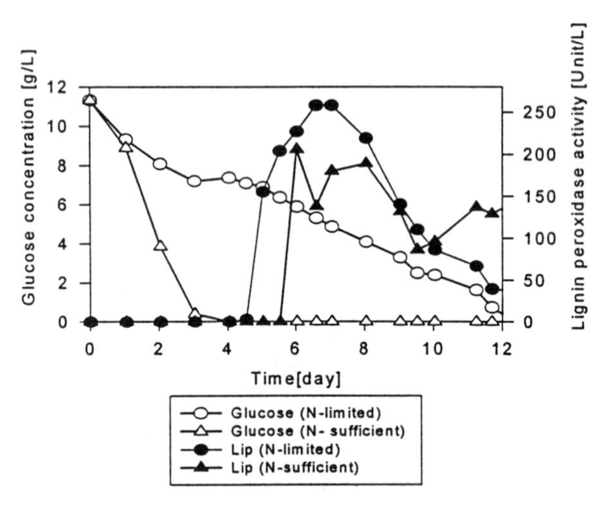

FIGURE 3. Production of lignin peroxidase in the different nitrogen concentration using Immobilized fungi

Biodegradation of PCP The white rot fungus *P. chrysosporium* has been shown to degrade a wide variety of environmental persistent organopollutants, including a number of organohalides. In this study, we have treated PCP with cell-free culture fluid. HPLC analysis of reaction mixtures containing PCP, culture fluid, and H_2O_2(or not) showed the appearance of unknown intermediates. The disappearance of PCP after 6 hour incubation of the reaction mixtures indicated that cell-free culture fluid can attack PCP in H_2O_2-added condition as well as in H_2O_2-omitted condition. The biodegradability was about 70 % from the HPLC assay based on the comparison with negative controls. In addition, intermediates formed were different in both cases ; one intermediate was formed in H_2O_2 containing mixture but two intermediates without H_2O_2 in the mixture(Table.1 Figure 4). Previously studies have shown that H_2O_2 is a

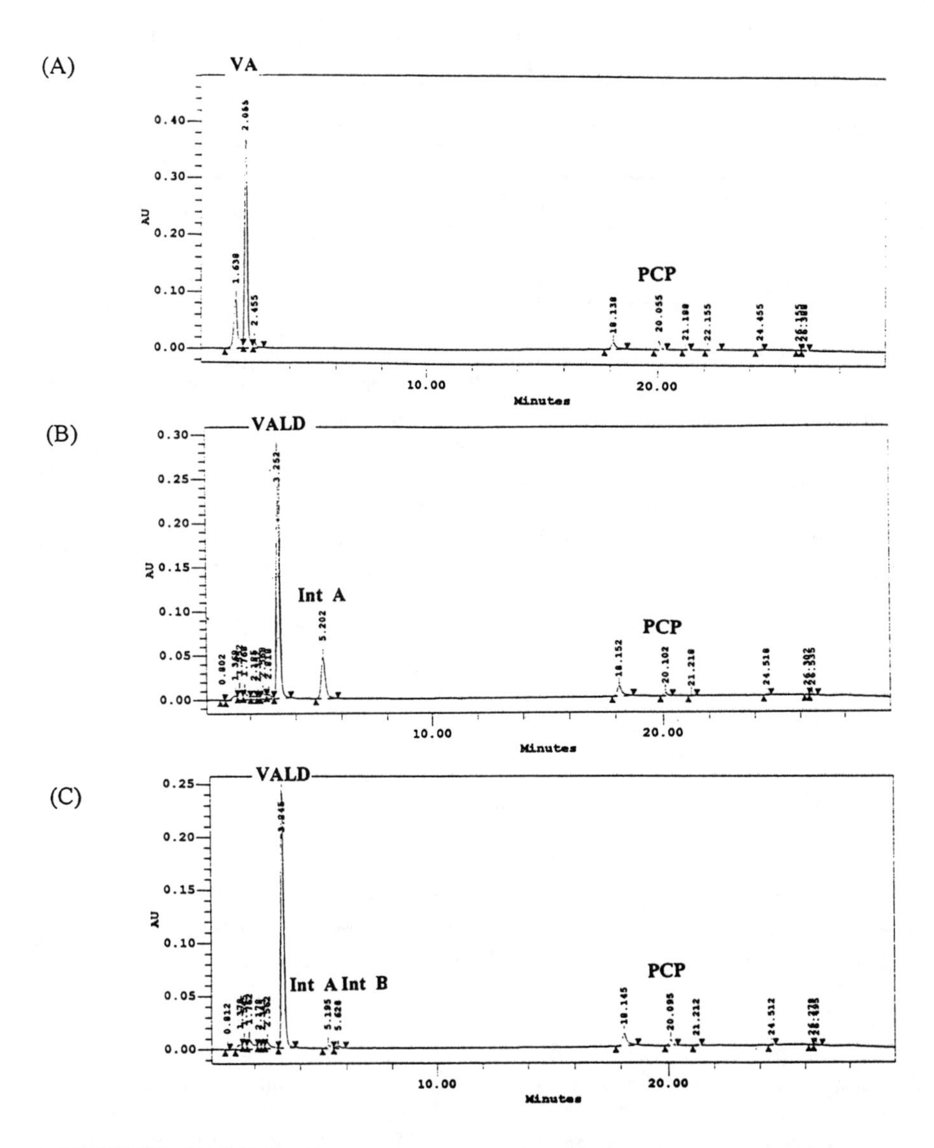

FIGURE 4. HPLC chromatogram of pentachlorophenol degradation and unknown product formation using culture fluid(pH 4.5) in nitrogen limited culture

A. Culture media(without LIP) + PCP(30 ppm),
B. Culture fluid(with LIP 40 unit/L) + PCP(30 ppm) + H_2O_2(0.5 mM),
C. Culture fluid(with LIP 40 unit/L) + PCP(30 ppm)

VA : veratryl alcohol, VALD : veratryl aldehyde, PCP : pentachlorophenol, Int A : Intermediate A, Int B : Intermediate B.

TABLE 1. Summary of PCP biodegradation in the culture fluid for 6 hours

	Culture fluid(LIP 40 unit/L)+ PCP(30 ppm) + H_2O_2(0.5 mM)	Culture fluid(LIP 40 unit/L) + PCP(30 ppm)
Removal % of PCP	73 %	68 %
Unknown intermediate	A	A, B

key component as an activation agent of lignin peroxidase in the oxidation reaction. However, in this study, PCP degradation was found to occur without H_2O_2 addition. There are two possibilities. The one is that another extracellular enzymes not related to H_2O_2 may be involved in the degradation of PCP. The other one is that H_2O_2 production by *P. chrysosporium* enough to initiate lignin peroxidase activation. This phenomenon is not yet scientifically proved. It could be most likely that *P. chrysosporium* is capable of degrading the PCP by producing extracellular enzymes other than ligninase and H_2O_2. Therefore cell-free culture fluids itself may be effective means for degrading recalcitrant chemicals including PCP.

REFERENCE

1. Cripps, C., J. A. Bumpus, and S. D. Aust. 1990. "Biodegradation of azo and heterocyclic dyes by *Phanerochaete chrysosporium*." *Appl. Environ. Microbial.* 56, 1114-1118.

2. Kennedy, D. W., S. D. Aust, and J. A. Bumpus. 1990. "Comparative biodegradation of alkyl halide insecticides by the white rot fungus, *Phanerochaete chrysosporium* (BKM-F-1767)." *Appl. Environ. microbial.* 56: 2347-2353.

3. Knapp, J. S. 1985. Biodegradation of cellulose and lignins in Comprehensive Biotechnology, Vol. 4, Murray Moo-Young ed. Permon Press.

4. Chung, N., and Steven D. Aust. 1995. "Veratryl Alcohol-Mediated Indirect Oxidation of Pentachlorophenol by Lignin peroxidase," *Arch. Biochem. Biophys.* 322: 143-148.

5. Kirk, T. K., E. Schultz, W. J. Connors, L. F. Lorenz, and J. G. Zeikus. 1978. "Influence of Culture Parameters on Lignin Metabolism by *Phenerochaete chrysosporium*," *Arch. Microbial.* 117: 277-285.

6 Kirk, M. and T. K. Kirk. 1988. "Lignin peroxidase of *Phanerochaete chrysosporium*," *Methods Enzymol.* 161: 238-249.

BIOTRANSFORMATION OF HEXACHLOROBENZEN BY ANAEROBIC ENRICHED CULTURES

In S. Kim (Kwangju Institute of Science and Technology, Kwangju, Korea)
Hirokazu Ishii (Japan Sewage Work Association, Tokyo, Japan)
Gregory D. Sayles (U.S. EPA, Cincinnati, Ohio)
Margaret J. Kupferle, and Tiehong L. Huang (University of Cincinnati
Cincinnati, Ohio)

ABSTRACT: The model compound, hexachlorobenzene (HCB), selected as highly chlorinated toxic organic hydrocarbon was degraded readily under anaerobic conditions using HCB acclimated mixed steady-state cultures. The rate of degradation increased as the initial HCB concentration increased. 1,3,5-TCB was the main product from the degradation of HCB. No significant biodegradation of 1,3,5-TCB under anaerobic conditions was observed. Three different sources of mixed cultures were investigated in our study. The HCB acclimated culture showed the highest HCB dechlorination reaction rate. Interestingly, ethanol-enriched cultures not acclimated to HCB dechlorinated HCB with a high activity as well. HCB was not dechlorinated by the municipal anaerobic digester sludge culture during the 37-day period of this study.

INTRODUCTION

Hexachlorobenzene (HCB) is used as fungicide and is toxic to microbial species. HCB is highly persistent in the environment, with a half-life of decades or more because of low solubility. The bulk of HCB released to the environment tends to remain in the solid phase. HCB is released to the aqueous phase slowly, thus becoming a long term constant source of groundwater contamination. Commonly used treatment processes containing acclimated aerobic microbial populations (e.g. activated sludge) generally do not biotransform HCB successfully. However, anaerobic microbial consortia have been shown to partially degrade highly chlorinated benzenes by means of reductive dechlorination (Mohn and Tiedje, 1992). Several factors such as the sources of biota, nature of inoculum, and the presence and type of cometabolites are known to be variables that affect the pattern of dechlorination (Liang and Grbic-Galic, 1991; Mohn and Tiedje, 1992; Mousa and Rogers, 1990). It is also reported that reductive dechlorination occurs more readily as the level of chlorination increase (Fathepure et al., 1988; Bhatnagar and Fathepure, 1991). However, after rapid initial dechlorination of highly chlorinated compounds, anaerobic biodegradation typically ends with less chlorinated benzenes and complete mineralization is difficult or impossible. Many researchers report that 1,3,5-trichlorobenzene is the most stable degradation product from HCB with a few exceptions (Bosma et al., 1988). On the other hand, these less chlorinated or non-chlorinated chemicals generally are degraded successfully in an aerobic environment. Thus, sequential anaerobic-aerobic treatment has the potential of providing relatively rapid complete destruction of a toxic contaminant such as hexachlorobenzene (Abramowicz, 1990; Avid et al, 1991; Armenante et al, 1992).

The objectives of this research were to determine the effect of different consortia and cometabolites on the anaerobic biotransformation rates for HCB and its intermediates identifying appropriate conditions for maximum biodegradation.

EXPERIMENTAL METHODS

Anaerobic Seed Cultures. Anaerobic cultures having identifiable and repeatable characteristics were developed and maintained in seed culture reactors so that subsequent biodegradation and dechlorination experiments conducted at different times would have a common microbial basis. Two reactors were prepared, one for unacclimated biomass as a control and one for HCB-acclimated biomass.

Eighteen liters of culture were continuously mixed with a magnetic stirrer in a 20-L glass carboy closed with a rubber stopper. The temperature was maintained at $35^{o}C$ and exposure of the cultures to light was minimized. Three glass tubes were inserted through the rubber stopper to provide access for feeding and withdrawing culture and for measuring gas production. Both seed cultures were fed one time per day on a draw-and-fill basis with a synthetic medium (Kim et al.,1994) designed to provide nutrients, buffering agents and trace minerals necessary to sustain anaerobic microbial growth. Both cultures also received ethanol (1g COD/L-day) as a carbon and energy source. The HCB-acclimated culture received 2mg/L-day HCB in addition to the ethanol. Nitrogen gas was used to pressurize the carboy and withdraw culture from the reactor as well as to anaerobically add the feed solution, maintaining a 20-day hydraulic (and solids) retention time.

Batch Degradation Studies. Batch degradation studies were done in two phases using serum bottle reactors. Phase 1 tests were conducted to investigate the effect of initial concentration on HCB degradation rate and to determine the pattern of transformation product formation. Batch dose of 5, 10, 20 and 50mg/L of HCB were tested in the concentration study. A replicate 10 mg/L treatment and an abiotic control dosed at 10mg/L were prepared as quality assurance measures. The concentration study was carried out in 500-mL serum bottles filled with HCB-acclimated culture from the laboratory seed culture reactor. Acetone was used as a solvent carrier for HCB when dosing the serum bottle, but the acetone was allowed to evaporate prior to addition of the culture to the bottle. The reactors were sub-sampled in 5-mL aliquots over a period of 33 days, and the samples were extracted and analyzed with gas chromatography. Subsequent experiments were done in 40-mL sacrificial vial reactors to minimize loss of the more volatile transformation products. For the transformation product study, a batch dose of 20mg/L HCB was spiked to the vials after HCB-acclimated culture had been added. The culture in half of the vial reactors was killed with mercuric chloride to serve as an abiotic control.

Phase 2 tests were conducted to investigate the extent of HCB degradation and the dechlorination pathways of microbial cultures from three different sources. Sacrificial 40-mL reactor vials were set up as described in the sections that follow according to the experimental design summarized in Table 1.

Table 1. Summary of experimental set-up for different mixed culture source.

Culture source	Materials added	No. Of vials	Function
	Seed + HCB (10 mg/L) + $HgCl_2$	15	Abiotic control
HCB acclimated	Seed + HCB (10 mg/L)	20	Test units
	Seed + HCB (10 mg/L)	20	Duplicates
	Seed + HCB (10 mg/L) + $HgCl_2$	15	Abiotic control
EtOH acclimated	Seed + HCB (10 mg/L)	20	Test units
	Seed + HCB (10 mg/L)	20	Duplicates
	Seed + HCB (10 mg/L) + $HgCl_2$	15	Abiotic control
Anaerobic digester sludge	Seed + HCB (10 mg/L)	20	Test units
	Seed + HCB (10 mg/L)	20	Duplicates

Initial HCB concentration in all treatments was fixed at 10 mg/L and the biomass concentration was adjusted to approximately 850 mg/L as volatile suspended solids (VSS). Culture acclimated to HCB (with ethanol as the primary substrate), culture acclimated to ethanol only, and anaerobic digester sludge from a municipal sewage treatment plant were selected for this test.

Analytical Methods. When a vial was sacrificed for analysis, 10 mLs of hexane and 2 mLs of acetone were added directly to the reaction vial and the vial was closed again tightly. The vials were mixed for 30 seconds using a vortex mixer, placed in a sonicator for 15 minutes, and centrifuged for 15 minutes at 1500 rpm. One mL of the hexane phase was transferred to a 2-mL glass vial sealed with a teflon-lined septum held in place with an aluminum crimp cap. A 10 (L aliquot of a 1000 mg/L 1,3,5-tribromobenzene stock solution was added to the extract as an internal standard. The extract was analyzed using gas chromatography.

Concentrations of HCB and penta-, tetra-, and trichlorobenzenes were determined using a 5890 Series II Hewlett Packard gas chromatograph equipped with HP Chemstation software®, a 30-m x 0.32-mm fused silica capillary column (SPB-1, Supelco, Inc.), and a ^{63}Ni electron-capture detector (ECD). The gas chromatograph equipped with the ECD detector was not sensitive enough to detect the di- and monochlorobenzenes, so the vials were immediately transferred to a 5890 Series II Hewlett Packard gas chromatograph equipped with a fused silica glass capillary column (30m x 0.25 mm ID, Quadex #202280), and a flame-ionization detector (FID). A 5890 Series II Hewlett Packard gas chromatograph equipped with a 2m x 2.0 mm glass column packed with Carbopack TM B-DA/

4% Carbowax 20 M (Supelco, Inc.) and a flame ionization detector was used to measure ethanol and volatile fatty acids concentrations.

RESULTS AND DISCUSSION

Phase 1 Test Results. Phase 1 tests were conducted to investigate the effect of initial concentration on HCB degradation rate and to determine the transformation pattern. Batch doses of 5, 10, 20 and 50 mg/L were tested. A replicate 10 mg/L treatment and an abiotic control dosed at 10 mg/L were prepared as quality assurance measures. Acetone was used as a solvent carrier for HCB and as an extranal cosubstrate in these experiments. The HCB-acclimated culture was the source of microorganisms. The rate of HCB degradation increased as initial HCB concentration increased (Figure1). No inhibition of culture activity was apparent at the HCB concentration tested. About 25 days were required to transform all of the HCB at the highest concentration (50 mg/L). The HCB concentration measured in the abiotic control remained constant at about 10-12 mg/L.

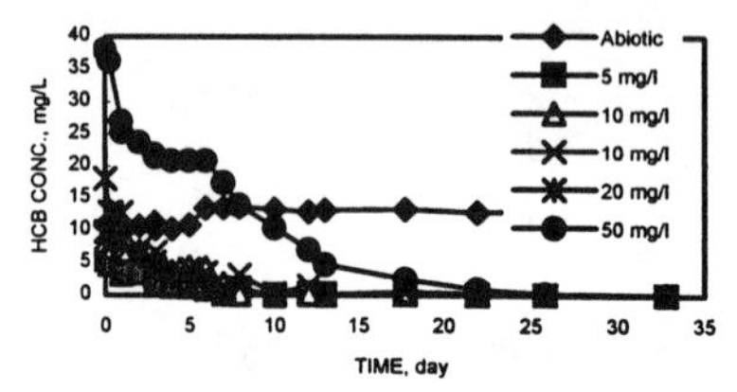

FIGURE 1. Anaerobic degradation of HCB in phase 1 test

The main product of HCB degradation was 1,3,5-TCB (Figure 2). The highest concentration of 1,3,5-TCB (molecular weight=181.5 g/mol) measured was 10 mg/L which represents about 79 % of the original 20 mg/L HCB (molecular weight = 285g/mol) on a molar basis. Other intermediates such as 1,4-dichlorobenzene (1,4-DCB), 1,2,4-trichlorobenzene (1,2,4-TCB), and 1,2,3,5-tetrachlorobenzene (1,2,3,5-TeCB) were detected at much lower concentration levels compared to 1,3,5-TCB(Figure 3). Pentachlorobenzene (PeCB) and 1,2,3,5-TeCB were detected at low concentrations during the first day of the experiment, but their concentrations decreased after one day of operation and disappeared completely after 10 days as they were transformed to less chlorinated compounds such as 1,2,4-TCB, 1,3,5-TCB and 1,4-DCB. Concentrations of 1,2,4-TCB, 1,3,5-TCB and 1,4-DCB increased continuously during the experiment.

Phase 2 Test Results. Phase 2 tests were conducted to investigate the extent of HCB degradation and the dechlorination pathways of microbial cultures from three different sources - a laboratory culture acclimated to HCB (with ethanol as the primary substrate), a laboratory culture acclimated to ethanol only, and anaerobic digester sludge from a municipal sewage treatment plant. The initial

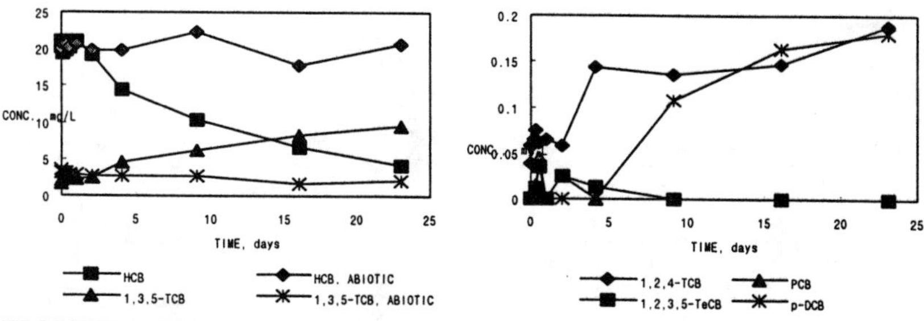

FIGURE 2. HCB degradation and its major intermediate, 1,3,5-TCB production in test unit

FIGURE 3. Concentrations of other minor intermediates in test unit

HCB concentration in all reactor vials was 10 mg/L, or 0.035 mM. The graphs shown in Figure 4 depict the conversion of HCB to less chlorinated benzenes on a molar basis over the course of 37 days. HCB was rapidly dechlorinated within a week by the HCB-acclimated culture(A), producing nearly equimolar amounts of 1,3,5-TCB with some production of PeCB and the two TeCB isomers initially and with a small amount of 1,2,4-TCB appearing later(A′). The 1,3,5-TCB was not transformed further. The pattern for the ethanol culture also exhibited significant dechlorination capacity preceded by a lag period of about a week(B). About 16 days were required for the complete transformation of HCB. As in the HCB-acclimated case, the major dechlorination product was 1,3,5-TCB(B′). Dichlorobenzenes were identified in samples collected near the end of the experiment using a GC with an ECD detector, but the amounts present were too near the method detection limit to be reliably quantitated. No significant amount of dechlorination was observed within 37 days when the culture source was minicipal anaerobic digester sludge(C). A small amount of 1,3,5-TCB was detected after several days, and it is possible that more dechlorination would have been observed if the experiment had been continued for a longer period. The results of this study indicate that acclimation to HCB improves rates of dechlorination.

CONCLUSIONS

The model compound, hexachlorobenzene (HCB), selected as highly chlorinated toxic organic hydrocarbon was degraded readily under anaerobic conditions using HCB acclimated mixed steady-state cultures. The rate of degradation increased as the initial HCB concentration increased. 1,3,5-TCB was the main product from the degradation of HCB. No significant biodegradation of 1,3,5-TCB under anaerobic conditions was observed.

Three different sources of mixed cultures were investigated in our study. The HCB acclimated culture showed the highest HCB dechlorination reaction rate. Interestingly, ethanol-enriched cultures not acclimated to HCB dechlorinated

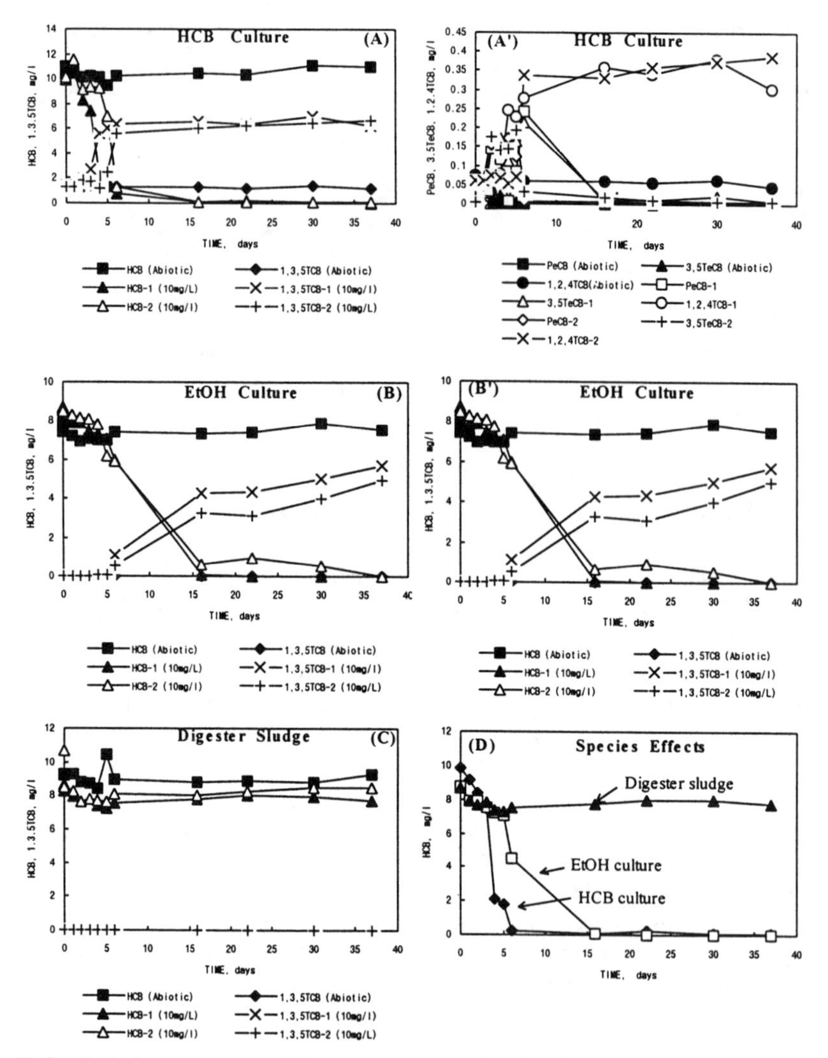

FIGURE 4. Effect of different source of mixed culture consortia on the dechlorination of HCB.

HCB with a high activity as well. HCB was not dechlorinated by the municipal anaerobic digester sludge culture during the 37-day period of this study.

The main degradation route observed in this study was HCB ---> PeCB ---> 1,2,3,5-TeCB ---> 1,3,5-TCB, which is similar to the pathway reported by Fathepure et al. (1988). Fathepure et al. reported that the major route of HCB dechlorination under their experimental conditions was HCB --->PeCB ---> 1,2,3,5-TeCB ---> 1,3,5-TCB and concluded that more than 90 % of the added HCB was recovered as 1,3,5-TCB. A minor route was HCB ---> PeCB --->

1,2,4,5-TeCB ---> 1,2,4-TCB ---> DCBs. Based on the results of our study, it may be more efficient to switch to an aerobic pathway for complete mineralization of accumulated 1,3,5-TCB, i.e.

anaerobic aerobic
HCB ----------> TCB ----------> $CO_2 + H_2O$ + Biomass

REFERENCES

Abramowicz.D.A. 1990. "Aerobic and Anaerobic Biodegradation of PCBs: a Review." *Crit. Rev. Biotechnol.* 10:241-251.

Armenante, P. M., D. Kafkewitz, G. Lewandowsky, C. Kung. 1992. "Integrated Anaerobic-Aerobic Process for the Biodegradation of Chlorinated Compounds. " *Environmental Progress*, 11(2): 113 -122.

Avid, P. J., L. Nies, and T. M. Vogel. 1991. *Sequential Anaerobic-Aerobic Biodegradation of PCBs in the River Model.*, In R. E. Hinchee and R. F. Offenbottel(ed.), In Situ Bioreclamation. Botterworth Heinemann, Boston. : 428-436.

Bhatnagar, L., and B. Z. Fathepure. 1991. *Mixed Cultures in Detoxification of Hazardous Wastes.* In G. Zeikus and E. A. Johnson (ed.), Mixed cultures in biotechnology, McGraw-Hill, Inc., New York. : 293-340

Bosma, T. N. P., J. R. van der Meer, G. Schraa, M. E. Tros, and A. J. B. Zehnder 1988, "Reductive Dechlorination of All Trichloro- and Dichlorobenzene Isomers. " *FEMS Microbial. Ecol.*, 53: 223-229.

Fathepure, B. Z., J. M. Tiedje, and S. A. Boyd. 1988. "Reductive Dechlorination of Hexachlorobenzene to Tri- and Dichlorobenzene in Anaerobic Sewage Sludge. " *Appl. Environ. Microbiol.*, 54: 327-330.

Kim, I. S., J. C. Young, and H. H. Tabak. 1994. "Kinetics of Acetogenesis and Methanogenesis in Anaerobic Reactions under Toxic Conditions." Water Environ. Fed. 66, No. 2, 119 - 132.

Liang, L., N., and D. Grbic-Galic. 1991. "Reductive Dechlorination of Hexa and Dichlorobenzene by Anaerobic Microcosms and Enrichments Derived from a Creosote-Contaminated Ground Water Aquifer. " *Abstr. Q-107, p. 294, Abstr. 91st Gen. Meet. Am. Soc. Microbiol. 1991. American Society for Microbiology*, Washington, D. C.

Mohn, W. W., and J. M. Tiedje. 1992. "Microbial Reductive Dehalogenation" *Microbiological Reviews*, 56(3) : 482 - 507.

Mousa, M. A., and J. E. Rogers. 1990. "Dechlorination of Hexachlorobenzene in Two Freshwater Pond Sediments Under Methanogenic Conditions. " *abstr. Q-45, P. 296. Abstr. 90th Annu. Meet. Am. Soc. Microbial. 1990. American Society for Microbiology, Washingto n, D.C.*

Reducing VOC Concentrations through Landfill Gas Removal and Cometabolic Degradation.

James D. Hartley (CH2M HILL, Sacramento, California)
Christopher M. Richgels (County of Sacramento, California)

ABSTRACT: At the Kiefer Landfill in Sacramento County, California, methods to reduce chlorinated ethylene concentrations in groundwater are being investigated. A monitoring well cross-gradient from the landfill became contaminated with chlorinated volatile organic compounds (VOCs) through landfill gas (LFG) migration. Subsequent removal of the LFG was accompanied by the reduction of VOC groundwater concentrations in this well. This study documents three months of testing at this well to ascertain the rate of LFG-borne VOC transfer to groundwater and VOC concentration reduction following LFG removal. Cyclic atmospheric pumping has also been employed to test the effect of aerobic conditions in the vadose zone. Data collected to date suggest that abiotic phase transfer, oxidation, and cometabolic dechlorination (using methane and toluene) are all significant in the observed vapor and groundwater concentration reductions.

INTRODUCTION

The Kiefer Landfill, owned and operated by the County of Sacramento for over 30 years, consists of two waste management units: the original 0.67 km^2 (165-acre) unlined module referred to as module M-1 and the modern 0.27 km^2 (67-acre) lined module M-1L (Figure 1). The subject of this paper is module M-1. This waste mass has historically exhaled LFG into the subsurface, and VOCs have been detected in the shallow aquifer beneath and downgradient of the site. A groundwater pump-and-treat facility was installed downgradient and has effectively contained the contaminant plume. At the same time, a cross-gradient monitoring well, MW-11A (Figure 1), displayed contamination loading characteristics that could not be explained by a hydraulic "leaky landfill" migration concept: the well was found to suddenly contain over 40 ug/L total VOCs following the completion of the adjacent lined module M-1L. Off-site field investigations verified a "cloud" of LFG in the overlying sand zone above the shallow aquifer that contained the VOCs also found in the groundwater plume.

Contaminant History of Monitoring Well MW-11A. Monitoring well MW-11A was installed in the spring of 1992. Initially it contained low concentrations of VOCs. At that time, MW-11A was approximately 200m (700 feet) from the limit of waste in module M-1, and the future location of the lined module M-1L was an open borrow pit.

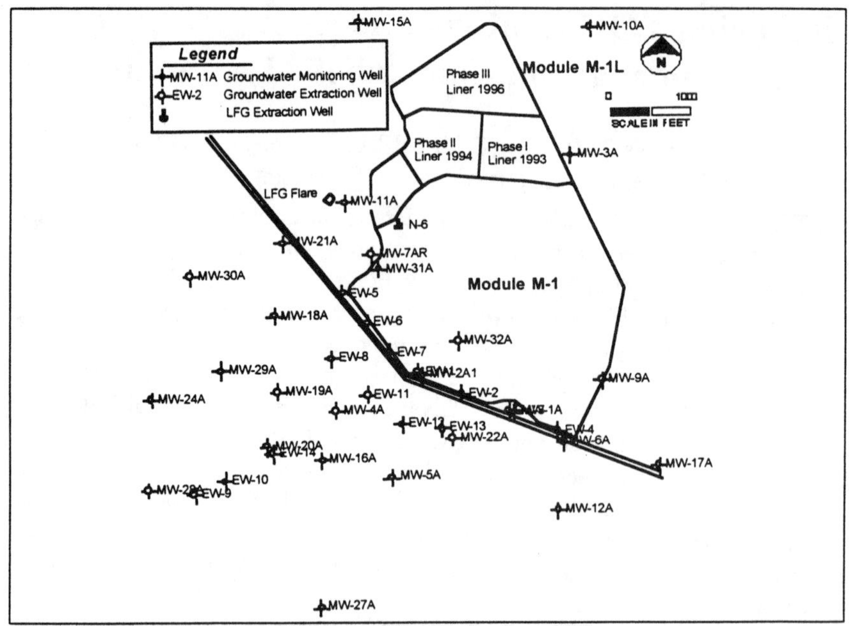

FIGURE 1. The Kiefer Landfill.

The screen of well MW-11A is located in a local geologic unit known as the Middle Sands of the Upper Mehrten Formation (referred to as Zone A). The overlying sand layer (the Upper Sands, Upper Mehrten) is in direct contact with the waste mass of module M-1. The well screen of MW-11A is within 6m of the upper sand layer and typical groundwater levels indicate approximately 2m of open well screen is above the water table (Figure 2).

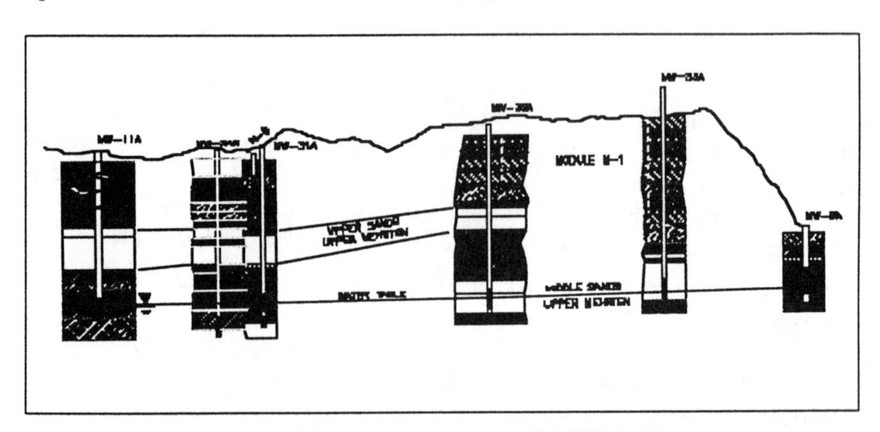

FIGURE 2. Monitoring well MW-11A.

Large magnitude swings in well concentrations were observed in MW-11A following the installation of module M-1L, beginning in 1993. After completion of the Phase I liner in September 1993, pressurized LFG was observed collecting beneath the exposed side slope liner. At the same time, the VOC concentration in MW-11A began to rise sharply (Figure 3) resulting in nearly a five to tenfold increase in VOCs over pre-liner levels. This elevated concentration continued until the LFG extraction system became operational in January 1997. In February 1997, MW-11A was sampled and analyzed for VOCs, revealing a significant decrease in levels from those found in the first half of 1996. This trend continued 6 months later when only a trace level of PCE was detected in MW-11A after nearly 10 months of LFG extraction system operation. These alternating responses of MW-11A to enhanced LFG migration (liner installations preventing ground surface venting) and LFG removal in module M-1 led to the current investigation.

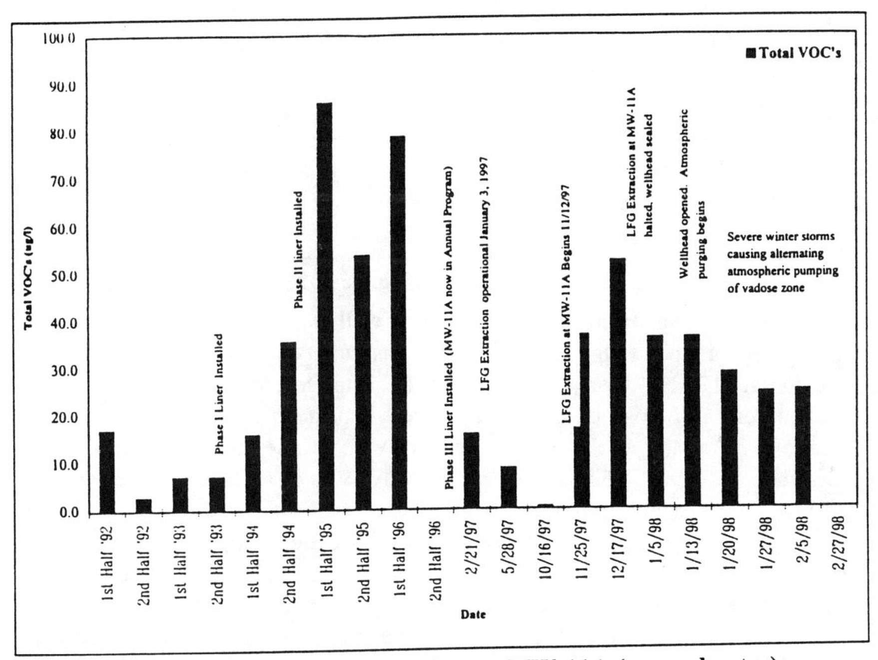

FIGURE 3. Historical total VOCs in MW-11A (groundwater).

Test LFG Extraction at MW-11A. MW-11A was constructed with a 12.7cm diameter PVC casing. The casing houses a dedicated 5cm purge pump, a bladder sampling pump, and a contained sounding pipe. A saddle tee connection point connected the well to a nearby manifold of the active LFG extraction system to provide vacuum for test extraction. Vacuum was first applied to MW-11A's casing on November 12, 1997. The extraction response (flow, pressure, O_2, and CH_4) is presented on Figure 4.

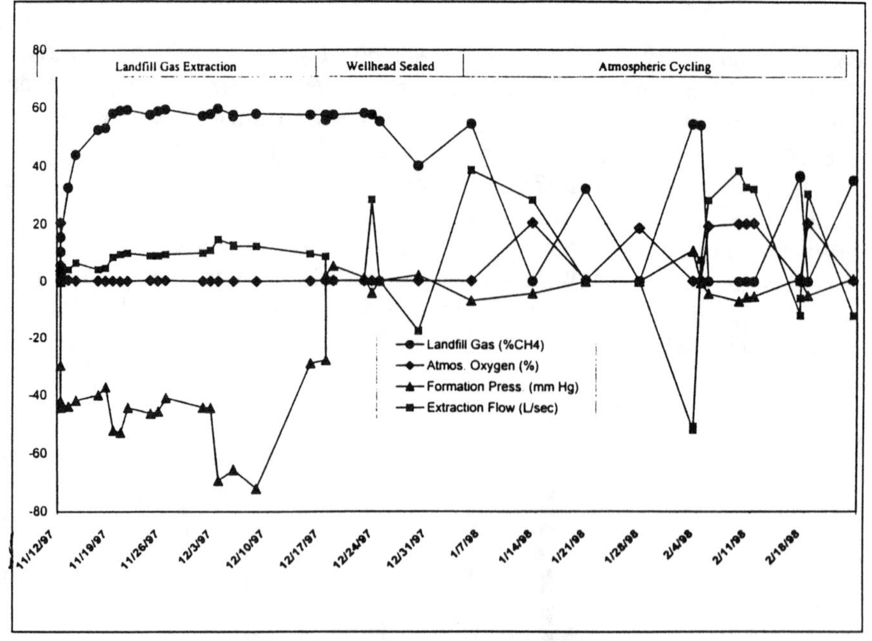

FIGURE 4. Extraction response of MW11A.

The initial vapor concentrations in this well were nearly atmospheric, but within a week after initiating extraction, the gas quality became essentially LFG. Extraction at MW-11A from November 12, 1997, to December 17, 1997, resulted in VOCs returning to the groundwater (see Figure 3, November 25, 1997). Extraction from the well was halted to discontinue further groundwater concentration increase. The LFG extraction valve was also closed to prevent any atmospheric recharge. On January 5, 1998, this well casing was vented to the atmosphere, and cyclic atmospheric "swabbing" of the vadose zone began (see Figure 4). Analytical data for both water and soil gas samples suggested a variety of degradation processes to be at work.

The sampling on October 16, 1997, revealed the water in this location was essentially clean relative to prior events. Reintroduction of LFG contact with the water table (November 12, 1997) recontaminated this well with predominately chlorinated ethylene compounds within 2 weeks. Once extraction from the well was concluded, the VOC concentrations began to decline. VOC and other concentrations declined further upon the cyclic introduction of atmospheric gases into the vadose zone between January 5 and February 5. The groundwater has been found to respond consistently with the vadose zone concentrations, though at a reduced rate.

INTERPRETATION OF DEGRADATION PROCESS

Vadose Zone. During atmospheric replenishment cycles, DO concentrations in the soil moisture are expected to have increased. This would stimulate the aerobic degradation of methane as well as the BTEX compounds and the lesser chlorinated ethylenes (DCEs, VC) in order of the oxidizing hierarchy of their compounds. In addition, the oxidation of toluene and (to a lesser extent) methane may also induce the cometabolization of TCE. PCE, which dechlorinates only under strongly reducing conditions, would not be expected to respond chemically to the conditions induced during atmospheric cycling.

These expected vadose zone trends were confirmed by vapor data collected during the two atmospheric replenishment cycles (Figure 5). BTEX compounds and DCEs (and VC to a slightly lesser extent) degraded rapidly as oxygen was reintroduced. With this cycling, the relatively persistent compound Freon 12 was found to return to approximately half its pre-cycling concentration. The response of PCE was similar, as would be expected (no strongly reducing conditions induced). TCE, which can be cometobolically degraded through the oxidation of toluene and methane, was found to be eliminated during atmospheric recharge and significantly suppressed during LFG replenishment, supporting the notion that aerobic conditions may have been sustained in the pore water during the replenishment of the LFG and LFG-borne substrates.

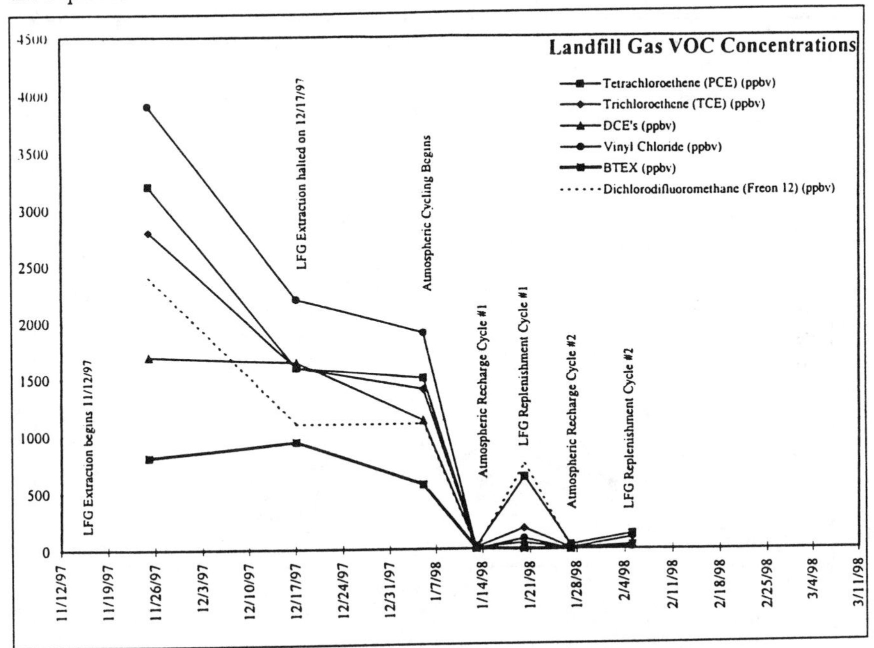

FIGURE 5. Landfill gas VOC concentrations

Groundwater. The concentration trends observed in the vadose zone track consistently with the most plausible processes. The more difficult interpretation is the effect vadose zone treatment has on concentrations of the same compounds in groundwater.

There are numerous factors to consider, and while the concentration data in the groundwater are specific, the processes responsible for trends in the data remain somewhat uncharacterized. For example, the actual depth and distribution of VOCs in the groundwater was not measured (no depth-specific sampling was performed), but given the brief duration of the exposure of the capillary fringe to LFG-borne VOCs and the relatively slow rate of aqueous diffusion, it is likely the newly supplied VOCs remained shallow. The actual depth of the water sampled by the dedicated pump system is also not known, and it is recognized that the pattern and rate of pumping can strongly influence the zones represented by the sample. What is available to interpret are the measured concentrations, which do suggest the following trends.

To the extent atmospheric cycling directly affected the vadose zone, the effect in groundwater should be damped (because of the access to groundwater only through aqueous diffusion) across the capillary fringe. As shown in Figure 6, this actually can be seen in the gradual and simple downward trend of the TCE concentrations. The concentrations of PCE, which are not judged to be treated except by phase transfer from the groundwater to the vadose zone, appear to respond with a slight lag to the atmospheric recharge cycles, when the vadose zone has been purged.

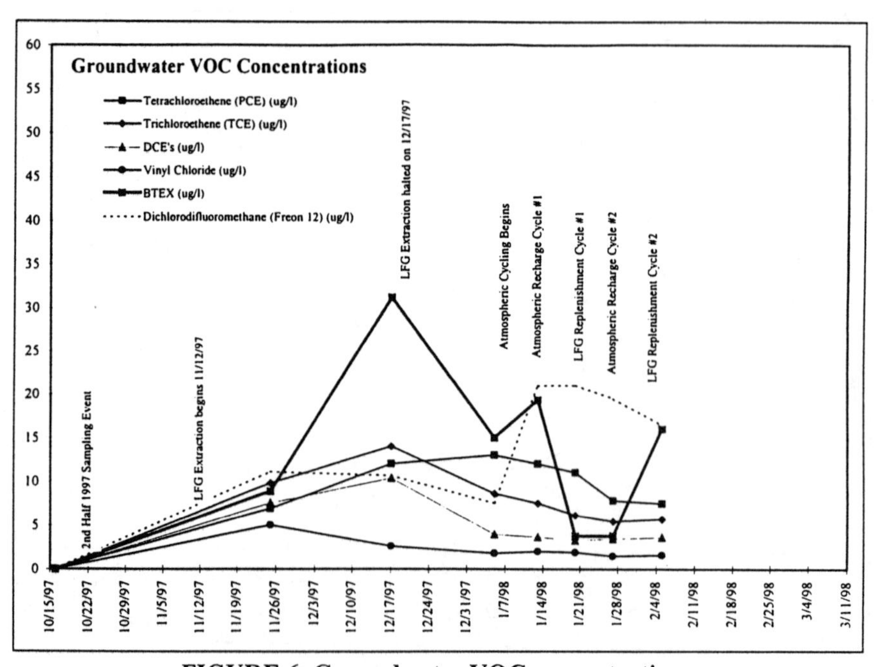

FIGURE 6. Groundwater VOC concentrations

As these trends continue, further conclusions will be drawn on the processes controlling TCE concentration reduction. As shown on Figure 6, TCE is responding less quickly than PCE to the removal of LFG, and is consistently declining while BTEX is present. In contrast, the DCEs and VC do not appear to be declining while BTEX is present. The BTEX shows a tendency to rapidly decline, except when recharged by a major LFG influx (cycle 2). These patterns will be investigated further, as will the post-BTEX rate of degradation of TCE, DCEs, and VC.

COMETABOLIC BIODEGRADATION OF *CIS*-1,2-DICHLOROETHENE BY ETHENE UTILIZING BACTERIA

Dieter Bryniok, Petra Koziollek, Susanne Bauer, and
Hans-Joachim Knackmuss (Fraunhofer IGB, Stuttgart, Germany)

ABSTRACT: Ethene was used as a novel substrate for cometabolic biodegradation of *cis*-1,2-dichloroethene (*c*DCE). Several ethene utilizing bacterial cultures were enriched that could efficiently cometabolize *c*DCE. One mixed culture was used for biodegradation experiments in continuous culture in a 1 liter stirred tank bioreactor and in a 40 l fixed biofilm reactor. The degradation processes were kept stable up to 6 months although it was operated under non sterile conditions. The specific transformation rates for *c*DCE were 8 to 9 $\mu mol \cdot min^{-1} \cdot g^{-1}_{protein}$. The maximum observed transformation yield was 0.51 mol_{cDCE}/mol_{ethene}.

INTRODUCTION

Tetrachloroethene (PCE) and trichloroethene (TCE) can be rapidly transformed to *cis*-1,2-dichloroethene (*c*DCE) by reductive dehalogenation (Vogel and McCarty, 1985, Wohlfarth and Diekert, 1997). The anaerobic transformation of *c*DCE to vinyl chloride (VC) and ethene is significantly slower. However, *c*DCE is readily cooxydized by aerobic bacteria harboring appropriate oxygenases.

Therefore, in close cooperation with the University of Stuttgart, Germany, we are developing an economic two stage process for the biodegradation of PCE and TCE in groundwater. The xenobiotics are reductively dehalogenated under strict anaerobic conditions in the first stage (Scholz-Muramatsu et al., 1995). Under these conditions *c*DCE accumulates. This metabolite is cometabolically mineralized in a subsequent aerobic stage (Figure 1).

Various auxiliary substrates like methane (Fogel et al., 1986, Little et al., 1988, Tsien et al. 1989), toluene (Wackett and Gibson, 1988), phenol (Nelson et al., 1987), cumene (Dabrock et al., 1992), and isoprene (Ewers et al., 1990) have been used for this cooxidative process. In practical application all these substrates caused serious difficulties due to the formation of highly reactive metabolites and suicide inactivation of the oxygenases (Little et al., 1988, Wackett and Householder, 1989, Anderson and McCarty, 1997). Additionally growth of contaminating microorganisms which utilized the auxiliary substrates caused decreasing chloroethene degradation rates. Therefore we used ethene as a structural analogue compound and novel auxiliary substrate for the enrichment of *c*DCE biodegrading bacteria.

MATERIALS AND METHODS

Ethene utilizing bacterial mixed cultures were enriched from the waste water treatment plant of a chloroethene producing facility in Germany. Ethene (3 to 20 % v/v) was fed over the gas phase as the sole source of carbon and energy. Cultures that were able to dechlorinate and mineralize *c*DCE completely were selected by determining the release of Cl^- with a silver precipitation technique (Ewers et al. 1990).

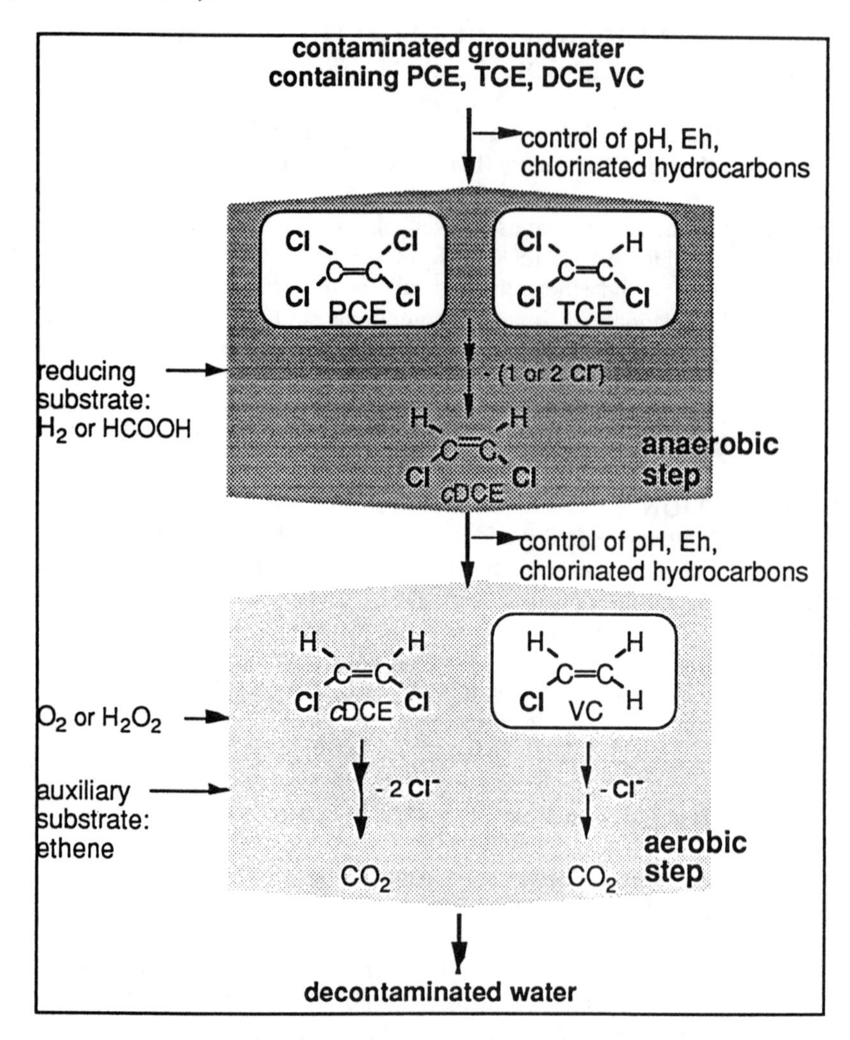

Figure 1. Two stage anaerobic/arobic process for the biodegradation of chloroethenes in groundwater

A mixed culture named K20 was used for experiments with a *c*DCE containing synthetic groundwater in a 1 liter stirred tank bioreactor, both in batch and continuous culture. Experiments with authentic *c*DCE contaminated ground water were conducted in a 40 liter fixed biofilm reactor with polyamide textile material as a support for the biofilm.

Ethene was supplied in the gas phase (0.9 to 1.9 % v/v) in all biodegradation experiments. The concentrations of *c*DCE, VC, and ethene in the gas phase were measured on line by gas chromatography with flame ionization detection (GC/FID). Liquid samples were determined by multiple headspace GC/FID analysis.

The concentration of Cl^- in water samples was measured with a chloride sensor. Dehalogenation rates were calculated as the ratio $\Delta[Cl^-] / 2\Delta[c\text{DCE}]$.

RESULTS AND DISCUSSION

Cometabolic Degradation of *c*DCE. The mixed culture K20 consists of five bacterial strains (Table 1). All five strains were highly tolerant of *c*DCE (up to 6 mM in the liquid phase) and able to degrade *c*DCE cometabolically in pure culture, but with considerably lower rates compared with the mixed culture.

K20 could degrade ethene [250 µM resp. 180 µM] and *c*DCE [800 µM] or vinyl chloride (VC) [70 µM] (Figure 2) simultaneously, both in batch or in continuous culture.

Table 1. Composition of the ethene and *c*DCE degrading bacterial mixed culture K20

strain	relative proportion in K20 [%]	cDCE degradation rate [$\mu mol \cdot min^{-1} \cdot g^{-1}_{prot.}$]	cometabolic transformation yield Ty
Mycobacterium sp. K1	79.0	5.0	0.43
Mycobacterium sp. K2	1.7	2.5	0.34
Corynebacterium sp. K3	15.9	7	0.48
Proteobacterium sp. K4	1.7	0.9	0.06
Proteobacterium sp. K5	1.7	0.5	0.02
mixed culture K20		9	0.51

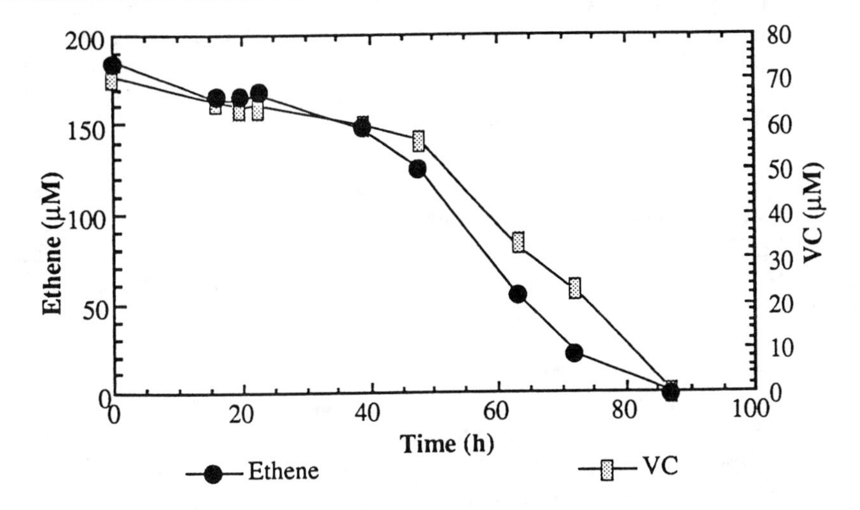

FIGURE 2. Simultaneous cometabolic biodegradation of ethene and VC by the mixed culture K20 during batch cultivation

Degradation of *c*DCE was not affected by the presence of TCE or PCE ($\leq$ 100 µM each). The specific transformation rate of *c*DCE (initial concentration $\leq$ 400 µM) was 8 to 9 $\mu mol \cdot min^{-1} \cdot g^{-1}_{protein}$.

The transformation yield T_y of the mixed culture K20 was high (0.51 mol_{cDCE}/mol_{ethene}) compared to the data (Tab 2) listed by Anderson and McCarty (1997).

TABLE 2. Transformation yield of the mixed culture K20 compared to reported data (laboratory scale experiments).

chloro-ethene	auxiliary substrate	culture	maximum observed T_y	reference
*c*DCE	methane	mixed culture	0.005	Dolan and McCarty, 1995
*c*DCE	methane/ formate	mixed culture	0.020	Dolan and McCarty, 1995
*c*DCE	methane/ formate	mixed culture	0.031	Chang and Alvarez-Cohen, 1996
*c*DCE	methane	mixed culture	0.058	Anderson and McCarty, 1997
***c*DCE**	**ethene**	**mixed culture K20**	**0.51**	
VC	methane	mixed culture	0.0072	Dolan and McCarty, 1995
VC	methane/ formate	mixed culture	0.013	Dolan and McCarty, 1995
VC	methane/ formate	mixed culture	0.058	Chang and Alvarez-Cohen, 1996
VC	methane	mixed culture	0.20	Anderson and McCarty, 1997
VC	**ethene**	**mixed culture K20**	**0.61**	

Chemostat experiments. Lab scale biodegradation experiments with K20 in continuous culture up to 20 weeks were carried out in a 1 liter stirred tank bioreactor. The *c*DCE transformation activity (7 $\mu mol \cdot min^{-1} \cdot g^{-1}_{protein}$) was stable over the entire time period, although the process was run under non sterile conditions. Release of Cl^- showed, that the degradation of *c*DCE was complete when the feeding rate did not exceed 18 $\mu M \cdot^{-1}$.

Higher feeding rates could not be applied, since the culture was substrate limited by ethene. The ethene feeding rate was 1.7 mg/l·h via gas phase at aeration rates of 1.2 l/h to 2.4 l/h and ethene concentrations of 1.8 to 0.9 % (v/v). Higher ethene feeding rates resulted in a high loss of ethene by stripping and, at higher aeration rates also *c*DCE appeared in the waste air.

Bioremediation of authentic groundwater. Optimization of cometabolic *c*DCE biodegradation was performed in a 40 liter fixed biofilm reactor with polyamide texile material as support for the biomass. The reactor was fed with authentic *c*DCE (60 µM resp. 500 µM) contaminated groundwater, which contained only negligible amounts of PCE and TCE. *c*DCE was continuously degraded to concentrations below 10 µg/l. The degradation rates (average 5.6 µmol/l·h) were stable over 6 months under non sterile conditions.

Determination of dehalogenation rates was not possible in authentic groundwater, since the background chloride level was approximately a thousandfold higher than the maximum chloride release from chloroethene biodegradation.

CONCLUSIONS

The advantages of ethene as auxiliary substrate are:

1) high transformation yield (0.51 mol_{cDCE}/mol_{ethene}),
2) simultaneous degradation of *c*DCE and ethene (ratio ≥ 4:1),
3) high tolerance to *c*DCE (≤ 6 mM)
4) high catabolic performance during continuous cultivation under non sterile conditions, and
5) low cost compared to other substrates.

The advantages and the experimental results suggest an economic process for biotreatment of groundwater and in-situ bioremediation of soil, which are contaminated with chloroethenes.

The cooxidative process is currently combined with an initial anaerobic step, that partially dechlorinates PCE and TCE. The two step anaerobic/aerbic process will allow the mineralization of mixtures of all chlorinated ethenes in groundwater. A pilot scale experiment at a former industrial production site, which is contaminated with PCE, TCE, and *c*DCE is planned.

ACKNOWLEDGEMENT

This project was supported by the Deutsche Bundesstiftung Umwelt, Osnabrück, Germany
We thank Brigitte Höhl, Peter Schlenk, and Joachim Regel for valuable technical assistance.

REFERENCES

Anderson, J. E. and P. L. McCarty. 1997. "Transformation Yields of Chlorinated Ethenes by a Methanotrophic Mixed Culture Expressing Particulate Methane Monooxygenase." *Appl. Environ. Microbiol. 63:* 687-693.

Chang, H.-L. and L. Alvarez-Cohen. 1996. "Biodegradation of Individual and Multiple Chlorinated Aliphatic Hydrocarbons by Methane-oxidizing Cultures." *Appl. Environ. Microbiol. 62* 3371-3377.

Dabrock, B., J. Riedel, J. Bertram, and G. Gottschalk. 1992. "Isopropylbenzene (Cumene) - a New Substrate for the Isolation of Trichloroethene-degrading Bacteria." *Arch. Microbiol. 158:* 9-13

Dolan, M. E. and P. L. McCarty. 1995. "Methanotrophic Chloroethene Transformation Capacities and 1, 1-Dichloroethene Transformation Product Toxicity. *Environ. Sci. Technol. 29:* 2741-2747.

Ewers, J., D. Freier-Schröder, and H.-J. Knackmuss. 1990. "Selection of Trichloroethene (TCE) Degrading Bacteria that Resist Inactivation by TCE" *Arch. Microbiol. 154:* 410-413.

Fogel, M. M., a. T. Taddeo, and S. Fogel. 1986. "Biodegradation of Chlorinated Ethenes by a Methane Utilizing Mixed Culture." *Appl. Environ. Microbiol. 51* 720-724.

Little, C. D., A. V. Palumbo, S. E. Herbes, M. E. Lidstrom, R. T. Tyndall, and P. J. Gilmer. 1988. "Trichloroethylene Biodegradation by a Methane-oxidizing Bacterium." *Appl. Environ. Microbiol. 54:* 951-956.

Nelson. M. J. K., S. O. Montgomery, W. R. Mahaffey, and P. H. Pritchard. 1987. "Biodegradation of Trichloroethylene and Involvement of an Aromatic Biodegradative Pathway." *Appl. Environ. Microbiol. 53:* 949-954.

Scholz-Muramatsu, H., A. Neumann, M. Meßmer, E. Moore, and G. Diekert. 1995. "Isolation and Characterization of Dehalospirillum multivorans gen. nov., sp. nov., a Tetrachloroethene-utilizing, Strictly Anaerobic Bacterium." *Arch. Microbiol. 163:* 48-56

Tsien, H. C., G. A. Brusseau, O. Yagi, T. Tabuchi. 1989. "Biodegradationof Trichloroethylene by *Methylosinus trichosporium* OB3b." *Appl. Environ. Microbiol. 55* 3155-3161.

Vogel, T. M. and P. L. McCarty. 1985. "Biotransformation of Tetrachloroethene to Trichloroethene, Dichloroethene, Vinyl Chloride, and Carbon Dioxide under Methanogenic Conditions." *Appl. Environ. Microbiol. 49:* 1080-1083.

Wackett, L. P. and D. T. Gibson. 1988. "Degradation of Trichloroethylene by Toluene Dioxygenase in Whole-cell Studies with *Pseudomonas putida* F1." *Appl. Environ. Microbiol. 54:* 1703-1708.

Wackett, L. P. and S. R. Householder. 1989. "Toxicity of Trichloroethene to *Pseudomonas putida* F1 is Mediated by Toluene Dioxygenase." *Appl. Environ. Microbiol. 55:* 2723-2725.

Wohlfarth, G. and G. Diekert. 1997. "Anaerobic Dehalogenases." *Curr. Opin. Biotechnol. 8:* 290-295.

BIOTREATABILITY STUDIES FOR REMEDIATION OF TCE-CONTAMINATED GROUNDWATER

M.Eguchi, H.Myoga, S.Sasaki, and Y.Miyake
(ORGANO CORPORATION, Tokyo, Japan)

ABSTRACT: The Biotreatability was studied by vial tests and column tests to evaluate the potential for bioremediation of the trichloroethylene (TCE) contaminated site in Kimitsu city, Japan. In the vial tests, we examined TCE degradation tests using methanotrophic bacteria cultivated from the contaminated soil and groundwater. The methanotrophic bacteria cultivated from the contaminated soil and groundwater degraded 5 mg/l of TCE within one hour. The batch soil column tests were conducted using soil and groundwater collected from the contaminated site. Four soil columns were prepared to evaluate the influence of methane concentration and dissolved oxygen (DO) concentration on TCE degradation. A control column was prepared to evaluate abiotic losses of TCE. The TCE concentration entering each column was designed to be kept at 0.5 mg/l. The TCE removal rate was approximately 20 percent during one month, in the test column provided methane (5 mg/l) and DO (20 mg/l). The results from the vial tests and the column tests suggest that biostimulation using methane is applicable for the TCE contaminated site in Kimitsu City.

INTRODUCTION

Chlorinated organic compounds are common groundwater contaminants, in particular, TCE appears to be the most widespread contaminant in Japan. Generally, TCE is removed by air-stripping and activated carbon adsorption from groundwater, but these processes cannot decompose the contaminant.

Recently, it has been demonstrated that chlorinated organic compounds can be degraded to nontoxic products under aerobic conditions by cometabolism of microorganism oxidizing methane, propane, phenol, or toluene.

Because there are many houses near the contaminated site of the factory, the safety of primary substrates being used for *in situ* bioremediation is very important, especially in Japan. This led to the selection of methane as the primary substrate in this evaluation.

In this study, vial tests were investigated to evaluate the potential of TCE degradation of methanotrophic bacteria cultivated from the TCE-contaminated site in Kimitsu city, Japan. And column studies were investigated to evaluate the potential for bioremediation using methane and the effect of operating variables of the concentration of methane and DO on the degree of TCE degradation.

MATERIALS AND METHODS

Mixed Methanotrophic Bacteria Cultures. The soil sample (20 g) and the groundwater sample (180 ml) from Kururi-Ichiba site in Kimitsu city (Suzuki, 1996) were placed into 700 ml vials containing 20 ml of a sterile nitrate mineral salts (NMS) liquid medium (Sunghoon et al., 1984). The vials were sealed with butyl rubber stoppers and aluminum caps, and the headspace (500 ml) within the vials was altered to create 1:4 methane-air atmosphere. The vials were incubated on a shaker (100 rpm) at 30 ℃ for 2 weeks.

TCE Degradation Tests. Methanotrophic bacteria, after 2 weeks cultivation, were collected by centrifugation at 10,000 x g for 15 minutes at 10 ℃. The bacteria pellets were washed in the groundwater, and resuspended in the same groundwater to give an A_{600} reading of 1.0. The suspensions (10 ml) were transferred in 60 ml vials, and the vials were sealed with butyl rubber stoppers and aluminum caps. TCE degradation was initiated by adding TCE to each vial to a final concentration of 1, 5, 10, 25 and 50 mg/l. The vials were then incubated at 30 ℃ with shaking at 100 rpm. TCE degradation was monitored by measuring TCE concentration in the head space gas of each vial by gas chromatography.

Column Studies. Stainless columns (25 x 400 mm) were packed with sandy aquifer material obtained in August 1996 from Kururi-Ichiba site in Kimitsu City. A control column was prepared to evaluate abiotic losses of TCE.

The operational conditions at each soil column are listed in Table 1. The TCE concentration entering the column was designed to keep 0.5 mg/l. The column fluids were exchanged daily with 600 ml of the new feed solutions. The column studies were conducted at a fixed temperature (20 ℃).

TCE Analyses. TCE concentrations were determined by head space analysis using a gas chromatograph (GC311, HNU SYSTEMS, USA) equipped with a photon ionized detector. The injector, oven, detector temperatures were 110, 70, 110 ℃, respectively. Helium was used as the carrier gas.

TABLE 1. The operating conditions of the columns studies.

Column number	Substances in influent				Remarks
	TCE(mg/l)	CH_4(mg/l)	DO(mg/l)	NMS(%)	
1	0.5	0	15	0	Control column
2	0.5	3	10	10	Low CH_4 conc. and low DO conc.
3	0.5	6	18	10	
4	0.5	10	10	10	High CH_4 conc. and low DO conc. After 8 days, 3mg/l CH_4, 10mg/l DO

RESULTS AND DISCUSSIONS

TCE Degradation Tests. Figure 1 shows the results of TCE degradation at different concentrations up to 50 mg/l by the mixed methanotrophic bacteria culture from the soil and groundwater sample in Kururi-Ichiba site. For the blank test, the mixed methanotrophic bacteria culture was not inoculated into the serum bottle containing 1 mg/l of TCE. The mixed methanotrophic bacteria culture degraded about 99.9 % within 1 hour in the case of initial TCE concentrations up to 5 mg/l. Above 10 mg/l, the mixed methanotrophic bacteria could degrade TCE, but could not complete TCE degradation.

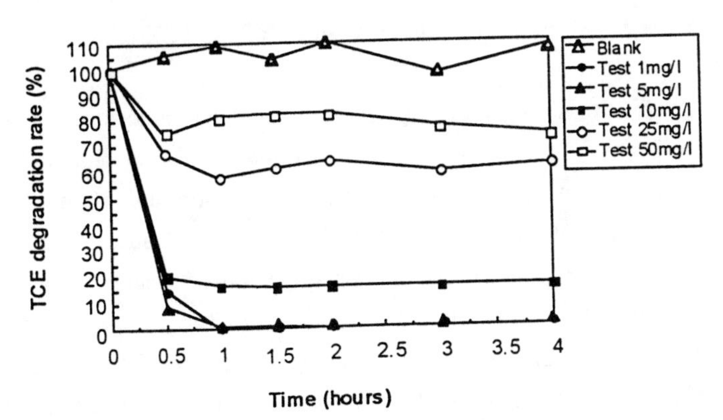

FIGURE 1. TCE degradation tests by the mixed methanotrophic bacteria culture at various concentrations.

It is well known that formate leads to increase TCE transformation capacity of methanotrophic bacteria. However, formate is difficult to use in the field of biostimulation, due to harm to the environment and high operating costs. From this point, formate was not added in this research.

This result suggest that the mixed methanotrophic bacteria cultivated from the soil and the groundwater in Kimitsu City have the TCE degradation ability and can degrade TCE at relatively high concentrations (10, 25, 50 mg/l) without being inhibited by high initial TCE concentrations.

Column Studies. Methane, DO, and nutrients (nitrate and phosphate) were consumed in all of the columns fed methane within 7 days (data not shown). Methanotrophic bacterial populations were increased from 10^2 cells/ml order to 10^6 cells/ml order within 14 days in all of the columns fed methane (data not shown).

Figure 2 shows the result of the TCE removal rate in each column. The TCE removal rate in the control column 1 fluctuated around zero. From this result, it was found that abiotic losses in the column was not significant in this study. Column 3 fed 6 mg/l methane, 20 mg/l DO, and nutrients showed 20% of TCE degradation within 10 days. After 10 days, the removal rate kept 20% of TCE degradation stably during 1 month. This is good agreement with the result of a field-scale biostimulation test using methane by Semprini et al. (1990). The guide for conducting treatability studies published by U.S. Environmental Protection Agency (1991) for applying bioremediation indicates that a contaminant removal rate of test column must keep 20% during 1 month. The result of column 3 offers encouragement for biostimulation using methane of the TCE contaminated site in Kimitsu City.

The effects of operating conditions, methane concentration and DO concentration, on TCE removal rate were evaluated. The column 2 fed low methane concentration (3 mg/l) and low DO concentration (10 mg/l) showed unstable TCE removal rate, and average TCE removal rate was about 10%. Lanzarone and McCarty (1990) discuss that there was no change in the degree of TCE degradation when the methane concentration (5.3 mg/l) was reduced to 2.7 mg/l. However, the TCE degradation was higher in column 3 fed 6 mg/l methane than in column 2 fed 3 mg/l methane in this column test. From this result, it appears that a major limitation on the biodegradation was the fed amounts of methane and DO. However, there is a no significant difference on methanotrophic bacteria populations between the column 3 and the column 2 fed low methane concentration.

The column 4 fed high methane concentration (10 mg/l) and low DO concentration (10 mg/l) showed no TCE degradation before 8 days. It was considered that DO level was insufficient to degrade TCE. Subsequently, when methane concentration was reduced to 3 mg/l, TCE degradation began and the degradation rate increased to about 35% temporarily. However, TCE degradation of the column 4 decreased to about 10% after 20 days.

It is well known that type II methanotroph possess a soluble methane monooxygenase (sMMO) that can degrade TCE, while type I methanotroph possess a particulate methane monooxygenase (pMMO) that oxidize TCE at a much lower rate than sMMO does. Results of competition experiments between type I and type II methanotrophs by Graham et al. (1993) suggest that type II methanotroph had a disadvantage when dissolved methane level was low.

Initially, the column 4 was operated to increase type II methanotroph by feeding high methane concentration. However, the operating condition, feeding high methane concentration (10 mg/l) for 1 week and then decreasing methane (3mg/l), had no advantage on the TCE removal rate after 8 days.

These column tests suggest that biostimulation using methane is applicable for the TCE contaminated site in Kimitsu City.

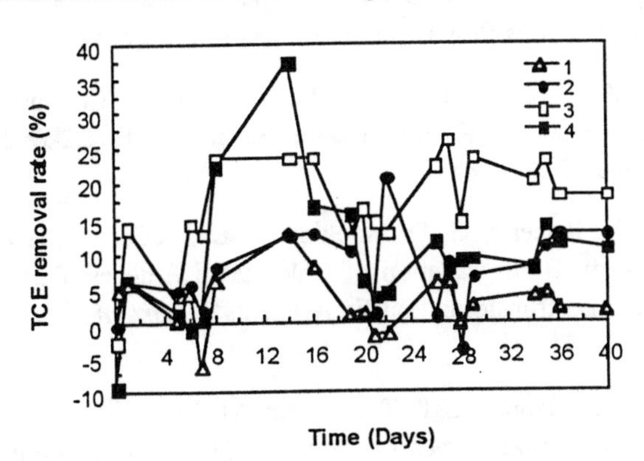

FIGURE 2. TCE removal rate in each column during the column tests.

CONCLUSIONS

The vial tests and the soil column tests were studied to evaluate the potential of bioremediation for TCE contaminated site in Kimitsu City, Japan. Our major findings are summarized as follows:

1. The mixed methanotrophic bacteria cultivated from the contaminated site had the ability of TCE degradation.
2. The mixed methanotrophic bacteria culture degraded about 99.9 % of TCE up to 5 mg/l within 1 hour.
3. In the column tests, column fed 6 mg/l methane, 20 mg/l DO, and nutrients showed 20% of TCE degradation and kept during 1 month.

ACKNOWLEDGMENTS

This study was conducted as one of the research and development activities for the bioremediation project which is handled by Research Institute of Innovative Technology for the Earth (RITE) in Japan, and funded by the Ministry of International Trade and Industry (MITI) through New Energy and Industrial Technology Development Organization (NEDO). Special thanks to Kimitsu City for their cooperation to make this study possible.

REFERENCES

U.S. Environmental Protection Agency. 1991. *Guide for Conducting Treatability Studies Under CERCLA: Aerobic Biodegradation Remedy Screening.* EPA/540/2-91/013A.

Graham, D. W., J. A. Chaudhary, R. S. Hanson, and R. G. Arnold. 1993. "Factors Affecting Competition Between Type I and II Methanotrophs in Two-organism, Continuous-flow Reactors." *Microbial Ecology 25*:1-17.

Lanzarone, N. A. and P. L. McCarty. 1990. "Column Studies on Methanotrophic Degradation of Trichloroethene and 1,2-Dichloroethane." *Ground Water 28*(6): 910-919.

Semprini, L., P. V. Roberts, G. D. Hopkins, P. L. McCarty. 1990. "A Field Evaluation of In-Situ Biodegradation of Chlorinated Ethenes: Part 2, Results of Biostimulation and Biotransformation Experiments." *Ground Water 28*(5): 715-727.

Sunghoon, P., M. L. Hanna, R. T. Taylor, and M. W. Droege. 1991. "Batch Cultivation of *Methylosinus trichosporium* OB3b. 1: Production of Soluble Methane Monooxygenase." *Biotechnology and Bioengineering 38*: 423-433.

Suzuki, Y., 1996. "National Project of Bio-remediation Technological Development for restoration of in situ Geo-pollution due to organochlorine compound." *Proceedings of the 6th Symposium on Geo-Environments and Geo-technics*, pp. 325-330.

FIELD APPLICATION OF IN SITU METHANOTROPHIC TREATMENT FOR TCE REMEDIATION

Robert Legrand and Andrew J. Morecraft (Radian International LLC, Austin, TX, and Morrisville, NC)
John A. Harju and Thomas D. Hayes (Gas Research Institute, Chicago, IL)
Terry C. Hazen (E. O. Lawrence Berkeley National Laboratory, Berkeley, CA)

ABSTRACT: In situ methanotrophic treatment technology (MTT) was evaluated at a site contaminated with chlorinated ethenes and hydrocarbons near a natural gas pipeline compressor station. The formation is characterized by approximately 50 ft (15 m) of saprolitic overburden above bedrock, and a depth to groundwater of 8 to 10 ft (2.4 to 3 m). The MTT system was automated and normally unattended. Air and methane were injected, along with nitrous oxide (N_2O) and triethylphosphate (TEP), as nutrient sources; TEP delivery was intermittent because of design issues. Trichloroethene (TCE) levels dropped from 2130 to 150 µg/L in the well initially exhibiting the highest concentration. The radius of influence of the air injection was approximately 30 ft (9 m). Methanotrophic bacteria increased over six orders of magnitude and eventually dominated the subsurface microbiota. The results indicate that, as long as nitrogen and phosphorus were reliably supplied, rapid (two to four weeks) growth of methanotrophs and associated oxidation of TCE followed. This pilot system was expanded to bioremediate the entire plume above bedrock; three additional injection wells were installed, along with observation wells, and a new TEP diffusion system was developed.

INTRODUCTION

Methanotrophic bacteria produce methane monooxygenase (MMO) to metabolize methane. This oxygenase is non-specific and oxidizes trichloroethene (TCE) to TCE epoxide, which breaks down rapidly into daughter products that are readily biodegradable (Wilson and Wilson, 1987). This is the principle behind methanotrophic treatment technology (MTT), as developed by the Gas Research Institute (GRI) in a multiyear, multidisciplinary research and development effort (Legrand, 1995). MTT was applied in situ at the U.S. Department of Energy's Savannah River site in 1993-94 to remediate a TCE plume. Air, methane, N_2O, and triethylphosphate (TEP) (nitrogen and phosphorus sources, respectively) were injected via a horizontal well to stimulate the growth of methanotrophs, achieving substantial growth of methanotrophs, production of MMO, and degradation of TCE (Hazen et al., 1997).

A gas transmission company and GRI contracted with Radian International LLC to evaluate in situ MTT at a natural gas pipeline compressor station, with subsurface contamination. The site is in rural Virginia. Depth to groundwater is 8 to 10 ft (2.4 to 3 m), and average groundwater velocity is 1.2 cm/day. The formation consists of approximately 50 ft (15 m) of saprolitic overburden (hydraulic conductivity: 3×10^{-4} cm/s) above bedrock.. The maximum concentration of chlorinated volatile organic compounds (VOCs)—mainly tetrachloroethene (PCE) and TCE—is approximately 2000 µg/L; some hydrocarbon contamination is also present. The contamination is found throughout the saturated saprolite and the upper fractured bedrock. The areal extent of the plume is around 1 acre (0.4 hectare).

Objective. The objective of this project was to evaluate the feasibility and effectiveness of in situ MTT at this site by showing substantial (one order of magnitude) and rapid (a few months) degradation of TCE, the main contaminant of concern. The process had to be simple to install and operate.

MATERIALS AND METHODS

Plan and cross-sectional views of the well locations can be found in Figures 1 and 2, respectively. The injection well (IW-1) was installed within 5 ft (1.5 m) of existing Well MW-7 to target a TCE hot spot and allow monitoring from MW-7. Additional observation wells (OW-1 through OW-4) were installed as indicated, based on an expected radius of influence (ROI) of less than 20 ft (6 m). The injection well was constructed of 1-in. (2.5-cm) diameter galvanized steel casing with 1 ft (30 cm) of 2-in. (5-cm) diameter stainless steel, wire-wrapped well screen. Clean uniform sand was emplaced to 6 in. (15 cm) above the screened interval, and a 2.5 ft (75 cm) bentonite seal was placed above this filter pack. The remaining annulus was grouted to land surface. The observation wells were constructed of 2-in. (5-cm) diameter Schedule 40 PVC casing, and 18 to 30 ft (5.5 to 9 m) of screen. Two soil vapor sampling points were also installed.

The MTT equipment was installed inside a shed and included the following:

- A 3/4 hp (0.56 kW) air compressor with a 30-gal. (114-L) air tank;
- A supply of methane and N_2O in laboratory cylinders, and liquid TEP with a metering pump;
- Piping and valving to successively inject methane, N_2O, and TEP in the air line; mass flow controllers (MFCs) for CH_4 and N_2O; and
- A lower explosive limit (LEL) detector to prevent the methane concentration from exceeding 80% of the LEL.

The airflow was initially 0.85 scfm (24 L/min), resulting in 20 psig (138 kPa) of pressure at the wellhead. The pressure dropped gradually during the course of the test run, requiring the flow rate to be increased to 1.24 scfm (35 L/min) during the last four weeks of the test. Methane and N_2O flow rates were controlled by MFCs to 4% and 0.02% of the airflow, respectively. TEP was metered in at approximately 0.24 mL gas/min. The system was unattended, although several site visits were made during the test run.

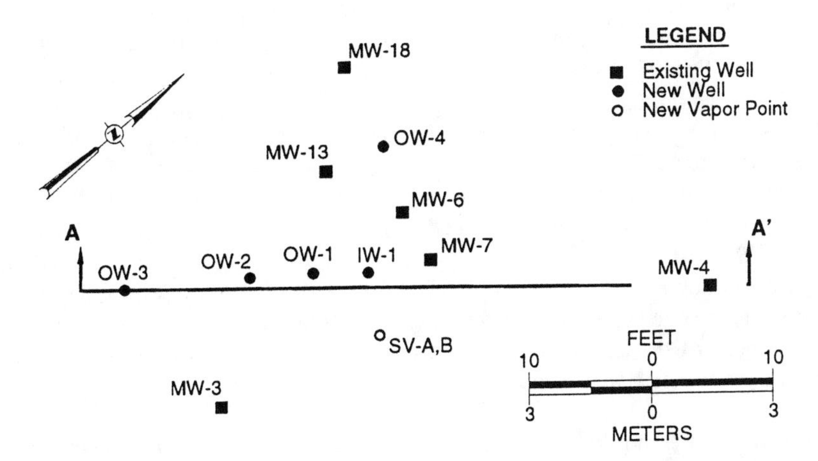

FIGURE 1. In situ MTT demonstration, plan view of wells

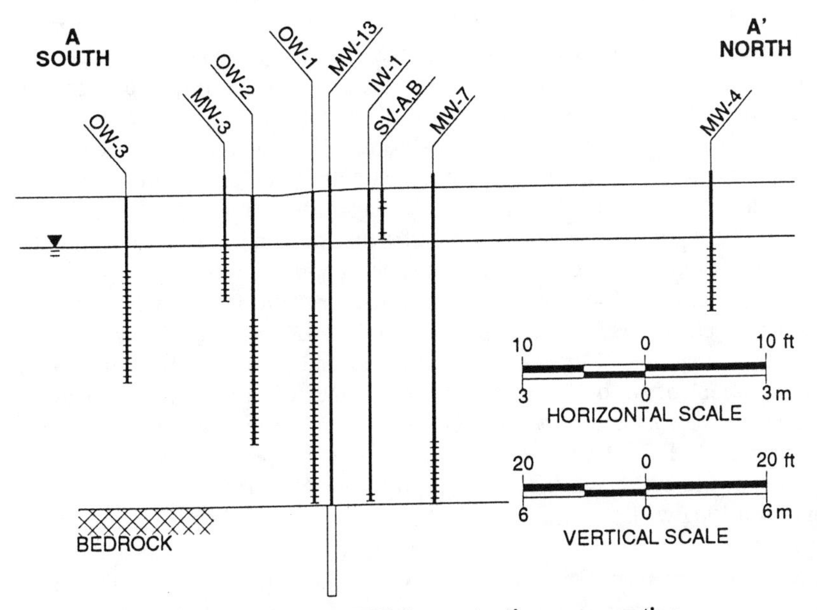

FIGURE 2. In situ MTT demonstration, cross section

RESULTS AND DISCUSSION

Operational history. The in situ MTT evaluation was operated for 139 days. At the beginning of the test, bubbling and pressure buildup were detected in Well OW-1, possibly due to short-circuiting from the nearby injection well. To prevent stripping of TCE, this well was tightly capped for the rest of the test run.

The methane feed was continuous for the first five weeks, then it was pulsed on a schedule of 8 hours on/16 hours off. The feed was interrupted by an unknown malfunction from Day 66 to Day 72, and by an electronic problem from Day 78 to Day 97. The original design called for metering in liquid TEP at a rate of 7 μL liquid/min, on the assumption that it would evaporate in the line. However, the project budget did not allow for the necessary metering hardware, so we settled for a flow rate one order of magnitude higher. Unfortunately, liquid TEP backed up into the system and disrupted the operation of the N_2O MFC; as a result, the flow of N_2O was interrupted from Day 51 to Day 111.

Several alternative systems for TEP delivery were evaluated, including a cylindrical bubble contactor, which failed to evaporate TEP at a measurable rate. Ultimately, we delivered an aqueous phosphate solution to the wells during the last four weeks of the test. In summary, phosphorus was delivered to the subsurface during the first few weeks and the last few weeks of the test.

Subsurface Chemistry. The copper concentration was below the detection limit of 1.4 μg/L, well below inhibitory concentrations for methanotrophs. Chloride was found at concentrations of 20 to 70 mg/L, precluding its use as a proxy for the degradation of approximately 2 mg/L of chlorinated ethenes. Nitrogen (nitrate-nitrite and total Kjeldahl) and phosphorus (orthophosphate and total phosphorus) were well below 1 mg/L, indicating that macronutrients would have to be added. Total organic carbon ranged from 1 to 10 mg/L, dissolved oxygen concentrations ranged from 0.1 to 2.8 mg/L, and pH ranged from 5.4 to 6.6.

At the end of the evaluation, total Kjeldahl nitrogen was measured at 0.4 to 4 mg/L, whereas phosphorus (orthophosphate and total phosphorus) was mostly below the detection limit of 0.01 mg/L; the highest values were 0.11 mg/L, in Well MW-7. These low concentrations soon after the addition of the phosphate solution confirm that the subsurface environment is phosphorus-limited.

Soil gas was also sampled and analyzed for methane and TCE. Background levels of methane were 2 to 5 ppmV, and no background TCE was detected (detection limit: 0.005 ppmV). On Day 21, and again on Day 48, methane concentrations of up to 4% by volume were found. We assume that disruptions in nutrient delivery limited carbon uptake, which allowed methane to build up in the subsurface. On Day 139, after four weeks of optimal operation, the methane levels had dropped to between 2 and 20 ppmV; 0.95 ppmV of TCE was measured in one vapor sampling well.

System Performance. Figure 3 depicts VOC concentrations in MW-7—the well with the highest initial TCE readings; note that TCE concentration is read on the right-hand *y*-axis. Starting around 2000 μg/L, the concentration of TCE declined dramatically in the first three weeks of the test. Despite the nutrient disruptions, it continued to decline, albeit more gradually, to 150 μg/L. PCE went from 50 μg/L to nondetect in three weeks; PCE is not subject to any known aerobic metabolism or cometabolism, but its degradation can be enhanced by nearby methanotrophic activity (Enzien et al., 1995). In two and a half months, benzene went from 17 μg/L to nondetect, and trans-1,2-DCE went from 25 μg/L to nondetect. Surprisingly, cis-1,2-DCE fluctuated but no clear decline is evident. DCA declined

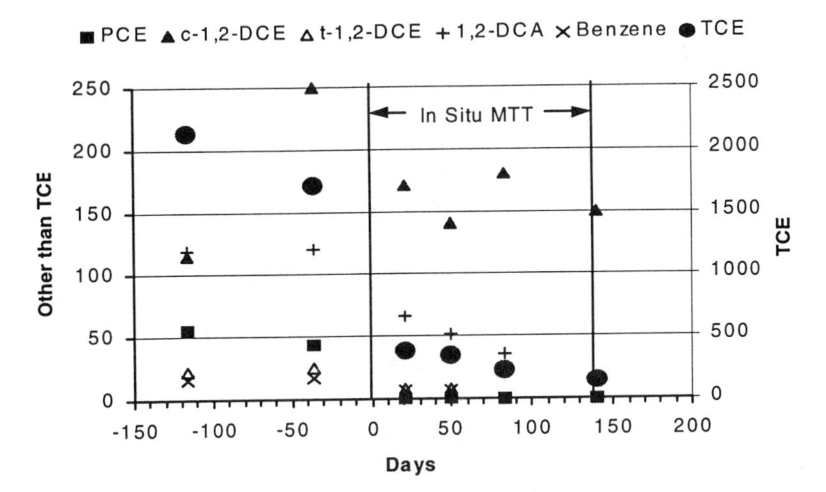

FIGURE 3. Impact of MTT on VOC concentrations (micrograms/L)

from 120 to 15 μg/L over the duration of the test.

There is some evidence of VOC displacement toward Wells OW-1 and OW-2. DCE and DCA increased slightly in OW-1, although the ultimate level was still below 100 μg/L; TCE increased from 200 to 660 μg/L in OW-1. We hypothesize that this displacement is due to mounding of the groundwater table caused by the air injection.

The pressure in the injection well was kept below 140% of the breakout pressure in order to minimize stripping and formation of preferential flow channels. The fact that some VOC levels remained constant or even increased in certain wells suggests that stripping was not a major factor in the VOC reductions achieved. That suggestion is reinforced by the fact that very little TCE was found in the soil gas at the end of the evaluation.

Radius of Influence. Oxygen saturation versus distance is shown in Figure 4. The bottom curve indicates that, initially, the groundwater was mostly anaerobic. The curves indicate that the ROI of the air injection eventually extended to approximately 30 ft (9 m). Note that the last set of data (Day 140) was obtained 24 hours after air injection was stopped, which explains the low saturation near the injection point. Helium tracing was conducted twice and supports the ROI estimate.

Growth of Methanotrophs. Methanotrophs were enumerated by using the most probable number procedure. On Day 94, very low counts were found (1 to 1500); note that there had been lengthy disruptions of the methane and N_2O delivery. On Day 140, however, methanotrophs in most wells exceeded the enumeration limit of 3.28×10^6, which demonstrates that, after methane, air, and nutrients are reliably supplied, massive methanotroph growth occurs in a matter of weeks. Total bacterial counts were between 2×10^6 and 3×10^7, indicating that methanotrophs had

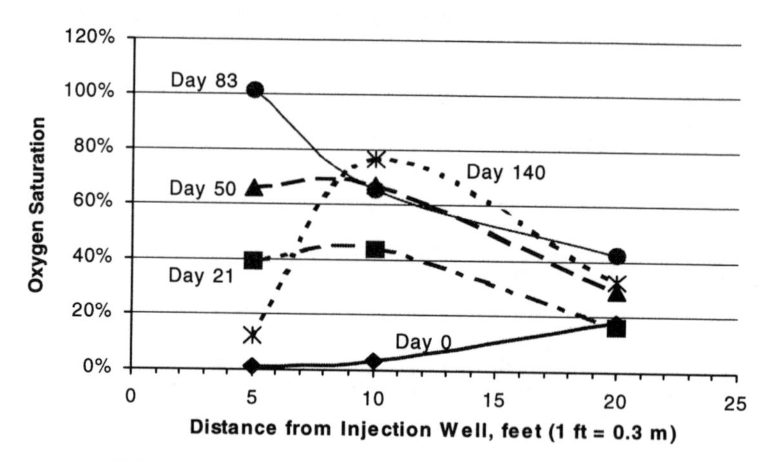

FIGURE 4. Oxygen saturation versus distance

become a dominant fraction of the microbiota.

Conclusions. In situ MTT using standard vertical wells is effective at rapidly cometabolizing TCE in the subsurface. An ROI of 30 ft (9 m) was achieved in this relatively tight formation. The pilot system described here was expanded to remediate the entire plume by the addition of two injection wells and two observation wells. The compressor, MFCs, and piping were upgraded to accommodate up to 5.3 scfm air (150 L/min) and correspondingly higher flow rates of other gases. A new TEP contactor was developed.

REFERENCES

Enzien, M. V., F. Picardal, T. C. Hazen, R. G. Arnold, and C. B. Fliermans. 1994. "Reductive Dechlorination of Trichloroethylene and Tetrachloroethylene Under Aerobic Conditions in a Sediment Column." *Appl. Environ. Microbiol. 60*(6): 2200-2204.

Hazen, T. C., K. H. Lombard, B. B. Lombard, M. V. Enzien, J. M. Dougherty, C. B. Fliermans, J. Wear, and C. A. Eddy-Dilek. 1997. *Full Scale Demonstration of In Situ Bioremediation of Chlorinated Solvents in the Deep Subsurface using Gaseous Nutrient Biostimulation.* Progress in Microbial Ecology (Symposium Proceedings, SBM Brazilian Society for Microbiology/ICOME- International Commission on Microbial Ecology) p. 597-604.

Legrand, R. 1995. *Methanotrophic Treatment Technology (MTT) - Final Report.* Prepared for Gas Research Institute, Chicago, IL; Contract No. 5091-253-2215.

Wilson, J. T., and B. H. Wilson. 1985. "Biotransformation of Trichloroethylene in Soil." *Appl. Environ. Microbiol. 49*(1): 242-243.

SUSTAINED BIODEGRADATION OF TRICHLOROETHYLENE IN A SUSPENDED GROWTH GAS TREATMENT REACTOR

Seung-Bong Lee, Jim P. Patton, Stuart E. Strand, and H. David Stensel
(University of Washington, Seattle, WA.)

ABSTRACT: A sparged shallow suspended growth gas treatment reactor, fed phenol for biomass growth, was investigated for the treatment of a trichloroethylene (TCE) contaminated gas. A fundamental mass transfer and biodegradation model that incorporated the effects of TCE metabolite toxicity was used to predict TCE removal efficiency in this treatment process. For a 40-cm deep laboratory scale reactor operated with a 6-day solids retention time (SRT), an applied phenol/TCE ratio of 2.41, and 515 mg/L biomass concentration, TCE degradation was sustained at a 70 percent removal efficiency for several consecutive operating SRTs. The model was able to predict the reactor performance and the model results suggested that the transformation capacity (Tc) of the test culture was about 2.0 mg TCE/mg biomass; well above most literature Tc values for both methane- and phenol-oxidizing cultures. Model simulations further showed that the gas treatment bioreactor performance and required applied phenol/TCE ratio is greatly effected by the transformation capacity.

INTRODUCTION

Air streams contaminated with trichloroethylene (TCE) can result from air stripping of contaminated groundwater or from soil vapor extraction of contaminated soil. Activated carbon is commonly used to treat such contaminated dilute gas streams, but at higher concentrations thermal oxidation may be cost effective. Aerobic biological treatment of contaminated gas streams is a more recent innovative alternative that has been demonstrated on a long-term basis for BTEX removal (Bielefeldt et al., 1997) and short-term for TCE degradation (Ensley and Kurisko, 1994; Landa et al., 1994).

In a biological suspended growth gas treatment reactor TCE-contaminated air is sparged through fine bubble diffusers which distribute the gas to the reactor and provide mixing of the suspended biomass. TCE is transferred from the gas phase to the liquid where it is cometabolically degraded by bacteria grown on another substrate; commonly phenol or methane. Ensley and Kurisko (1994) reported on a TCE degradation over a 14-day text period in a shallow sparged laboratory reactor. Phenol was used to support a *B. cepacia* G4 biomass and at an average applied phenol/TCE ratio of 30.1 about 95% TCE conversion was observed. Landa et al. (1994) also used a strain of toluene-fed G4 bacteria and operated a sparged suspended growth gas treatment reactor at variable TCE loadings over a period of 20 days to achieve TCE removals ranging from 7 to 90%. The toluene/TCE consumption ratio ranged from 15:1 to 67:1. It was apparent from their test that the growth substrate to TCE removal ratio was important in sustaining TCE degradation.

The efficiency of bacteria degrading TCE is limited by the production of a metabolite toxicity thought to be the epoxidation of TCE by oxygenase enzyme responsible for initiating TCE degradation (Alvarez-Cohen and McCarty, 1991). The term, transformation capacity (Tc) has been used by Alvarez-Cohen and McCarty (1991) to describe the inactivation of TCE-degrading bacteria in proportion to the amount of TCE degraded. Tc values in mg TCE degraded/mg biomass inactivated for methane oxidizing bacteria have ranged from 0.05 to 0.10 (Chang and Alvarez-Cohen, 1995) for methane-oxidizing bacteria and from 0.031 to 0.082 for phenol-oxidizing bacteria (Chang and Alvarez-Cohen, 1995). Greater amounts of TCE degradation per unit of biomass have been observed with other mixed cultures. Landa et al. (1994) degraded 0.12 mg TCE/mg biomass and Bielefeldt et al. (1995) reported 0.51 mg TCE degradation per mg biomass, for toluene and phenol-fed cultures, respectively. Chang and Alvarez-Cohen. (1997) also found that the Tc for a methane -oxidizing culture could be increased to 0.28 with formate addition. Lower Tc values would require more phenol addition to replace biomass inactivated during TCE degradation.

OBJECTIVES

The ability to sustain TCE degradation in biological gas treatment reactors is related to the ability of the culture to overcome the toxicity effect of TCE degradation intermediates. The purpose of this paper is to present a model that describes TCE removal in a suspended growth gas treatment reactor and to evaluate the reactor design parameters that affect sustained TCE removal efficiencies. The effect of the transformation capacity was incorporated into the bioreactor model to predict the mass feed phenol/TCE ratio needed to sustain reactor performance.

BIOREACTOR MODEL

Three basic equations used to describe the TCE removal in the gas treatment reactor are summarized in Table 1. They account of mass transfer of TCE from the gas to the liquid phase, the gas flow rate, the reactor volume, the TCE biodegradation kinetics, the biomass production from phenol consumption, the biomass loss due to endogenous decay and solids wasting, and biomass inactivated by using Tc to account for metabolite toxicity from TCE degradation. The table includes parameter values used in model simulation runs to investigate the effect of key operating conditions on TCE removal efficiency.

REACTOR OPERATION VERSUS MODEL PREDICTION

TCE removal was sustained in a lab-scale shallow sparged suspended growth reactor. The lab reactor contained a 2-liter liquid volume at a 0.4-meter depth, with a diffuser stone located at the bottom of the column. A syringe pump fed pure TCE into the air stream which was sparged through a water cylinder to stabilize the influent TCE concentration to the bioreactor.

TABLE 1. Model equations, parameters, and parameter values used for model simulations.

Model Equations	
$C_{g,e} = H \cdot C_L + (C_{g,0} - H \cdot C_L) \cdot \exp\left(\frac{-k_L a \cdot D}{Q_{gas} / A \cdot H}\right)$	(1)
$\frac{K \cdot C_L \cdot X}{K_S + \cdot C_L} = Q_{gas}(C_{g,0} - C_{g,e})$	(2)
$V\frac{dX}{dt} = Y \cdot \Delta P - K_d \cdot X - \frac{Q_{gas}(C_{g,0} - C_{g,e})}{T_C} - \frac{X}{SRT}$	(3)

Operation parameter		Value
$C_{g,0}$	= influent TCE gas concentration, mg/L	0.5
$C_{g,e}$	= effluent TCE gas concentration, mg/L	-
C_L	= liquid TCE concentration, mg/L	-
H	= dimensionless Henry's constant for TCE	0.42
$k_L a$	= mass transfer coefficient for TCE, d^{-1}	128
Q_{gas}	= gas flow rate, L/d	300
D	= liquid depth in reactor, m	0.4
A	= surface area in reactor, m^2	0.005
K	= TCE specific degradation rate, mgTCE/mgVSS-d	0.25
K_S	= half-velocity constant, mg/L	0.25
K_d	= endogenous decay constant, d^{-1}	0.06
X	= biomass, mg/L	-
Y	= yield, mg VSS/mg phenol	0.6
ΔP	= phenol feed rate, mg/-d	333
SRT	= sludge retention time, days	6

The reactor influent and effluent gases were automatically measured using SRI8610B gas chromatography (GC; Restek DB-624 megabore capillary column, oven 130°C, FID). Gas sampling lines were wrapped with heating tape. Gas standards were prepared by injecting weighted TCE pure liquids into 160-mL serum vials with 80mL deionized water. Headspace TCE concentrations were based on the liquid volume added and the Henry's constant. Reactor temperature was kept at 25±1°C. The reactor feed solution contained 1,000 mg/L phenol and the following nutrient concentrations: 700 mg/L KH_2PO_4, 1000 mg/L K_2HPO_4, 200 mg/L NH_4Cl, 30 mg/L $MgSO_4$, 66.5 mg/L $CaCl_2 \cdot 2H_2O$, 300 mg/L $NaHCO_3$, 55 μg/L $CuCl_2 \cdot 2H_2O$, 150 μg/L $ZnCl_2$, 20 μg/L $NiCl_2 \cdot 6H_2O$, 880 μg/L $FeSO_4 \cdot 7H_2O$, 135 μg/L $Al_2(SO_4)_3 \cdot 18H_2O$, 280 μg/L $MnCl_2 \cdot 4H_2O$, 55 μg/L $CoCl_2 \cdot 6H_2O$, 30 μg/L $NaMoO_4 \cdot 2H_2O$, and 50 μg/L H_3BO_3. The feed solution was fed every 2 hours with a masterflex pump to minimize competitive inhibition of phenol with TCE.

The reactor operating conditions for reactor results that were used to compare the observed effluent TCE concentrations to the model predictions are shown in Table 2 and represent 44 days of sustained TCE treatment. The model adequately predicted the effluent gas TCE concentration and percent removal efficiency, but a very high Tc value of 2.0 mg TCE/mg biomass had to be assumed. This suggests that the TCE degradation did not produce metabolite toxicity in the reactor culture or it was at a sufficiently low level to have no affect on the TCE-degrading culture. Further work is on-going to investigate the culture Tc characteristics.

TABLE 2. Reactor operating conditions and model fit.

Operating parameter	**Value**
Influent TCE, mg/L	0.446
TCE Loading, mg/mgVSS-d	0.134
SRT, days	6
Phenol/TCE Feed Ratio	2.41
Reactor Biomass, mg/L	515
Temperature, °C	25±1
pH	6.8±0.2
Reactor Depth, cm	40
Air flow rate, ml/min	215
Operating time, days	44
Model fit result	**% Removal**
Model run	68 %
Experimental result	70 %

MODEL SIMULATION RESULTS

Figure 1 shows the importance of the reactor feed Phenol/TCE (P/T) ratio and Tc value on the reactor performance (a low Tc value is indicative of a greater amount of biomass inactivation during TCE degradation). At common literature Tc values of 0.05 to 0.10 the TCE removal efficiency is only 28 to 50% at an applied P/T ratio of 10. If none or little metabolite toxicity exist (Tc at 1000) a higher TCE removal efficiency can occur and a relatively low applied P/T ratio can be used. Lower P/T ratios reduce the cost of the biological treatment system.

Model simulation results shown in Figures 2 and 3 were based on a Tc value of 0.5 mg/mg. The results in Figure 2 show that at a given feed P/T ratio the TCE removal efficiency can be improved by increasing the SRT. The SRT effect is less at higher P/T ratios. This result is due to the higher biomass concentrations that can be maintained at longer SRTs or due to feeding more phenol.

Figure 3 shows that a critical depth is reached where the TCE removal efficiency is no longer improved by adding more reactor depth. Beyond this depth (0.8 meters for the reactor operating conditions assumed) the TCE removal is no longer mass transfer limited and is controlled by the biodegradation capacity. At higher P/T ratios more biomass is present to improve the TCE-degrading capacity.

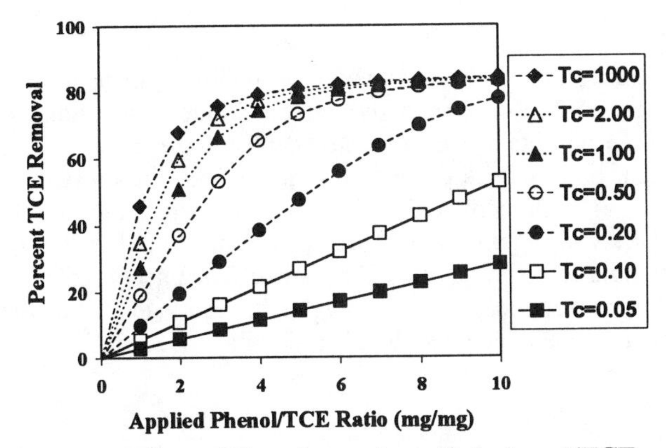

FIGURE 1. Effect of Tc value and applied phenol/TCE ratio on % TCE removed.

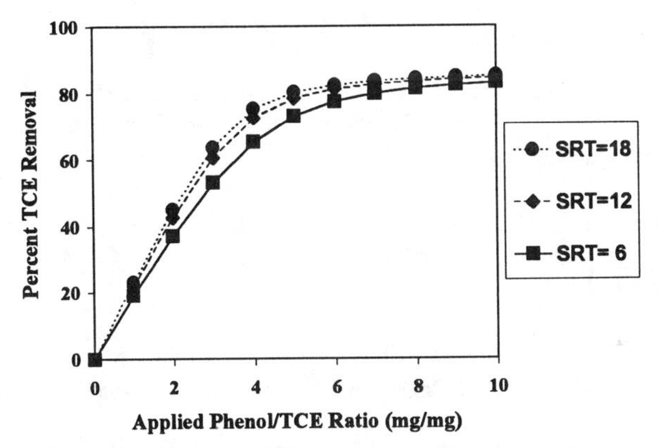

FIGURE 2. Percent TCE removal efficiency is increased at higher reactor SRTs.

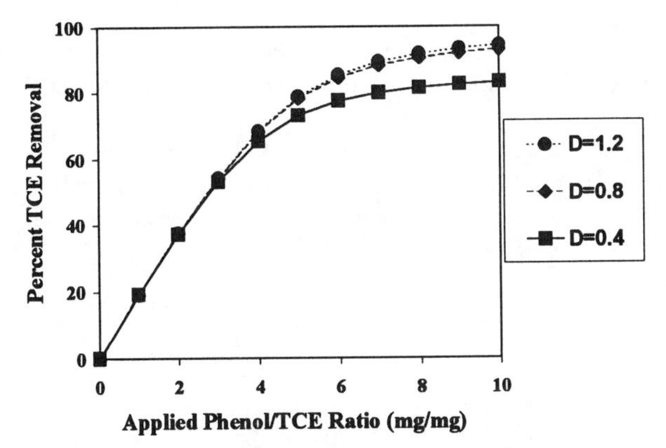

FIGURE 3. Increased reactor depth to 0.8 meter improves TCE removal efficiency.

SUMMARY

TCE degradation can be sustained in a shallow suspended growth reactor for the treatment of TCE-contaminated gases. Model simulations suggest that 90% removal can be obtained at a reactor depth of 0.8 meters. The ability to sustain treatment performance depends on the applied phenol/TCE ratio. TCE-degrading cultures with commonly observed Tc values of 0.05 to 0.10 mg TCE/mg biomass will require very high applied phenol/TCE ratios (greater than 10) to achieve significant levels of TCE removal. The phenol fed TCE-degrading culture used in this study may have a Tc value in the range of 2.0 which resulted in a modest operating applied phenol/TCE ratio of about 2.5.

REFERENCES

Alvarez-Cohen, L. and P.L. McCarty. 1991. "Effects of Toxicity, Aeration, and Reductant Supply on Trichloroethylene Transformation by a Mixed Methanotrophic Culture." *Appl. Environ. Microbiol.* 57:228.

Bielefeldt, A.R. and H.D. Stensel. 1997. "Biodegradation of BTEX-Contaminated Gas in a Sparged Shallow Liquid Reactor." *In Situ and On-Site Bioremediation: Volume 5*. p.37 Battelle Press.

Bielefeldt, A.R., H.D. Stensel, and S.E. Strand. 1995. "Cometabolic Degradation of TCE and DCE without Intermediate Toxicity." *Journal of Environ. Engr. Div., Amer. Soc. of Civil Engrs.*, 121:791.

Chang, H.L. and L. Alvarez-Cohen. 1995. "Model for the Cometabolic Biodegradation of Chlorinated Organics." *Environ. Sci. Technol.* 29:2357.

Chang, H.L. and L. Alvarez-Cohen. 1997. "Two-Stage Methanotrophic Bioreactor for the Treatment of Chlorinated Organic Wastewater." *Wat. Research.* 31:2026.

Ensley, B.D. and P.R. Kurisko. 1994. "A Gas Lift Bioreactor for Removal of Contaminants from the Vapor Phase." *Appl. Environ. Microbiol.* 60:285.

Landa, A.S., E.M. Sipkema, J. Weijma, A.A.C.M. Beenackers, J. Dolfing, and D.B. Janssen. 1994. "Cometabolic Degradation of Trichloroethylene by *Pseudomonas cepacia* G4 in a Chemostat with Toluene as the Primary Substrate." *Appl. Environ. Microbiol.* 60:3368.

TRICHLOROETHYLENE BIOREMEDIATION BY *METHYLOSINUS TRICHOSPORIUM* OB3B IMMOBILIZED IN A FIBROUS-BED BIOREACTOR

Amy L. Kneidel, ***Hojae Shim***, and Shang-Tian Yang
(The Ohio State University, Columbus, Ohio)

ABSTRACT: A laboratory scale, novel fibrous-bed, immobilized cell bioreactor was developed for bioremediation of trichloroethylene (TCE). *Methylosinus trichosporium* OB3b mutant PP358 was employed for aerobically cometabolizing TCE using the soluble methane monooxygenase. The fibrous-bed bioreactor system operated in recycle batch mode successfully degraded TCE. As the initial TCE concentration increased, the degradation rate increased. Maximum degradation rates were achieved for 10-15 mg/L TCE. At high TCE concentration (~23 mg/L), however, rates were lower, probably due to TCE toxicity. Kinetic study showed that higher TCE degradation rate (up to 32 times or 84.77 mg/L/d), tolerance of higher TCE concentrations (up to 11 times or 22.6 mg/L), and longer operational stability (over one year), were achieved for this unique bioremediation system, compared to free cell and other immobilized cell systems.

INTRODUCTION

TCE is commonly used in industry as a cleaning solvent and degreaser; however, it has been shown to be toxic to humans at high exposures, carcinogenic in mice, and is a suspected carcinogen in humans. The drinking water standard for TCE has been set at 5 μg/L since TCE is one of the most frequently detected organic contaminants in water supplies from groundwater sources in the United States (Montgomery, 1996). Another prevalent site for TCE contamination is U.S. Air Force bases where TCE is used as a cleaning solvent and for removal of paint. Therefore, the U.S. Air Force Office of Scientific Research has listed TCE as one of the most desirable chemicals to be treated at Air Force installations.

Bioremediation transforms toxic contaminants to nontoxic products and also has proven to be a cost-effective technology. The Department of Energy predicts that bioremediation of groundwater can reduce clean-up costs by 50 to 75 % in comparison to the traditional method of pump and treat. Also, the costs to bioremediate soils can be up to 10 times cheaper than the conventional dig and burn technologies (U.S. EPA, 1995).

Microorganisms are capable of degrading TCE under both anaerobic and aerobic conditions. Under anaerobic conditions, chlorine atoms are sequentially removed to form less chlorinated organics by reductive dechlorination. Unfortunately, usually only partial dechlorination occurs allowing products, like dichloroethylene and vinyl chloride which are even more toxic than TCE, to accumulate. However, under aerobic conditions, TCE is oxidized to form TCE epoxide which is rapidly converted to glyoxylic acid, and finally to CO_2. Thus, the aerobic process has proven to be advantageous since it forms nontoxic products and occurs more rapidly.

TCE cannot be used by microorganisms as their sole source of carbon and energy; it appears to be biodegradable only through cometabolism. Previous studies have identified several cosubstrates including methane, phenol, and toluene, for TCE cometabolism. Dealing with a cometabolic process presents a challenge because the enzyme responsible for catalyzing the degradation of TCE also catalyzes the degradation of the energy source. Thus, a competition exists between TCE and the energy source which is more easily and preferentially degraded.

Since the efficiency of *in situ* TCE treatment methods is limited, recent effort has concentrated on developing and optimizing bioreactors for more efficient biodegradation. More specifically, for biotreatment of TCE several bioreactor types such as hollow fiber, attached film, bubble column, packed bed, and stirred tank have been well studied. These bioreactors can be classified as either free cell or immobilized cell system. Traditionally most previous research has explored free cell systems. Recently attention has been given to immobilized systems because immobilization of cells within a porous matrix or on a solid surface can have many advantages over free cell systems (Enzien et al., 1994; Friday and Portier, 1991; Miller et al., 1990; Shuler and Kargi, 1992). Immobilization methods previously investigated include: biofilms (Speitel and Segar, 1995), attached films (Fennell et al., 1993), resting cell filters (Hanna and Taylor, 1996), calcium alginate gel beads (Uchiyama, 1992; 1995), and sediment columns (Enzien et al., 1994). Despite their potential advantages, however, several common problems were also encountered by these immobilized cell systems.

To overcome the problems with the above mentioned immobilized cell bioreactors, a novel fibrous-bed bioreactor was developed. This bioreactor contains a convoluted fibrous matrix for cell immobilization. The cells are immobilized in a spiral-wound, porous, fibrous matrix, which provides a large surface area for cell attachment and a large void space for cell entrapment. There are built-in vertical gaps between the spiral-wound layers of the fibrous material to allow excess or dead cell biomass to fall off to the bottom of the reactor, gases such as air to flow freely upward and escape from the top of the reactor, and liquid medium to be pumped through the packed bed without substantial pressure drop. Mass transfer limitations in the bioreactor can be controlled by using an appropriate thickness of fibrous packing and by allowing sufficient liquid medium to flow through the porous fibrous matrix where most cells are present. Therefore, this new reactor packing design provides many superior qualities including well-defined hydrodynamic features that allow for easy operation and scale-up, high reaction rate and loading capacity, and long-term stability, and also is advantageous for carrying out multiphase reactions that take place in biodegradation and biofiltration of hazardous chemicals such as TCE.

Preliminary studies of this bioreactor were conducted to analyze its potential for treatment of groundwater and air contaminated with benzene, toluene, ethylbenzene, and xylenes (BTEX) (Shim, 1997). In comparison to free cell systems, immobilized cells in a liquid-continuous fibrous-bed bireactor tolerated higher concentration (>1,000 mg/L) of benzene and toluene, and gave at least 16-fold higher degradation rates for benzene, ethylbenzene, and *ortho*-xylene, and a 9-fold higher degradation rate for toluene. In addition, the bioreactor showed a long-term stability of more than a year.

Objective. The overall goal of this research was to develop and evaluate the fibrous-bed, immobilized cell bioreactor for its ability and efficiency in treating TCE-contaminated water. A methanotroph, *Methylosinus trichosporium* OB3b mutant PP358 with antibiotic resistance, was selected as the microorganism to aerobically cometabolize TCE, because it had previously proven to give faster degradation rates than most other microorganisms commonly used (Hanna and Taylor, 1996). Specific objectives used to accomplish the overall goal included: 1) studying TCE biodegradation kinetics in a free cell system, 2) studying TCE biodegradation kinetics in an immobilized cell system, and 3) studying the bioreactor long-term performance.

MATERIALS AND METHODS

Medium. *M. trichosporium* OB3b mutant PP358 was grown in a nitrate minimal salts (NMS) medium containing $NaNO_3$ (0.85 g/L), KH_2PO_4 (0.68 g/L), Na_2HPO_4 (0.71 g/L), $MgSO_4 \cdot 7H_2O$ (0.25 g/L), $CaCl_2 \cdot 2H_2O$ (1.47 mg/L), $Na_2MoO_4 \cdot 2H_2O$ (2.42 mg/L), $MnSO_4 \cdot H_2O$ (0.17 mg/L), $CoCl_2 \cdot 6H_2O$ (0.24 mg/L), $ZnSO_4 \cdot 7H_2O$ (0.29 mg/L), H_3BO_3 (0.06 mg/L), and $FeSO_4 \cdot 7H_2O$ (1.12 mg/L). The medium was prepared with deionized water and supplemented with antibiotics, streptomycin sulfate and nalidixic acid to minimize bacterial contamination, and cycloheximide to prevent fungal growth. These supplements were added to the medium at 20 mg/L each.

Free Cell Culturing. To culture free cells, a 2-L glass flask filled with 1,400 mL NMS medium, continuously sparged with methane and air, was used. Methane at a flowrate of 20 mL/min and air at a flowrate of 150 mL/min allowed a high cell density. Free cells were tested for their ability to degrade TCE after growth on both methane and methanol. For cells grown on methane, after TCE was added to the flask, the gas outlet and methane and air supply lines were closed to prevent loss of gaseous TCE and to eliminate soluble methane monooxygenase (sMMO) enzyme competition between TCE and methane. The TCE concentration was detected by gas chromatography (GC). To verify that TCE was biodegraded, the chloride ion concentration was determined by ion chromatography (IC). After TCE was degraded, methane and air were again continuously supplied to rejuvenate the cells and enzyme. For cells grown on methanol, the cell suspensions removed from the serum bottle cultures were centrifuged to remove any methanol present in the liquid to prevent the enzyme competition, then cell pellets were resuspended in fresh medium. The cell suspension aliquots were placed in serum tubes (25 mL) sealed with aluminum crimped Viton stoppers, with 15 mL headspace. Various concentrations of TCE were injected and the tubes were placed in the shaker (25 °C, 200 rpm).

Bioreactor Construction and Start-Up. The bioreactor was constructed from a glass column (5 cm i.d. and 36 cm height) containing a fibrous packing material (100 % cotton terry cloth) affixed to a stainless steel wire screen. The wire screen and terry cloth were wound into a spiral configuration leaving 5 mm gaps between the spiral layers for liquid flow. The bottom of the reactor was packed with ceramic saddles to support the fibrous bed and to ensure proper liquid distribution. Figure 1 shows a schematic of the fibrous-bed bioreactor system for TCE bioremediation in recycle batch mode.

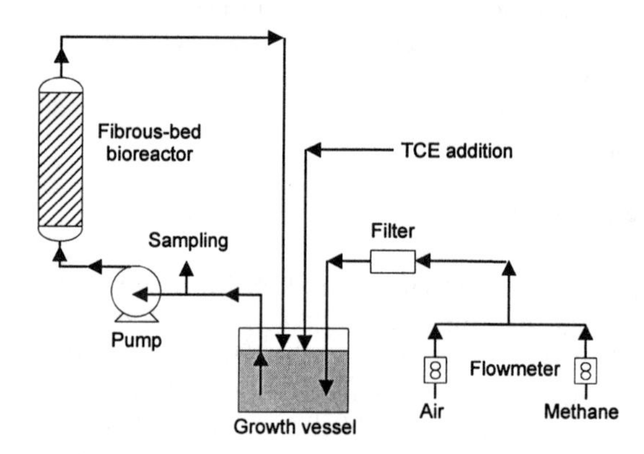

FIGURE 1. Fibrous-bed bioreactor system for TCE treatment with recycle batch operation.

After construction, ~1 L of cell suspension of *M. trichosporium* OB3b mutant PP358, pregrown on methane, was pumped into the growth vessel and then circulated through the bioreactor. The bioreactor system was maintained at ambient temperature, ~23 °C, and the cells were allowed to grow and immobilize in the reactor.

Recycle Batch Operation. Two modes of recycle batch operation were employed: growth/rejuvenation mode and TCE degradation mode. During the growth/rejuvenation mode, cells were allowed to grow on methane and synthesize sMMO. During the TCE degradation mode, the sMMO was allowed to degrade TCE. It was necessary to alternate between these two modes of operation because when methane is present TCE cannot be degraded since methane is preferentially degraded by sMMO. However, cells require methane as their energy source and therefore rejuvenation periods are required between TCE degradation periods to sustain cell viability and sMMO production. For the growth/rejuvenation mode, the cell suspension was continuously circulated through the reactor and air and methane were continuously supplied (Fig. 1). For the TCE degradation mode, air and methane were no longer supplied to avoid competition between methane and TCE for the enzyme. Liquid TCE was added to the growth vessel at a desired concentration and then circulated through the bioreactor. To track the TCE degradation, the TCE concentration was determined by GC and the chloride ion concentration by IC. Every two weeks fresh medium was added to restore the liquid level in the growth vessel. Various concentrations (2.5-25 mg/L) of TCE were also tested to determine what concentrations caused toxic effects.

Long-Term Operation. A fed batch system was employed to determine the long-term stability of the bioreactor. After a high cell density in the reactor was achieved, the cells were exposed to intermittent additions of TCE.

Analysis of TCE, Methanol, and Chloride Ion Concentrations. The concentrations of TCE and methanol in the liquid phase were determined using GC. Two μL sample was injected to Varian 3400 gas chromatograph equipped with a flame ionization detector and a glass column packed with Alltech Carbograph 2, 80/100. Nitrogen was used as the carrier gas at a flow rate of 30 mL/min. The temperatures for column, injector, and detector were 220, 220, and 250 °C, respectively. Five mL liquid sample of cell suspension taken from the growth vessel and centrifuged for 20 min at 4,000 rpm was injected into Dionex ion chromatograph equipped with an ED40 electrochemical detector, GP40 gradient pump, AS40 automated sampler, IonPac column AG14, and anion self-regenerating supressor-I, to determine the chloride ion concentration as TCE was biodegraded.

RESULTS AND DISCUSSION

TCE Degradation Kinetics with Free Cells. TCE degradation kinetics were studied with free cells to determine their growth rate and their ability to degrade TCE. Based on the cell density changed with time during growth on methane, the specific growth rate was determined to be 0.013 h^{-1} (data not shown). The TCE degradation rate of free cells was determined by tracking the TCE concentration with time. For instance, for an initial TCE concentration of 11.88 mg/L, the degradation rate was 12.30 mg/L/d (data not shown).

TCE Degradation Kinetics with Immobilized Cells. This was studied in the bioreactor operated in recycle batch with intermittent additions of TCE. Figure 2 shows typical data obtained for the degradation of TCE. As the TCE concentration decreased, the chloride ion concentration increased, confirming that TCE was actually biodegraded rather than simply adsorbed or lost by leakage.

The effect of initial TCE concentration on degradation rate was also evaluated by measuring the change in TCE concentration with time at various initial TCE concentrations in the range of 5-21 mg/L. Figure 3 shows the change in TCE concentration with time for 3 different initial TCE concentrations along with their respective degradation rates. Figure 4 summarizes the result of TCE degradation at various initial concentrations with immobilized cells. In general, optimal degradation rates were achieved for TCE concentrations in the range of 10-15 mg/L and the degradation rate increased with increasing TCE concentration up to ~10 mg/L. Above 10 mg/L, the rate appeared to reach a constant value (~50 mg/L/d). At high TCE concentration (~23 mg/L), however, the degradation rate was lower (~30 mg/L/d), indicating TCE toxicity effects.

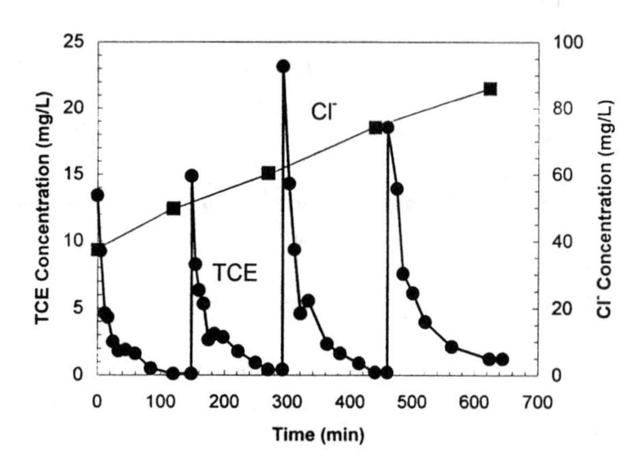

FIGURE 2. Kinetics of TCE degradation by immobilized cells of *Methylosinus trichosporium* OB3b mutant PP358.

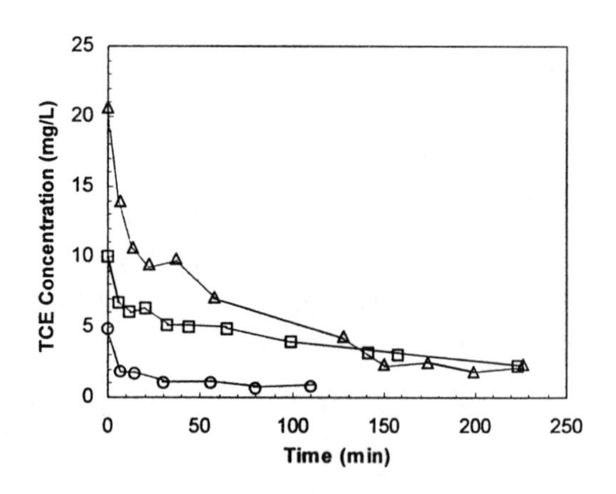

FIGURE 3. Degradation of TCE by immobilized cells at various initial concentrations.

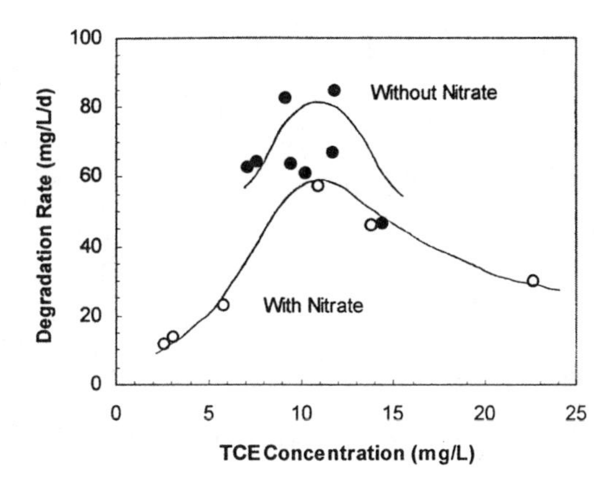

FIGURE 4. Effect of TCE concentration on degradation by immobilized cells.

It was also observed that much higher TCE degradation rates (up to 84.77 mg/L/d) were achieved without the presence of nitrate in the medium (Fig. 4). Figure 5 shows TCE degradation rate as a function of nitrate concentration in the medium. The highest degradation rate occurred when nitrate was depleted. When *M. trichosporium* OB3b is completely deprived of a combined nitrogen source, it is capable of fixing nitrogen. During nitrogen fixation, the microorganism can maintain high growth yields; however, it grows at a much lower growth rate, compared to an optimal nitrate concentration for growth (Bowman and Sayler, 1994). But, when fixing nitrogen, some NADH is utilized by the nitrogenase, making it unavailable for use in sMMO catalyzed reactions. Bowman and Sayler (1994) found that under nitrate depleted conditions, the sMMO activity in growing cells was 10 % of the activity of control cultures with nitrate present. They found that between 0 and 124 mg/L nitrate, sMMO activity steadily increased. The effect of nitrate on resting (immobilized) cells during TCE degradation, however, has not previously been studied. These two effects of nitrate on growth rate and sMMO activity in growing cells can be useful for developing a bioreactor system for TCE degradation. *M. trichosporium* OB3b requires methane to maintain growth and sMMO activity; however, in the presence of methane, TCE will not be degraded. The sMMO preferentially degrades methane, rather than TCE. This competition problem has made the design of continuous TCE-treatment bioreactors very challenging. The trend shown in Figure 5 offers a solution to this difficulty. It suggests that when deprived of nitrate, the cells are growing at a significantly lower rate and thus not utilizing all of the sMMO present for methane oxidation. Instead, even in the presence of methane, the residual sMMO is being used for TCE degradation. Therefore, the competition for the enzyme can be regulated by controlling the nitrate concentration.

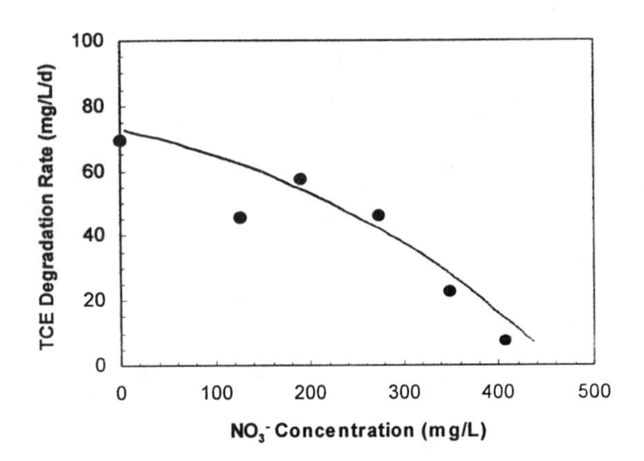

FIGURE 5. Effect of nitrate concentration on degradation of TCE (10 mg/L) by immobilized cells.

Reactor Long-Term Stability. The fibrous-bed bioreactor was operated over a year alternating between TCE degradation and cell rejuvenation modes. The long-term stability of the bioreactor system seemed to be good (data not shown). There was no clogging of the packed column and not much accumulation of biomass. As a result, there was no deterioration in the reactor flow distribution and the immobilizing matrix was not disrupted. There did not appear to be diffusional limitations since the amount of dissolved methane and oxygen available to the cells was adequate to support growth.

ACKNOWGEDGMENTS

Authors would like to thank Ms. Carol E. Aziz and Dr. George Georgiou from the University of Texas at Austin, for providing the culture of *Methylosinus trichosporium* OB3b mutant PP358.

REFERENCES

Bowman, J. P. and G. S. Sayler. 1994. "Optimization and Maintenance of Soluble Methane Monooxygenase Activity in *Methylosinus trichosporium* OB3b." *Biodegradation. 5*: 1-11.

Enzien, M. V., F. Picardal, T. C. Hazen, R. G. Arnold, and C. B. Fliermans. 1994. "Reductive Dechlorination of Trichloroethylene under Aerobic Conditions in a Sediment Column." *Appl. Environ. Microbiol. 60*: 2200-2204.

Fennell, D. E., Y. M. Nelson, S. E. Underhill, T. E. White, and W. J. Jewell. 1993. "TCE Degradation in a Methanotrophic Attached-Film Bioreactor." *Biotechnol. Bioeng. 42*: 859-872.

Friday, D. D. and R. J. Portier. 1991. "Development of an Immobilized Microbe Bioreactor for VOC Applications." *Environ. Prog. 10*: 30-39.

Hanna, M. L. and R. T. Taylor. 1996. "Attachment/Detachment and Trichloroethylene Degradation-Longevity of a Resting Cell *Methylosinus trichosporium* OB3b Filter." *Biotechnol. Bioeng. 51*: 659-672.

Miller, G. P., R. J. Portier, D. G. Hoover, D. Friday, and J. L. Sicard. 1990. "Biodegradation of Chlorinated Hydrocarbons in an Immobilized Bed Reactor." *Environ. Prog. 9*: 161-164.

Montgomery, J. H. 1996. *Groundwater Chemicals Desk Reference.* 2nd ed., pp. 866-1003. CRC Press, Inc., New York.

Shim, H. 1997. "BTEX Degradation by a Coculture of *Pseudomonas putida* and *Pseudomonas fluorescens* Immoblized in a Fibrous-Bed Bioreactor." Ph. D. Dissertation, The Ohio State University, Columbus, OH.

Shuler, M. L. and F. Kargi. 1992. *Bioprocess Engineering.* pp. 61-78 and 148-161. Prentice Hall, Englewood Cliffs, NJ.

Speitel, G. E. and R. L. Segar. 1995. "Cometabolism in Biofilm Reactors." *Wat. Sci. Technol. 31*: 215-225.

Uchiyama, H., K. Oguri, M. Nishibayashi, E. Kokufuta, and O. Yagi. 1995. "Tricholoethylene Degradation by Cells of a Methane-Utilizing Bacterium, *Methylocystis* sp. M, Immobilized in Calcium Alginate." *J. Ferment. Bioeng. 79*: 608-613.

Uchiyama, H., K. Oguri, O. Yagi, and E. Kokufuta. 1992. "Trichloroethylene Degradation by Immobilized Resting-Cells of *Methylocystis* sp. M in a Gas-Solid Bioreactor." *Biotechnol. Lett. 14*: 619-622.

U. S. Environmental Protection Agency. 1995. *Bioremediation in the Field.* EPA/540/N-95/500.

COMETABOLISM OF CHLORINATED VOCs DOWNGRADIENT OF A FUEL HYDROCARBON SOURCE

Peter I. Dacyk (Parsons Engineering Science, Inc., Cincinnati, Ohio)
William D. Hughes (Parsons Engineering Science, Inc., Cincinnati, Ohio)

ABSTRACT: Cometabolism of chlorinated volatile organic compounds (VOCs) was observed in an aquifer contaminated with chlorinated VOCs by a tetrachloroethene (PCE) release from an industrial laundry facility. The cometabolism occurred when fuel hydrocarbons migrated into the aquifer from an upgradient bulk fuel storage terminal. The presence of trichloroethene (TCE) and cis-1,2 dichloroethene (1,2-DCE) suggested a mechanism was operating within the aquifer which allowed biodegradation of the PCE. An examination of groundwater data revealed fuel hydrocarbons were present within the aquifer. The aerobic nature of the aquifer beneath the laundry facility precluded the likelihood of reductive dechlorination. Fortuitous degradation of the chlorinated compounds through cometabolic processes was the most likely influence in biodegrading PCE to its lesser-chlorinated degradation byproducts. Recent work in the area of *insitu* remediation of chlorinated VOCs has advocated the injection of fuel hydrocarbons to enhance degradation of chlorinated VOCs. A proposal was presented to allow migration of the fuel hydrocarbons from the fuel terminal. Regulatory concerns of off-site fuel migration and liability concerns with the bulk fuel terminal owners prevented the cessation of active remediation at the fuel terminal. A field-scale pilot test will be performed at the laundry facility utilizing propane as an alternative carbon source to stimulate cometabolic degradation of the chlorinated solvents.

INTRODUCTION

The solvent release site is a commercial laundry located in southwestern Ohio. The laundry facility sits atop a buried valley aquifer composed primarily of glacial sand and gravel deposits with interbedded clay, silt, and sand stringers. The buried valley aquifer typically contains low dissolved oxygen (D.O.) concentrations (< 1 mg/L) and is reducing in nature. The aquifer is the primary potable water source for the surrounding communities. A groundwater production well field is located approximately 1070 meters north (downgradient) of the site.

The laundry facility is located immediately downgradient of a bulk fuel storage complex consisting of four bulk storage terminals. Each of the four fuel terminals has reported hydrocarbon releases. A variety of remediation systems are in use at the terminals. The fuel terminal which is immediately upgradient of the laundry facility uses pump and treat technology coupled with soil vapor extraction as its primary remediation technology. Fuel hydrocarbons from the terminal were detected in groundwater sampled at the laundry facility during the early stages of remediation monitoring. The fuel hydrocarbons (benzene, toluene, ethylbenzene, and xylenes -BTEX) diminished to non-detect levels in the laundry facility's monitoring wells in late 1992; however BTEX concentrations at the fuel terminal's downgradient perimeter wells persisted into mid-1996.

Dry cleaning operations at the facility used tetrachloroethene (PCE) as the primary cleaning solvent through 1986. A release of PCE occurred in 1985 when a 1000 gallon above ground storage tank (AST) ruptured during refilling. The released solvent flowed into a nearby storm sewer which empties into an on-site infiltration pond. An additional chlorinated compound release of undetermined quantity may have occurred from underground storage tanks (USTs) located adjacent to the AST. Active remediation of the chlorinated VOCs began in 1989 with the installation of groundwater extraction wells and an air stripper to treat the extracted groundwater. Soil vapor extraction (SVE) wells were added later to strip VOCs from contaminated vadose zone soils. Groundwater treatment and SVE were subsequently terminated in the mid 1990's due to poor VOC removal rates. The groundwater recovery wells were retained to provide hydraulic control of the solvent plume. An air-sparge system with vapor extraction was installed and began operation in 1997. VOC removal rates improved but initial results indicate that mechanical removal of the chlorinated VOCs by air sparging and SVE will not decrease VOC concentrations to drinking water standards. Figure 1 is a site plan showing the solvent source area, groundwater recovery wells, and select monitoring wells.

SITE CHARACTERIZATION

Groundwater sampling at the laundry facility revealed BTEX compounds to be migrating from the fuel terminal onto the laundry facility property. PCE as well as TCE and cis 1,2-dichloroethene (1,2-DCE) were detected in monitoring wells located down and cross-gradient from the solvent release point. No chlorinated hydrocarbon, other than PCE, was known to have been released at the site; however, the commercial-grade PCE may have contained impurities which include TCE and 1,2 (DCE). The lesser chlorinated TCE and 1,2-DCE hydrocarbons are typical byproducts of PCE degradation (Guest et al., 1995). The presence of both TCE and 1,2-DCE in concentrations greater than PCE was the first indication that the PCE was undergoing dechlorination to it's lesser - chlorinated degradation byproducts Additionally, several groundwater samples revealed the presence of minor (< 5 µg/L) concentrations of vinyl chloride (VC).

Dissolved oxygen (D.O.) data collected during groundwater sampling events showed the aquifer beneath the laundry facility to have D.O. concentrations generally greater than 1 mg/L and concentrations of 3 mg/L were not uncommon. The high D.O. concentrations show the aquifer beneath the laundry facility to be aerobic rather than anaerobic. Lower D.O. (< 1 mg/L) conditions prevailed at the facility concurrent with BTEX detections in the upgradient fuel terminal's perimeter groundwater monitoring wells (pre-1996). Dissolved oxygen data collected at the site prior to 1994 show unusually high (> 3 mg/L) D.O. concentrations which were determined to be questionable based on unreliable instrumentation used when these data were collected.

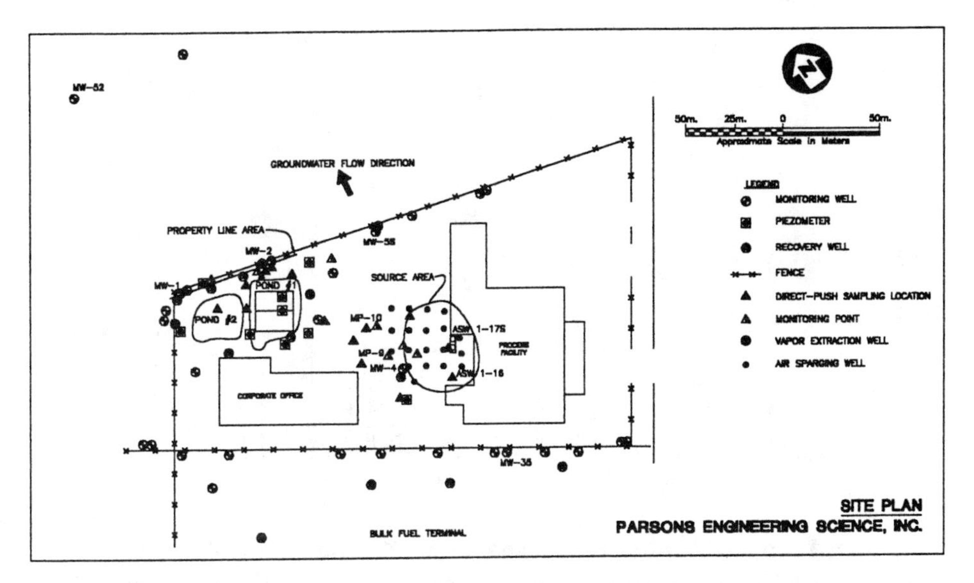

FIGURE 1. Site plan of industrial laundry facility showing location of solvent source area, groundwater recovery wells, and select monitoring wells.

EVIDENCE OF COMETABOLISM

Aerobic conditions within the aquifer beneath the laundry facility preclude the occurrence of reductive dechlorination of PCE. The presence of the less chlorinated TCE and 1,2-DCE degradation byproducts suggest degradation of PCE was occurring under aerobic conditions, which is uncommon for highly chlorinated compounds such as PCE (Weidemeier et al., 1996). Figure 2 represents groundwater analytical results taken from a monitoring well (MW-4) located near the solvent release area. Groundwater BTEX (dissolved) data collected from a monitoring well (MW-35) located on the downgradient edge of the fuel terminal is superimposed on the Figure 2 data. The degradation of PCE to less chlorinated degradation byproducts may have occurred as a fortuitous byproduct of the biodegradation of the BTEX compounds present in the aquifer beneath the laundry facility. The degradation of PCE may have been catalyzed by an enzyme or cofactor released by the microbes during degradation of the BTEX (McCarty and Semprini, 1994). No groundwater or soil analyses were performed to determine the quantity of native organic carbon available for biodegradation. Groundwater analytical results from the laundry facility taken primarily in 1992 show a TCE/PCE ratio greater than 1. This ratio corresponds to the appearance of BTEX compounds in the laundry facility's wells and in the high BTEX concentration in monitoring wells adjacent to the laundry facility (MW-35 on Figure 2). Post 1992 groundwater results show a decreasing TCE/PCE ratio with PCE becoming the dominant chlorinated compound after 1994. A similar trend is

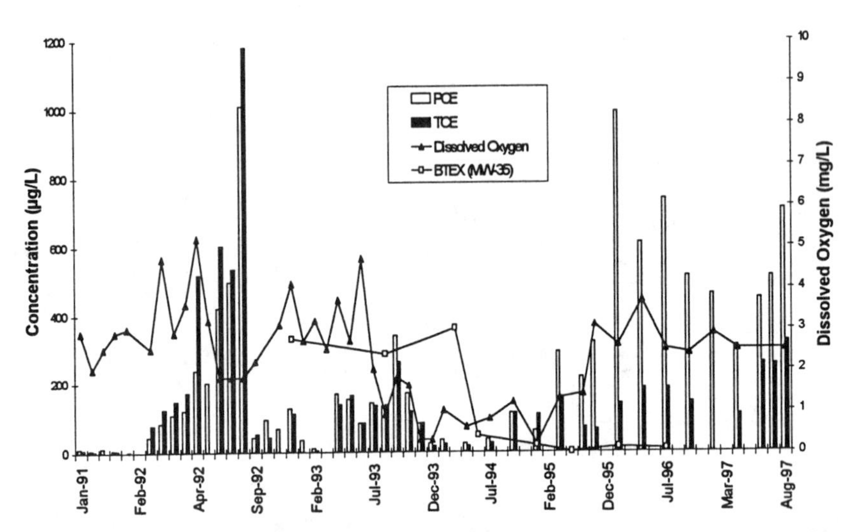

FIGURE 2. PCE ,TCE, BTEX, and dissolved oxygen concentrations in groundwater sampled from monitoring wells MW-4 and MW-35 and located near the solvent source area and immediately up-gradient of the laundry facility.

observed in the 1,2-DCE/PCE ratio. This trend has remained to the present, even with overall decreasing concentrations of PCE through mechanical removal. The reduction in TCE and 1,2-DCE concentrations relative to PCE concentrations corresponds to the disappearance of BTEX compounds from groundwater sampled from the fuel terminal's downgradient perimeter wells and a corresponding increase in D.O. concentrations within the laundry facility's groundwater. Minor favorable fluctuations in the PCE/TCE ratio (increasing TCE concentrations) are observed in several wells at the site after disappearance of the fuel hydrocarbons. The increase in TCE (and 1,2-DCE) and decrease in PCE correlates to low D.O. concentrations measured in these wells. The low D.O. may have produced transient anaerobic conditions which stimulated reductive dechlorination. The PCE to TCE and PCE to 1,2-DCE ratio increased once the aquifer returned to aerobic conditions.

INTRODUCTION OF ANTHROPOGENIC CARBON

The favorable TCE/PCE ratio observed in groundwater samples concurrent with BTEX detections indicates conditions at the site may favor cometabolism of PCE when sufficient carbon is available for microbial biodegradation. An attempt was made to convince local regulatory agencies and the operators of the adjacent fuel terminal to cease active remediation at the fuel terminal. A case was made which argued that discontinuing remediation at the terminal would allow fuel hydrocarbons to reach the laundry facility and stimulate cometabolism of the hydrocarbons to reach the laundry facility and stimulate cometabolism of the chlorinated hydrocarbons. The proximity of the terminal and laundry facility to the production well field and liability concerns with the terminal owners precluded shut-down of remediation at the terminal. As an alternative to discontinuing active groundwater remediation at the fuel terminal, a pilot test evaluating the injection of propane into the aquifer at the solvent release area will be performed in the near future. The propane will be injected through the existing air sparge system. Propane will be a source of anthropogenic carbon (electron donor) for microbial degradation which should cometabolize the PCE and TCE. If the pilot test shows favorable distribution of propane into the aquifer and consequent reduction in PCE and TCE concentrations, a full scale system will be designed using the existing sparge injection wells. Preliminary design planning predicts a propane injection ratio of approximately 100:1 (air to propane). The propane will be injected in pure form over a period of several minutes followed by several hours of air injection. Extracted vapors will be monitored continuously by a lower explosive limit (LEL) detector for safety purposes. Continued groundwater and soil vapor monitoring will be used to evaluate whether introduction of the anthropogenic carbon has stimulated degradation of the chlorinated hydrocarbons under aerobic conditions.

CONCLUSIONS

Degradation of PCE to its less chlorinated TCE and 1,2-DCE degradation byproducts was occurring in a predominantly aerobic aquifer. The degradation of the PCE was occurring in the presence of fuel hydrocarbons. The degradation of the PCE was probably the result of cometabolic processes as microbes were degrading fuel hydrocarbons introduced into the aquifer from fuel releases at an adjacent, upgradient fuel terminal. During the suspected cometabolic PCE degradation period, the TCE to PCE ratio generally remained above 1. Once the fuel hydrocarbons were depleted, the TCE to PCE ratio decreased to less than 1 and PCE has become the dominant chlorinated hydrocarbon present within the site's groundwater. Present removal of PCE and the other chlorinated hydrocarbons appears to be limited to mechanical processes active at the site. Anthropogenic carbon, in the form of propane, will be introduced into the aquifer to stimulate a resumption of cometabolism of the PCE and its degradation byproducts.

REFERENCES

Guest, P. R., Benson, L. A., and Rainsberger, T. J., 1995. *Inferring Biodegradation Processes for Trichloroethene from Geochemical Data*. Presented at Battelle International Symposium on In Situ and On-Site Bioreclamation. April 24-27. San Diego California.

McCarty, P.L. and L. Semprini. 1994 "Ground-Water Treatment for Chlorinated Solvents" In Norris, R.D. (Ed.) Handbook of Bioremediation. Lewis Publishers, Boca Raton, Fl. Pp. 87-116.

Wiedemeier, T. H., Swanson, M. A., Moutoux, D. E., Wilson, J. T., Kampbell, D. H., Hansen, J. E., and P. Haas. 1996. *Overview of the Technical Protocol for Natural Attenuation of Chlorinated Aliphatic Hydrocarbons in Groundwater Under Development for the Air Force Center for Environmental Excellence.* Presented at Symposium on Natural Attenuation of Chlorinated Organics in Groundwater. September 11-13. Dallas, Texas.

COMETABOLIC BIOFILTRATION OF TCE USING BIOLUMINESCENT REPORTER BACTERIA

Chris D. Cox, Kevin G. Robinson, Hae-Jin Woo, Charles Leon Wright, and John Sanseverino (University of Tennessee, Knoxville, TN, USA)

ABSTRACT

Biofiltration of TCE was studied using *Pseudomonas putida* strains F1 and TVA8. TVA8 is a genetically-engineered bioluminescent reporter bacterium for induction of the *tod* (toluene dioxygenase) operon. TCE removal in excess of 95% was achieved when toluene was fed intermittently to the biofilter compared to about 30% removal when toluene was fed continuously. Oxidation of TCE resulted in a decrease in the whole-cell Tod enzyme activity and resultant TCE degradation. Bioluminescence of TVA8 increases during toluene loading, indicating induction of the Tod enzyme system.

INTRODUCTION

Solvent-contaminated gas streams requiring treatment prior to atmospheric release originate from many industrial processes and from remediation process such as air sparging and soil-vapor extraction. Biofiltration is a process in which contaminated gas streams are passed though a bed of porous media supporting a biofilm. Contaminants are transferred from the gas stream to the biofilm and are subsequently biodegraded. In the case of chlorinated aliphatic compounds, such as TCE, aerobic degradation occurs almost exclusively by cometabolic processes. Competitive inhibition, NADH limitation, and product toxicity often limit the ability to sustain high removal rates in cometabolic processes (Chang and Alvarez-Cohen, 1995). The goal of this research is to incorporate an understanding of the rate of enzyme synthesis and inactivation into biofiltration design and operation and to investigate the use of bioluminescent reporter bacteria to determine TCE degradation potential within biofilters.

MATERIALS AND METHODS

Pseudomonas putida F1 was a gift from David T. Gibson (University of Iowa). *Pseudomonas putida* TVA8 is a genetically-engineered bioluminescent reporter bacterium for *tod* (toluene dioxygenase) activity. It was constructed by inserting a *tod-lux* fusion into the chromosome of F1 (Applegate et al., 1997).

P. p. strains F1 and TVA8 were inoculated from frozen stock cultures and grown in 500 ml of Hunters medium. The phosphate buffer of Hunters medium was replaced with KH_2PO_4 (2.15 g L^{-1}) and K_2HPO_4 (5.3 g L^{-1}). The final pH was 7.0. Liquid toluene was suspended in a glass bulb in the head space of the flask. The cultures were harvested by centrifugation at 1000 rpm for 15 min and transferred into fresh medium containing diatomaceous earth pellets (Celite) in a 4 L flask. The flask was then placed on a shaker to provide an opportunity for the organisms to attach to the diatomaceous earth pellets; a low mixing intensity was used to minimize erosion. The inoculated pellets were transferred to the biofiltration column.

A schematic diagram and detailed description of the biofiltration apparatus used for the experiments with F1 have been given elsewhere (Cox et al., 1998). A similar apparatus was used for the experiments with TVA8. The F1 and TVA8 columns measured 65 and 25 cm in length and 10 and 3.2 cm in diameter, respectively. Both column were constructed of glass with sampling ports sealed with mininert caps distributed along the column length. The columns were packed with Celite Biocatalyst Carrier R-635 (Manville, Denver, Colo.) to provided physical support for biofilm. The flow rate of toluene- and TCE-saturated air were varied using mass flow controllers to establish the desired composition of inlet gas to the column. Nutrients and moisture were added to the column via an aqueous solution introduced at the top of the column.

Whole-cell Tod activity, NADH, and protein concentration were determined for the immobilized cells in the biofiltration reactor. Two pellets taken from each stage of the biofiltration column were mixed with 2 ml of sterile 0.1 % $Na_2P_2O_7$ and vortexed to detach immobilized cells from pellets. The cell mixture was centrifuged and washed with phosphate buffer solution (pH=7.2). The final cell pellets were resuspended with 5 ml of fresh medium. The cells were dispersed by sonicating them for 3 second, followed by 30 seconds of shaking; this cycle was repeated three times. The prepared cell culture was used for measurement of the normalized enzyme activity and concentrations of NADH and protein.

A fluorescence-based method of relating whole-cell Tod enzyme activity to indoxyl formation has been reported elsewhere (Woo et al., 1998). Briefly, indole is added to cells prepared as described above and allowed to react for two minutes. The cells were centrifuged and the fluorescence of the supernatant was determined and compared to a blank containing no cells. The excitation and emission wavelengths were 365 nm and 470 nm, respectively. The rate of increase in the fluorescence of the sample can be related to the rate of indigo formation; the whole-cell Tod activity is normalized to the protein concentration of the sample and reported as mg indigo min^{-1} (mg protein)$^{-1}$.

Optical density (OD_{600}) was measured on a Beckman Spectrophotometer (Model DU-70; Fullerton,CA) and related to protein concentration using Bovine serum albumin (BSA) as a protein standard. NADH was measured by fluorescence (Model 240, Perkin-Elmer, Inc., Buckinghamshire, England). Excitation and emission wavelengths were 340 nm and 425 nm, respectively. The normalized whole-cell NADH concentration was expressed as μmol NADH (mg protein)$^{-1}$.

Bioluminescence was measured using a photomultiplier tube (PMT). Light was collected and sent to the PMT using a liquid light pipe secured perpendicular and flush to the column using a collar. The signal from the PMT was recorded in real time using a digital display (model 7070, Oriel, Stratford,

CT) connected to a PC. The entire biofilter was enclosed in a light-tight box during bioluminescent monitoring.

RESULTS AND DISCUSSION

TCE was fed continually in all biofiltration experiments. Two different feeding schemes were used for toluene: continuous and intermittent feeding. A typical experiment of each type will be described. In the continuos feeding experiment, *P. p.* F1 was used and the emphasis of discussion will be the role of Tod enzyme activity on biodegradation of TCE. In the intermittent feeding experiment using *P. p.* TVA8, the emphasis will be on interpretation of bioluminescent data.

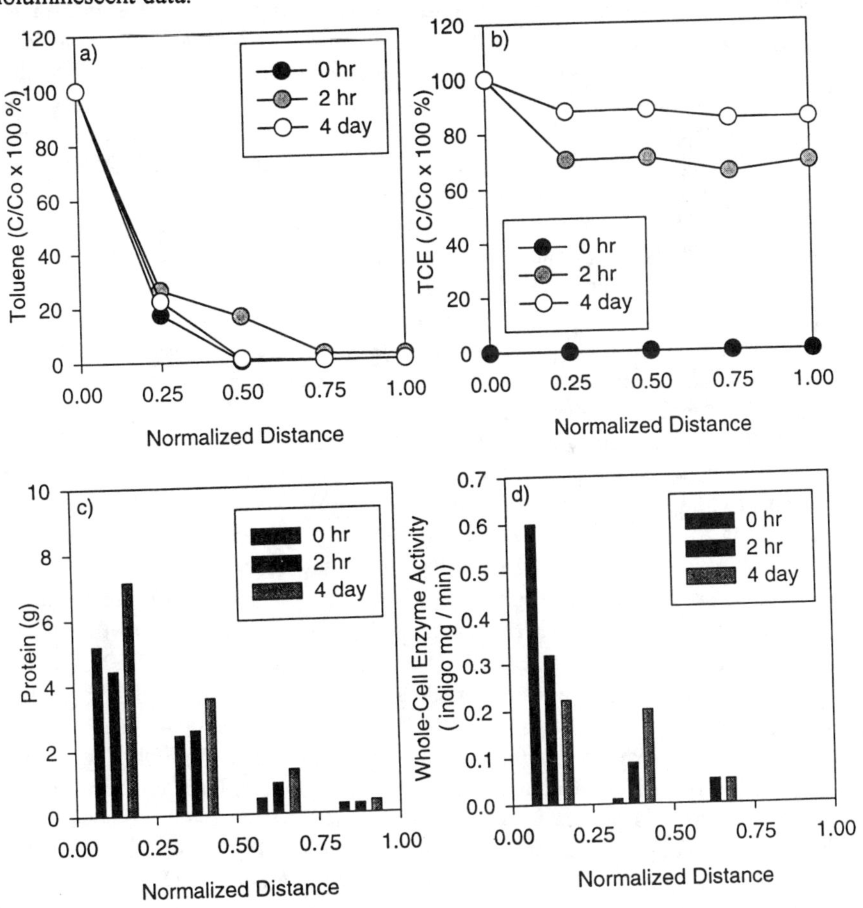

FIGURE 1. Biofiltration of TCE by strain F1 with continuous feeding of toluene: a) toluene removal profile, b) TCE removal profile, c) protein profile, d) whole-cell Tod activity profile.

In the first experiment, both toluene and TCE were fed to the column simultaneously at concentrations of 1684 and 170 μg/L, respectively. The gas flow rate to the column was 500 mL/min yielding an empty-bed contact time (EBCT) of 10 minutes. The loading rates of toluene and TCE were about 0.084 and 0.0085 μg (mg protein)$^{-1}$ min^{-1}, respectively. Profiles of toluene, TCE, biomass (measured as cell protein) and whole-cell Tod enzyme activity through the column during four days of operation are shown in Figure 1. The data for time = 0 represents conditions just before initiating TCE flow after feeding toluene alone for an extended period of time. Toluene is completely degraded within the first half of the column (Figure 1a). Biomass (Figure 1c) and whole-cell Tod enzyme activity (Figure 1d) are also proportionally higher near the inlet of the column where the loading of toluene is greatest. A higher toluene loading would result in an increase in the biological activity of the biofilter, but would also increase competitive inhibition. About 30% removal of TCE is achieved during the first 2 hrs of feeding. Toluene loading is about an order of magnitude greater than TCE loading, resulting in significant competitive inhibition of TCE and poor removal efficiency. TCE removal decreases to about 15% after 4 days of operation indicating that low levels of TCE removal can be sustained for long periods of time with continuos feeding of toluene. The decrease in whole-cell Tod activity that occurs after the initiation of TCE feed to the biofilter can be attributed to product toxicity. The capacity of the biofilter to remove TCE decreases with time as the extent of enzyme deactivation increases. Under these conditions, NADH levels in the column remained constant across the column at about 0.45 μmol/mg protein and were not limiting.

The experiment above illustrates the difficulty in achieving high TCE removal rates when toluene is present. For this reason, it is advantageous to feed toluene to the biofilter intermittently. A typical experiment with intermittent feeding of toluene using TVA8 bioluminescent reporter bacteria is shown in Figure 2. Toluene, at a concentration of 12 mg/L, was fed at a rate of 40 mL/min, corresponding to a toluene loading rate of 3.7 μg (mg protein)$^{-1}$ min^{-1}. A high toluene loading rate is advantageous to maximize protein growth and Tod activity within the filter. TCE was fed at a concentration of 600-950 μg/L at a rate of 5 mL per minute, corresponding to a loading rate of 0.035 μg (mg protein)$^{-1}$ min^{-1}. The EBCT during TCE feeding was about 40 min. At the beginning of the experiment, toluene is fed to the biofilter and virtually no TCE degradation occurs due to competitive inhibition. The effluent concentration of toluene during this time remained below 0.5 mg/L. The flow of toluene to the biofilter was interrupted after about 200 minutes and TCE removal in excess of 95% was maintained for over 100 minutes. Despite a higher TCE loading to the filter, the fraction removal of TCE was much greater with intermittent feeding of toluene compared to continuous feeding. Additional experiments indicated that TCE-degrading capability of the column can be regenerated over an indefinite number of cycles by feeding toluene for approximately twice as long as the period of TCE feeding alone. This suggests that three biofilters in rotational duty, one actively treating TCE and two being regenerated, may be used to continuously degrade TCE to a high extent.

The bioluminescent reporter bacteria *P.p.* TVA8 was evaluated as an indicator of TCE degradation potential. Bioluminescence is observed to increase steadily during toluene feeding, indicating the induction of Tod enzyme. The momentary sharp increase in bioluminescence that is observed upon cessation of toluene flow is probably the result of increased oxygen availability to the luciferase enzyme. After the flow of toluene is interrupted, bioluminescence decreases steadily; however, attempts to correlate this decrease with the duration of viable TCE degradation were unsuccessful. Within the biofilter, the rate of decrease in bioluminescence was found to be independent of TCE concentration. Once luciferase is fully induced, bioluminescence is controlled by the availability of the long chain fatty aldehyde tetradecanal and/or reduced riboflavin phosphate ($FMNH_2$) (Meihgen, 1991). The fact that these factors are different than those controlling TCE cometabolism may explain the inability to correlate a decrease in bioluminescence to a decrease in TCE degradation. Addition of acetate to the column stimulates bioluminescence, but does not significantly enhance or prolong TCE degradation.

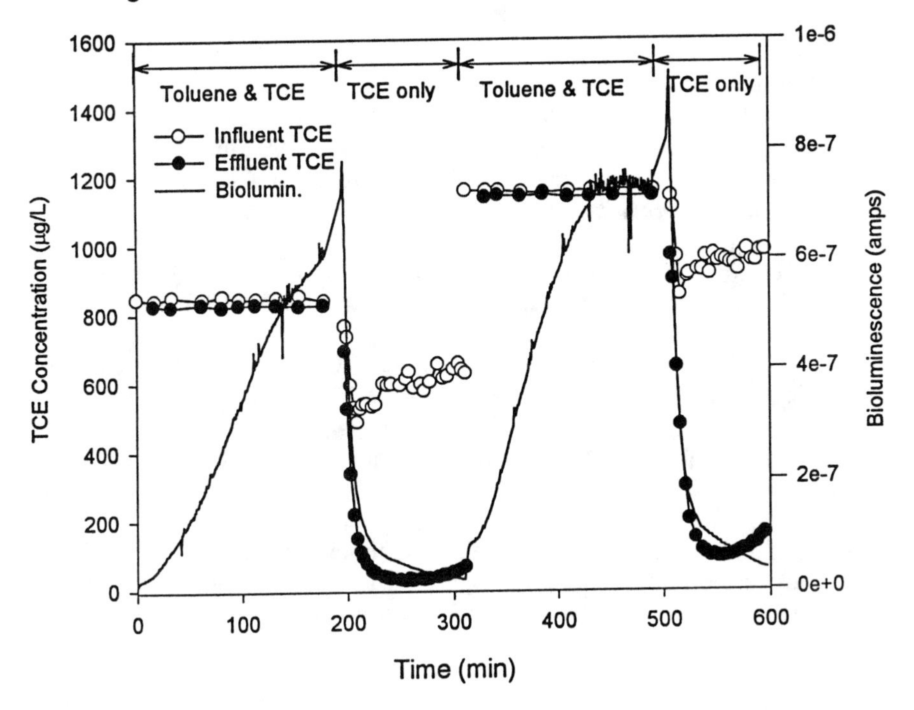

FIGURE 2. TCE removal and bioluminescence during biofiltration of TCE by strain TVA8 with intermittent feeding of toluene.

CONCLUSIONS

Whole-cell Tod enzyme activity within the biofilm can be measured and used as an indicator of TCE biodegradation potential. A quantitative understanding of the factors controlling Tod activity in cells during TCE degradation may yield improved design and operational methodologies for cometabolic processes. High removal rates of TCE can be sustained by feeding toluene to the biofilter intermittently. The induction of Tod enzyme can be detected using the bioluminescent reporter strain TVA8. One promising application of these organisms is *in situ* subsurface remediation, where the spatial distribution of injected toluene is uncertain. The reporter bacteria could be used to provide online monitoring of the spatial and temporal variation in oxygenase activity during injection of primary substrate to the subsurface.

ACKNOWLEDGMENTS

This research was funded through the Waste Management Education and Research Institute of the University of Tennessee and through a cooperative agreement between the Department of Energy, The University of Tennessee, and Tennessee State University.

REFERENCES

Applegate, B., S. Kehrmeyer, and G. Sayler. 1998. "A Modified Mini-Tn5 System for Chromosomally-Introduced *lux* Reporters for Chemical Sensing." *Applied Environ. Microbiol.* Submitted.

Chang, H.L. and L. Alvarez-Cohen. 1995. "Model for the Cometabolic Biodegradation of Chlorinated Organics." *Environ. Sci. Technol.* 29, 2357-2367.

Cox, C.D., H.J. Woo, and K.R. Robinson. 1998. "Cometabolic Biodegradation of TCE in the Gas Phase." *Water Sci. and Technol.* In press.

Meighen, E.A. 1991. "Molecular Biology of Bacterial Bioluminescence." *Microbial Rev.* 55(1) 123-142.

Woo, H.J., J. Saneverino, C.D. Cox, K.G. Robinson, and G.S. Sayler. 1998. "The Measurement of Toluene Dioxygenase Activity in Biofilm Culture of *Pseudomonas putida* F1." *J. Microbial Methods.* Submitted.

COMETABOLIC BIOVENTING OF CHLORINATED SOLVENTS AT A FORMER WASTE LAGOON

Evan E. Cox, Todd A. McAlary, and David W. Major
(Beak International Incorporated, Guelph, Ontario, Canada)
Jason Allan and Leo Lehmicke
(Beak International Incorporated, Kirkland, Washington)
Scott L. Neville (Aerojet Propulsion Systems, Sacramento, California)

ABSTRACT: Field and laboratory studies and a pilot test have been conducted to evaluate cometabolic bioventing to remediate chlorinated volatile organic compounds (VOCs) in the vadose zone at a former septic waste lagoon (former lagoon) at a Superfund site in California. An initial soil gas investigation at the former lagoon indicated that significant biodegradation of the chlorinated VOCs is occurring naturally under a combination of anaerobic and aerobic redox conditions. Specifically, the parent solvents disposed of at the former lagoon (namely trichloroethene [TCE], 1,1,1-trichloroethane [1,1,1-TCA] and chloroform [CF]) are being reductively dechlorinated under strongly methanogenic conditions within remaining buried septage material while dechlorination products diffusing from the buried septage material (namely 1,2-dichloroethene [1,2-DCE], vinyl chloride [VC], 1,1-dichloroethane [1,1-DCA] and dichloromethane [DCM]) are aerobically biodegrading to CO_2 through a combination of direct mineralization (for VC and DCM) and cometabolic biodegradation processes. Laboratory microcosm studies confirmed the intrinsic biodegradation processes and indicated that natural degradation rates could be increased to up to 5.2%/day through oxygen and methane addition. A pilot test of cometabolic bioventing was conducted at the site, and confirmed that degradation of selected chlorinated VOCs could be enhanced in situ through continuous addition of oxygen and methane. Further studies are expected and full-scale application of the technology is being considered.

SITE BACKGROUND

The former lagoon had a surface area of approximately 550 m^2 and an approximate depth of 6 m. It received waste from septic tanks at the facility from the early 1960's until 1979. Laboratory chemical wastes were likely incorporated into the waste as sinks and floor drains at the facility were typically plumbed into the septic tank systems. In 1979, the lagoon was filled with dredge tailings from the surrounding area, and graded flat. The dredge tailings are remnants of historic gold dredging operations that have disturbed alluvial sediments to depths of up to 20 m below ground surface (bgs). A septage (sludge) layer remains at depths of up to 9 m bgs. A perched watertable exists at approximately 12 m bgs.

Chlorinated VOCs have been previously detected in the soil gas near the former lagoon, with total VOC concentrations ranging up to 57,600 µg/L. Historical VOC concentrations in the soil gas suggest that residual pure phase VOCs are present in the remaining septage layer. VC, an anaerobic biodegradation product of TCE, comprises a significant portion (as much as 92%) of the VOCs in the soil gas, indicating that TCE is anaerobically biodegrading in the remaining septage material. Methane has also been detected in the soil gas at the former lagoon, with concentrations ranging as high as 142,000 μg/L.

APPROACH AND METHODS

Intrinsic Biodegradation Survey. A baseline soil gas survey was initially conducted to assess the nature and extent of intrinsic biodegradation of chlorinated VOCs occurring in the remaining septage and the surrounding vadose zone. Soil gas probes and wells were installed within the remaining septage material and along a transect extending from the boundary of the septage material to the boundary of VOC distribution (approximately 20 m). Samples were collected for analysis of VOCs by EPA Method TO-14 and natural gases (e.g., oxygen , methane, carbon dioxide, ethene) by ASTM method 1945.

Laboratory Microcosm Studies. Microcosm studies using site soil were conducted to confirm intrinsic degradation processes that were hypothesized to be occurring in situ. Microcosms were constructed using 100 mL serum bottles containing 20 grams of homogenized soil. Treatment and sterile control microcosms were amended with selected chlorinated VOCs and cometabolites (e.g., methane or toluene) known to promote aerobic cometabolic biodegradation; active control microcosms were amended with selected chlorinated VOCs but no cometabolite. Microcosms were incubated in the dark at room temperature (approximately 22°C) for a period of approximately 14 weeks. The rate and extent of oxygen and cometabolite consumption and chlorinated VOC biodegradation (direct mineralization and/or cometabolism) were monitored over time.

Pilot Testing. A pilot test of cometabolic bioventing was conducted to demonstrate that VOC degradation could be enhanced in situ through the addition of oxygen and methane to the vadose zone. The pilot test consisted of two main tasks, which were respirometry testing and biostimulation testing.

Respirometry tests were conducted at each of the soil gas wells on the transect to assess the range of methane and oxygen respiration rates in the pilot test area and to refine design and operations of subsequent biostimulation trials. For respirometry testing, a mixture of air containing methane (2.5%) and helium (1%) was injected at each well at a flow rate of between 0.5 and 1 standard cubic foot per minute (scfm) for a period of 2 hours. Soil gas samples were then collected from each well over an expanding time scale for analysis of methane, oxygen, carbon dioxide and helium. Methane and oxygen data were compared to helium tracer data to assess relative rates of diffusive losses. Diffusive losses were then subtracted from observed losses to estimate methane and oxygen respiration.

Biostimulation trials were conducted by delivering oxygen and methane into the vadose zone to enhance both direct mineralization and/or cometabolic (methanotrophic) biodegradation of the target VOCs. Mineralization was expected to be the dominant biodegradation process for VC and DCM, whereas cometabolism was expected to be the dominant biodegradation process for TCE, cis-1,2-DCE, trans-1,2-DCE, 1,1-DCE, 1,1-DCA, and CF. For 1,1,1-TCA, both abiotic transformation to 1,1-DCE and possibly cometabolism (to CO_2) was expected. PCE and CT (present at trace concentrations) were not expected to biodegrade under aerobic conditions in the pilot test. The fate of Freon-113 (present throughout the pilot test area) was uncertain.

Figure 1 provides a schematic of the vapor process control system used for the biostimulation injections. The process involved extracting soil vapor containing target VOCs from the vadose zone, amendment of the extracted soil vapor stream with atmospheric oxygen, helium (up to 1% v/v) and methane (up to 1.5% v/v), and recirculation of the resultant vapor stream to the vadose zone via an injection well to enhance VOC biodegradation. Helium breakthrough was monitored at each well using a real-time helium detector, and this data was used to assess travel/residence times for the purposes of calculating degradation rates. Soil gas samples were collected from the injection well (to establish initial [T_0] concentrations to compare against) and from each of the wells where 100% helium breakthrough was observed. Resulting VOC concentration data were corrected relative to the helium tracer to account for breakthrough. VOC concentration data were also corrected relative to VOC-specific retardation factors to account for apparent VOC losses due to gas-phase VOCs partitioning into the soil moisture and sorption of gas-phase VOCs onto the soil.

RESULTS AND DISCUSSION

Intrinsic Biodegradation of VOCs in the Vadose Zone. Results of the soil gas survey indicate that chlorinated VOCs are intrinsically biodegrading through a sequential combination of anaerobic and aerobic processes. The anaerobic biodegradation processes are occurring within the septage layer itself. TCE is being reductively dechlorinated to 1,2-DCE (predominantly the cis-isomer) and VC; 1,1,1-TCA is being reductively dechlorinated to 1,1-DCA and CA; and CF is being reductively dechlorinated to DCM. To some extent these dechlorination daughter products are being further anaerobically biodegraded to non-chlorinated end products such as ethene (for VC), ethane (for CA), or CO_2 (for DCM). However, the rate of degradation of the dechlorination daughter products, particularly VC, is slower than their rate of diffusion from the septage layer to the surrounding vadose zone (which is aerobic). Methane is being produced in the septage and is also diffusing from the septage into the surrounding aerobic vadose zone. Once VOCs diffuse into the surrounding aerobic vadose zone, they appear to be biodegrading to CO_2 through a combination of direct mineralization (e.g., for VC, CA, DCM) and cometabolic biodegradation processes.

Laboratory Microcosm Results. Results of the microcosm studies confirmed that the indigenous microorganisms can biodegrade most of the chlorinated VOCs in the soil gas at the former lagoon to CO_2, through a combination of anaerobic and aerobic (direct mineralization and cometabolism) processes. Under simulated in situ conditions (Active Control, no cometabolite added), degradation rates for 1,2-DCE, 1,1-DCE, VC, 1,1,1-TCA, CF and DCM ranged up to 3.6 %/day, with higher rates generally reported for the chlorinated ethenes. These rates were increased (e.g., up to 5.2%/day) through addition of oxygen and methane.

Field Respirometry Results. Results of respirometry testing indicate that the indigenous microorganisms in the vadose zone at the former lagoon are acclimated for methane metabolism (i.e., methanotrophic bacteria appear to be present and active). Calculated methane half-lives ranged from 4 to 11 days. Calculated oxygen half-lives ranged from 4 to 12 days. The fastest methane and oxygen respiration rates were observed at well SGW-2 (location shown in Figure 1), located at a distance of approximately 10 m from the boundary of the remaining septage. Slower methane consumption rates were measured at wells located near the boundary of the remaining septage (SGW-1 and E1), where oxygen demand appeared to relate to the presence of organic chemicals other than the added methane (e.g., materials related to the septage).

Field Biostimulation Trials. Significant VOC degradation was observed during two short-term continuous injection trials. Concentrations of total chlorinated ethenes (TCE, cis-1,2-DCE, trans-1,2-DCE, 1,1-DCE and VC) decreased by approximately 70-80% in the two trials, whereas the concentrations of total chlorinated ethanes (1,1,1-TCA, 1,1-DCA and Freon 113) and total chlorinated methanes (CF and Freon 11) decreased by 30-35%. Specific mass losses for target VOCs, corrected to helium tracer breakthrough and for chemical-specific retardation, were: TCE 12-70%; cis-1,2-DCE 82-90%; 1,1-DCE 40-55%; VC 40-67%; 1,1,1-TCA 60-70%; 1,1-DCA 11-45%; and Freon 113 26-30% (Figure 2). Little to no mass loss was observed for PCE, CT and Freon 11. Mass loss rates and extents generally increased over successive injections, suggesting increased microbial activity (possibly microbial growth) related to the prior exposure to methane and possibly VOCs (e.g., toluene and VC).

In both trials, oxygen concentrations dropped significantly (more than 70%) across the test area, with corresponding increases in CO_2 concentrations (Figure 2). By comparison, methane concentrations decreased by less than 20% (Figure 2). Toluene, present at low concentrations (<20 μg/L) in the injection stream, was rapidly depleted over the test area and may have stimulated some cometabolism of the chlorinated ethenes by aromatic oxidizers (Figure 2). Additional studies are expected to better understand the relationship between methane use oxygen consumption/demand in this enhanced system, and to more fully evaluate VOC degradation over a longer period of operation.

CONCLUSIONS

The results of the studies to date provide encouraging evidence that microorganisms are capable of intrinsically biodegrading chlorinated VOCs in vadose zone systems, either through direct mineralization (for VC and DCM) or cometabolism in the presence of methane (and possibly toluene). At this site, bioventing may be a feasible remediation technology, pending results of additional longer-term studies.

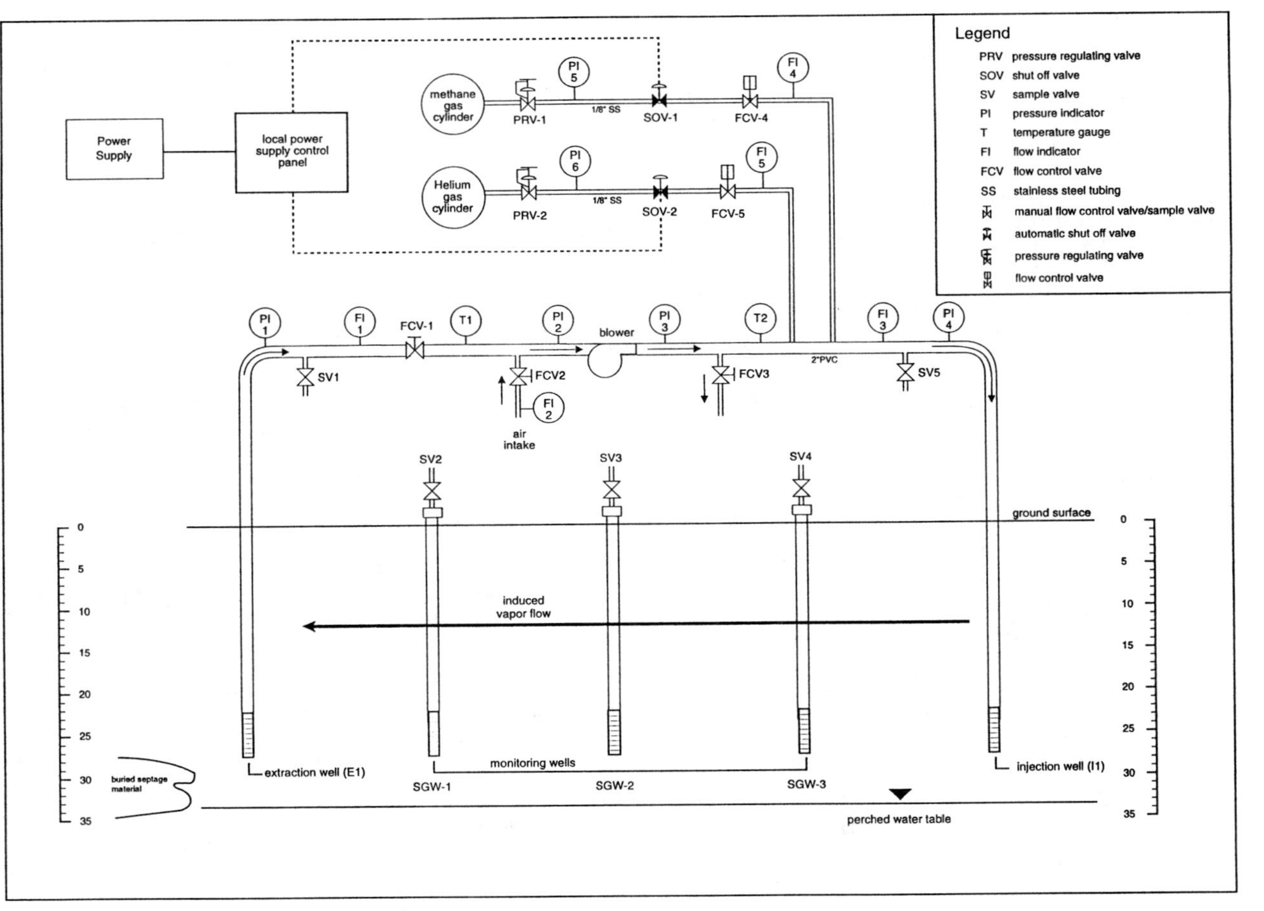

Figure 1: Schematic of the Cometabolic Bioventing Process Control System

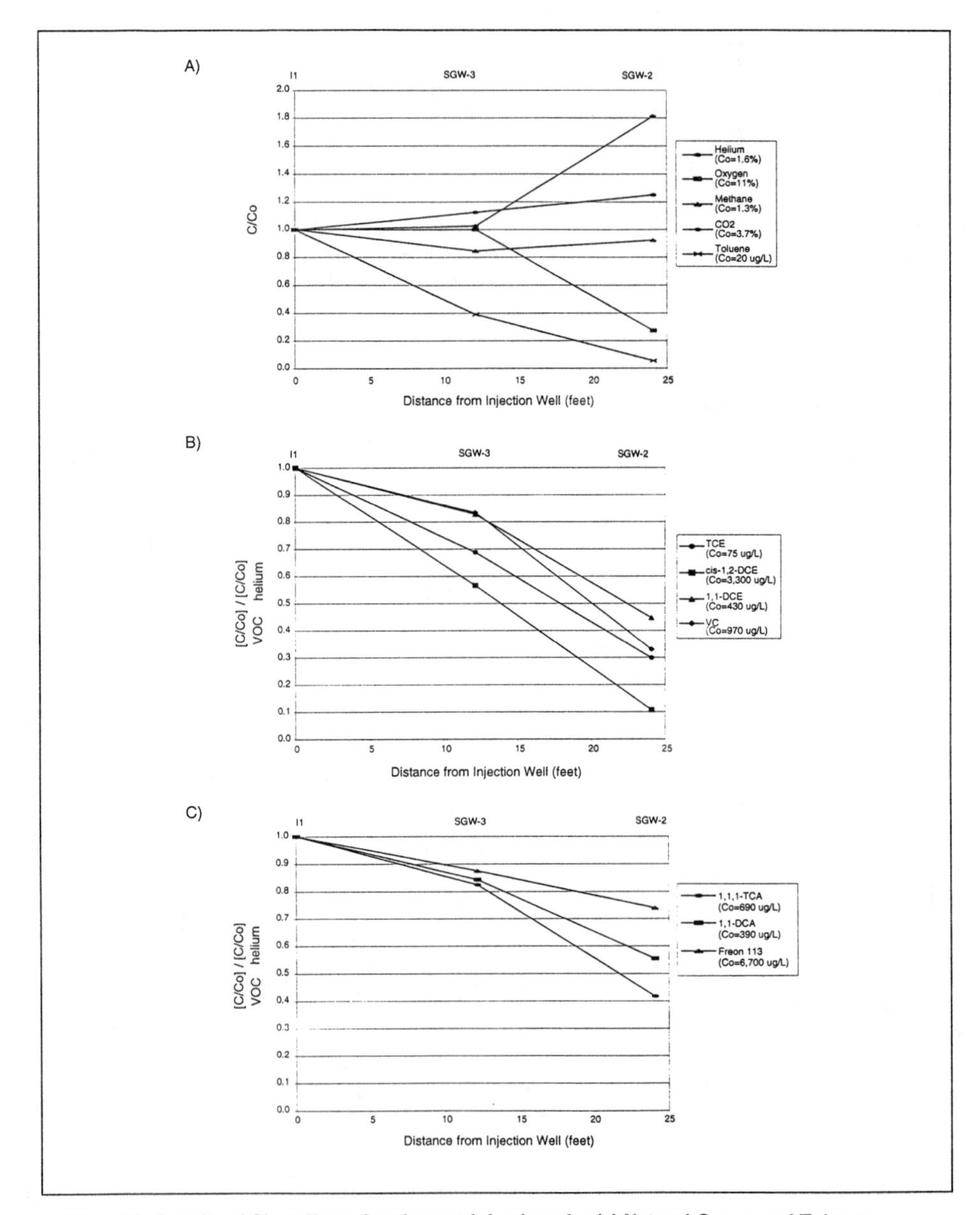

Figure 2: Results of Short Term Continuous Injections for (a) Natural Gases and Toluene, (b) Specific Chlorinated Ethenes, (c) Specific Chlorinated Ethanes

BIODEGRADATION OF TETRACHLOROETHYLENE AND TRICHLOROETHYLENE USING MIXED SPECIES MICROBIAL MATS

Walter O'Niell, Valentine Nzengung and John Noakes (Univ. of Georgia, Athens, GA)
Judith Bender and Peter Phillips (Clark Atlanta University, Atlanta, GA)

Abstract: Experiments were conducted to investigate the degradation of tetrachloroethylene (PCE) and trichloroethylene (TCE) by mixed-species microbial mats. Batch experiments were conducted to evaluate the kinetics of PCE and TCE degradation by mat material in liquid growth media. Additional experiments were conducted to investigate PCE degradation in a plant growth chamber that simulated natural environmental conditions, and samples were analyzed for transformation products. Rapid uptake followed by slower degradation of both PCE and TCE was observed in the experiments. These results indicate that microbial mats have potential for application to remediation of media contaminated by these recalcitrant compounds.

INTRODUCTION

Tetrachloroethylene (PCE:C_2Cl_4) and trichloroethylene (TCE:C_3HCl_3) are common xenobiotic environmental contaminants. Widespread use of these volatile solvents in industrial applications has led to extensive contamination of soils, groundwater and surface waters (Westrick et al., 1984). Both compounds are recalcitrant to natural degradation. Removal of these compounds from contaminated sites using currently available technologies may require decades to centuries. Therefore, more efficient and cost-effective technologies to remediate soils and water contaminated by these chemicals are needed (Newman et al., 1997).

Microbial mats have excellent potential for application to biodegradation of organic compounds because of their limited growth requirements, ability to support a mixed bacterial community, close proximity of aerobic and anaerobic zones within the mat, and survival capability in a wide range of environments. The mats, which are primarily composed of cyanobacteria (*Oscillatoria* spp.) and related heterotrophic bacteria, are self-sustaining photoautotrophs. The mats are stratified, with a lower layer containing anaerobic bacteria that can biodegrade chlorinated organics through reductive dehalogenation. The near surface layers of the mat are photosynthetic and may degrade halogenated compounds through oxidative pathways.

Objective. The goals of this study were three-fold: (1) determine the sorptive capacity of microbial mats for PCE and TCE, (2) measure the transformation rates of PCE in batch samples (static transformation) and in a simulated aquatic environment (growth chamber), and (3) identify PCE transformation products.

The mats used in our experiments were provided by Drs. Judith Bender and Dr. Peter Phillips of Clark Atlanta University and were grown using a patented technique developed by Drs. Bender and Phillips.

RESULTS AND DISCUSSION

Sorption Studies. Batch samples of microbial mat and Allen/Arnon (AA) growth media were placed in glass vials, dosed with either PCE or TCE, and sealed. Control samples which did not contain any mat were also prepared. All samples were maintained at 20° C under continuously lighted conditions. PCE and its metabolites were extracted from the liquid and mat phases of the samples using hexane or a hexane/methanol mix and analyzed using a gas chromatograph with an electron capture detector (GC/ECD).

The concentration of solvent sorbed to the mat was plotted versus the liquid-phase concentration, as shown for PCE in Figure 1. This uptake was described by a linear sorption model, and a partitioning coefficient was determined by calculating the slope of each line. A partition coefficient of 6.3 ml/g was determined for PCE while the partition coefficient for TCE was calculated at 2.9 ml/g. These relatively large partitioning coefficients are attributed to sorption of PCE and TCE to lipids in the microbial mats.

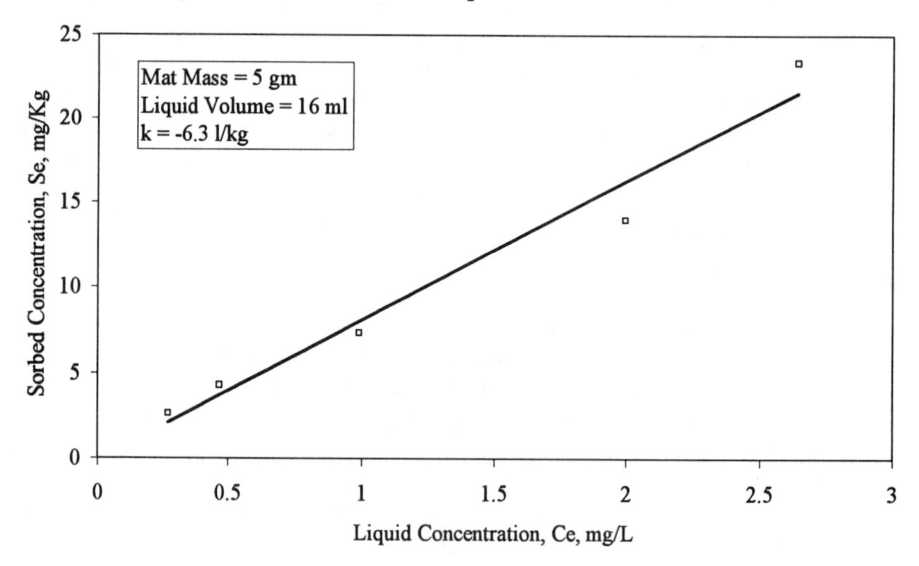

Figure 1. Sorption of PCE to Microbial Mat (6 Hour Incubation)

Degradation Kinetics in Static Transformation Experiments. Batch samples of mat and AA growth media were prepared, placed in glass vials, dosed with PCE, and sealed. Control samples were prepared using AA media without mat.

All samples were maintained at 20° C under 12 hour light/12 hour dark conditions. PCE was extracted from the liquid and mat phases of the samples at set intervals and the concentration was determined by GC/ECD analysis.

Representative results of the batch transformation experiments are shown in Figure 2. These results indicate that the liquid-phase PCE concentration decreased quickly (15 minutes or less). This decrease of dissolved-phase PCE concentration is attributed to partitioning from the liquid to the mat phase. The rapid uptake by the mat is followed by a slower degradation and transformation of the PCE. The half-lives of PCE in the batch samples varied from 18.5 to 27.9 days. TCE, a reductive dehalogenation product of PCE, was generally detected in the samples in one week or less.

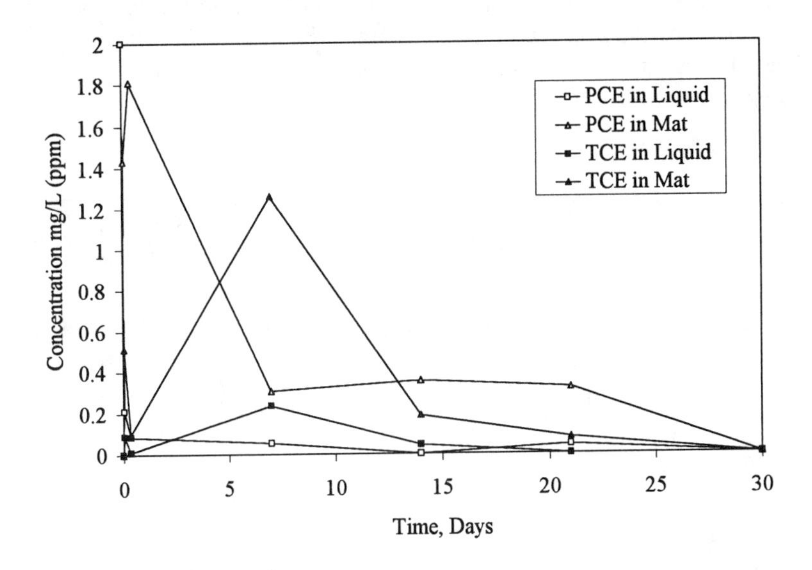

Figure 2. Batch Transformation Experiment Results, 8 grams of Mat Dosed with 2 mg/L PCE

Degradation Kinetics in Continuous Air Flow Growth Chamber Experiments. The growth chamber experiments were conducted using 5 liter glass chambers with air inlets and outlets. The chamber was filled with 4 liters of liquid AA growth media and 300 grams of mat. A stock solution of PCE or TCE in methanol was added to the chamber to make a liquid-phase concentration of 10 mg/l (ppm). Medical-grade air was provided to the chamber from a pressurized tank. The air was moistened before entering the chamber to minimize volatilization, and the experiment was conducted at 20° C under a 12 hour light/12 hour dark cycle to simulate natural growth conditions for the microbial mat. Three traps containing ethyleneglycolmonomethylether were installed at the chamber outlet to capture vapor-phase PCE and its metabolites.

The liquid-phase was sampled immediately after dosing, and at 24 to 48 hour intervals thereafter. After 24 hours, the initial concentration in the mat chambers decreased 10 to 20 percent, which is attributed to sorption and volatilization. After the initial 24 hour period, the PCE and TCE concentrations decreased exponentially (Figure 3). TCE was detected in the liquid-phase of the chamber dosed with PCE after approximately 5 days, and the TCE concentration increased with time. The lag period between dosing and the detection of TCE may correspond to the time required for the mat to establish a well-developed anaerobic zone. The half-life of PCE in the chamber averaged 4.5 days, and the half-life of TCE averaged 8.25 days. In estimating the half-lives, losses due to volatilization were not accounted for. Therefore, half-lives may be greater than the values reported.

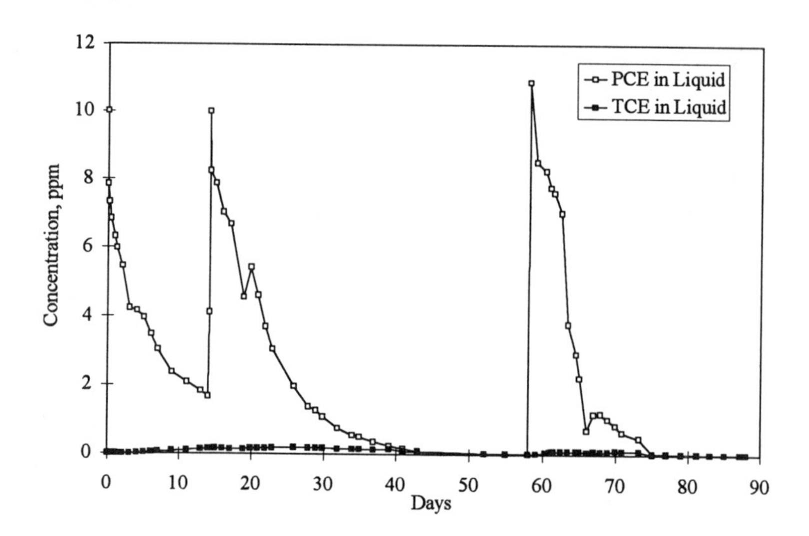

Figure 3. PCE Transformation in a Continuous Flow Growth Chamber

Products Testing. Liquid-, mat-, and vapor-phase samples from the batch and growth chamber experiments have been analyzed for dichloroethylene (DCE) and vinyl chloride (VC) using a GC/ECD, a GC equipped with a flame ionization detector, and a GC equipped with a mass spectrometer. DCE and VC were of particular interest, as they would be likely transformation products if PCE was primarily degraded through reductive dehalogenation. Neither of these compounds has been detected to date indicating that the PCE may be transformed through more than one pathway in these experimental systems.

Samples are currently being analyzed for chloroacetic acids using GC/ECD and GC/MS analysis of methyl tert-butyl ether derivatized extracts. The detection of these acids may be indicative of an oxidative transformation pathway in these experimental systems.

CONCLUSIONS

The results of these experiments indicate that microbial mats sequester and metabolize PCE and TCE. In all experiments, rapid sorption of the PCE to the mat was observed, and the large partition coefficients indicate that sorption is a significant uptake process for both PCE and TCE. The partition coefficient of TCE is approximately half that of the value calculated for PCE. This difference is attributed to the greater aqueous solubility of TCE relative to PCE.

The rapid sorption of the PCE and TCE observed in the experiments was followed by slower uptake of these compounds from the liquid phase. In samples dosed with PCE, TCE typically appeared in the liquid-phase and mat-phase of the samples in one week or less. The appearance of TCE indicates that the PCE is being dehalogenated, but the absence of DCE and VC suggest that PCE degradation is not following a complete dehalogenation pathway. The subsequent decrease in TCE concentration indicates that this transformation product is being degraded as well. PCE had a shorter half-life in the growth chamber than in the sealed batch vials, although the mat-to-liquid ratio was smaller in the chamber than in the batch samples. This result suggests that the mat was able to more effectively degrade the PCE under the more natural conditions of the growth chamber.

Additional experimentation is needed to better understand the uptake and degradation of PCE and TCE by microbial mats, but the results indicate that the mats have potential for application to the remediation of media contaminated by PCE and TCE.

REFERENCES

Newman, L.A., S.E. Strand, N. Choe, J. Duffy, G. Ekuan, M. Ruszaj, B.B. Shurtleff, J. Wilmoth, P. Heilman and M.P. Gordon, 1997. "Uptake and Biotransformation of Trichloroethylene by Hybrid Poplars" *Environmental. Science & Technology* 31:4. 1062-1067.

Westrick, J.J., J.W. Mello and R.F. Thomas, 1980. "The groundwater supply survey" *Journal of the American Water Works Association* 76. 52-59.

MODELING PHYTOREMEDIATION OF LAND CONTAMINATED BY HYDROCARBONS

M.Yavuz Corapcioglu, Robert L. Rhykerd, Clyde L. Munster, Malcolm C. Drew, and Kijune Sung (Texas A&M University)
Yoon-Young Chang (Korea Inst. of Sci. and Technol.)

ABSTRACT: In recent years, phytoremediation, i.e., the use of plants to clean up soils contaminated with organics, has become a promising new area of research, particularly for *in situ* cleanup of large volumes of slightly contaminated soils. A model that can be used as predictive tool in phytoremediation operations was developed to simulate the transport and fate of a residual hydrocarbon contaminant interacting with plant roots in a partially saturated soil. Time-specific distribution of root quantity through soil, as well as root uptake of soil water and hydrocarbon, was incorporated into the model. A sandy loam, which is dominant in soils of agricultural importance, was selected for simulations. Cotton, which has well-documented plant properties, was used as the model plant. Model parameters involving root growth and root distribution were obtained from field data reported in the literature and ranges of reported literature values were used to obtain a realistic simulation of a phytoremediation operation. Following the verification of the root growth model with published experimental data, it has been demonstrated that plant characteristics such as the root radius are more dominant than contaminant properties in the overall rate of phytoremediation operation. The ability to simulate the fate of a hydrocarbon contaminant is essential in designing technically efficient and cost-effective, plant-aided remedial strategies and in evaluating the effectiveness of a proposed phytoremediation scheme. The model presented can provide insight into the selection and optimization of a specific strategy.

INTRODUCTION

The use of plants to stimulate the remediation of a soil widely contaminated with an organic chemical at low soil concentrations represents a potential low-cost and effective alternative for waste management (Jones, 1991). The application of various plants to contaminated sites with various organic pollutants has been well documented for both laboratory and field studies and reviewed in a number of published papers (Corapcioglu, 1992; Matso, 1995; Shimp et al., 1993).

Before implementation of phytoremediation is attempted, the governing mechanisms should be identified and quantified. Mathematical modeling is an economical approach to obtaining this information. Although a number of developed conceptual models are in existence, they generally have focused on the fate of ionic nutrients and some metals in the root-soil zone (Bresler, 1987; Tracy and Marino, 1989). Such models incorporate the transport mechanisms of the nutrients and other solutes to the surface of roots and penetration of ions into the roots. Presently, several theoretical models for the effect of vegetation on the fate of organic chemicals in a hazardous waste site have been published (Corapcioglu, 1992; Davis et al., 1993; Bell, 1992). Although currently in a developmental stage for the simulation of the fate of organic contaminants in the root-soil zone, these models have shown the possibility of being predictive tools.

The objective of this study was to validate a mathematical model for understanding and predicting the effect of plants on the remediation of large volumes of soil slightly contaminated with nonvolatile hydrocarbons. Specifically the model was tested to determine the effect of root absorption on movement of hydrocarbons and hydrocarbon uptake by roots.

MATERIALS AND METHODS

The phytoremediation model has been developed to predict plant aided remediation of hydrocarbon contaminated soils and parameters to validate the model have been identified (Chang and Corapcioglu 1998). The parameters examined to determine their influence on phytoremediation are described in Table 1.

Table 1. Parameters used to examine the influence of plants on the fate of hydrocarbons in soil.

Symbol	Description	Units	Source
L_{Td}	root density	cm cm^{-3}	Taylor and Keppler 1975
R_r	root radius	cm	Nye and Tinker 1977
t_T	constant time of maturity	days	simulation
R_{cf}	root concentration factor	none	Bell 1992
q_{av}	mean daily uptake rate	cm^3 cm^{-1} d^{-1}	Taylor et al. 1992
T_{scf}	concentration movement into plant shoots	none	Bell 1992

These parameters were incorporated into the phytoremediation model to determine the influence of root absorption on the movement of hydrocarbons and hydrocarbon uptake by plants.

RESULTS AND DISCUSSION

Effect of Root Absorption on the Movement of a Hydrocarbon: Migration of a hydrocarbon, with infiltrating water following irrigation, may be retarded by root absorption of a hydrocarbon through a root zone. The extent of this retardation depends upon plant properties, soil conditions, and chemical characteristics of the hydrocarbon. Figure 1 shows the simulation results for the effects of the root absorption on the movement of a hydrocarbon for various input parameters that represent plant and hydrocarbon properties such as L_{Td}, R_r, t_T, and R_{cf} at 2000 h simulation. Any processes of hydrocarbon degradation in soils are neglected in the simulation in order to better observe the effects of plant roots on the transport of the hydrocarbon. Figure 1a shows sensitivity of the simulation

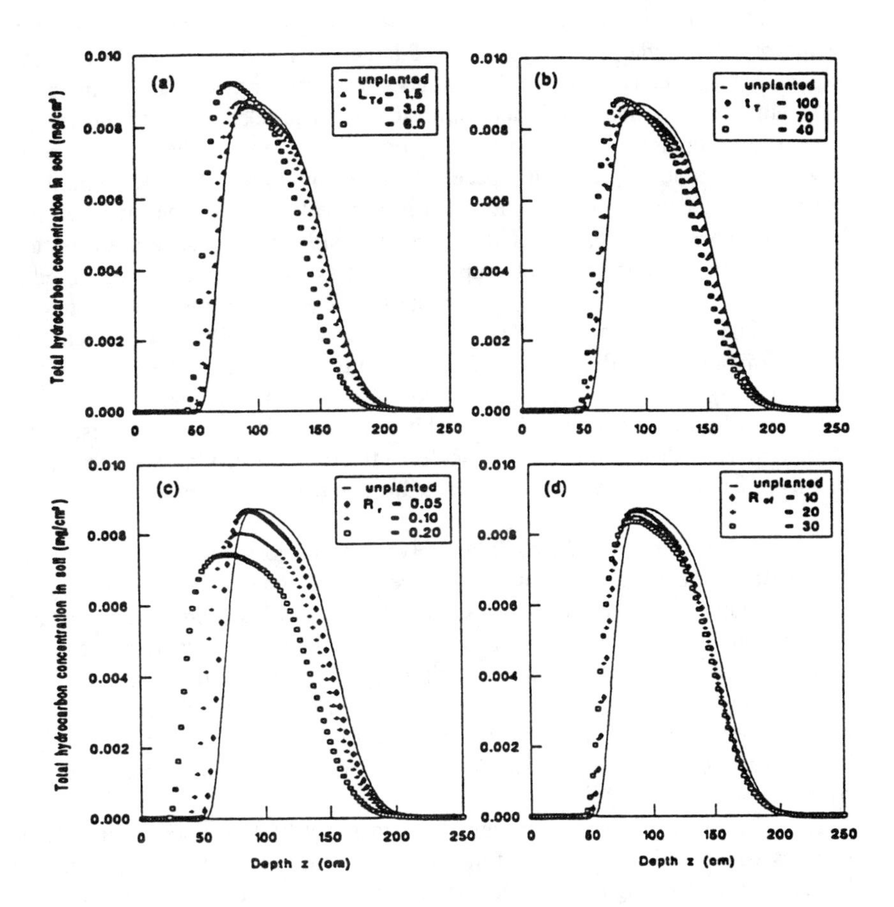

Figure 1. Effect of root absorption on hydrocarbon migration in unsaturated soil for different values of the model parameters (a) L_{Td} (cm cm^{-3}), (b) t_T (day) (c) R_r (cm) (d) R_{cf} at 2000 h simulation.

model to changes of L_{Td}, (cm cm^{-3}) from 1.5 to 3.0 and 6.0. Presence of plant roots with large value of L_{Td}, which represent available absorption sites, clearly shows the effect of retardation of the organic plume by root organic absorption under the simulated condition. The retardation by plant roots is seen to be carried out by reduction of the organic dispersion as well as convection producing relatively narrow concentration profiles in Figure 1a. As shown in Figure 1b, the more rapidly the root reaches its mature size, the more plant roots influence the migration of the hydrocarbon for the same reasons that variation in the hydrocarbon concentration profiles with change of L_{Td} is observed. Figure 1c illustrates high sensitivity of the model to the root radius R_r. Increase of R_r in a range of the most likely values for crops exhibits apparent decrease of migration of the organic plume due to build-up of larger root surface area. In addition, increase of R_r at constant t_T and L_{Td} induces relatively intense gradient in spatial distribution of root quantity in the topmost soil. The relatively rapid downward movement of organic plume in less dense rooting zone leads to the large spreading of the concentration profile in space, as shown in Figure 1c. The simulated profiles of organic plume for various R_{cf} in Figure 1d, on the other hand, exhibit little difference in the extent of the organic retardation by the root absorption, while evident influence of plant properties such as R_r, t_T, and L_{Td} is observed in Figure 1a, b and c respectively. The values of R_{cf} used in this simulation were estimated based on a range of literature values for commonly known hydrocarbons in contaminated sites (Bell, 1992). Under conditions examined, the retardation of the hydrocarbon corresponding to the root absorption shows small sensitivity to changes in R_{cf}, as seen in Figure 1d. This indicates that plant characteristics are more important than organic properties for effective intercepting and retardation of a hydrocarbon by slowly growing roots through the root-soil zone.

Hydrocarbon Uptake by Roots: Hydrocarbon losses into shoots with the transpiration stream of a plant was also investigated numerically. In addition the model sensitivity to plant properties such as q_{av}, t_T, and L_{Td} is presented. Organic degradation by biological activity is not included in the simulation for better understanding of the effect of the root uptake processes. For this analysis, T_{scf} is set at 1, which indicates no rejection of solutes in the root uptake process. This allows a better analysis of the model sensitivity to model parameters. In general T_{scf} is less than 1, therefore actual uptake of a hydrocarbon by roots may be overestimated.

The extent of organic matter loss into shoot by root uptake depends on plant parameters. The influence of the plant properties on the fate of hydrocarbon is demonstrated in Figure 2, which shows the normalized mass in the organic plume as a function of time for different values of q_{av}, L_{Td}, and t_T. Because of linear relationships of root organic uptake rate the normalized mass versus time curves in Figure 2a and b show identical trends. The rate of mass loss linearly increases over time with increasing q_{av} and L_{Td} after 1000 h. Although variation in t_T in a range of the study shows little influence in change of the rate of mass loss, as shown in Figure 2c, the effect of root uptake on organic mass loss is relatively early observed for a rapid growing root system which has small t_T. Although playing a minor role in organic mass loss in the simulation, organic uptake by root may be important in the use of resistant plants to toxic chemicals.

CONCLUSIONS

This model can identify the dominant parameters and be a very effective tool in optimizing data collection efforts. However, it should be noted that due to plant-, contaminant-, and soil-specific nature of various model parameters, the

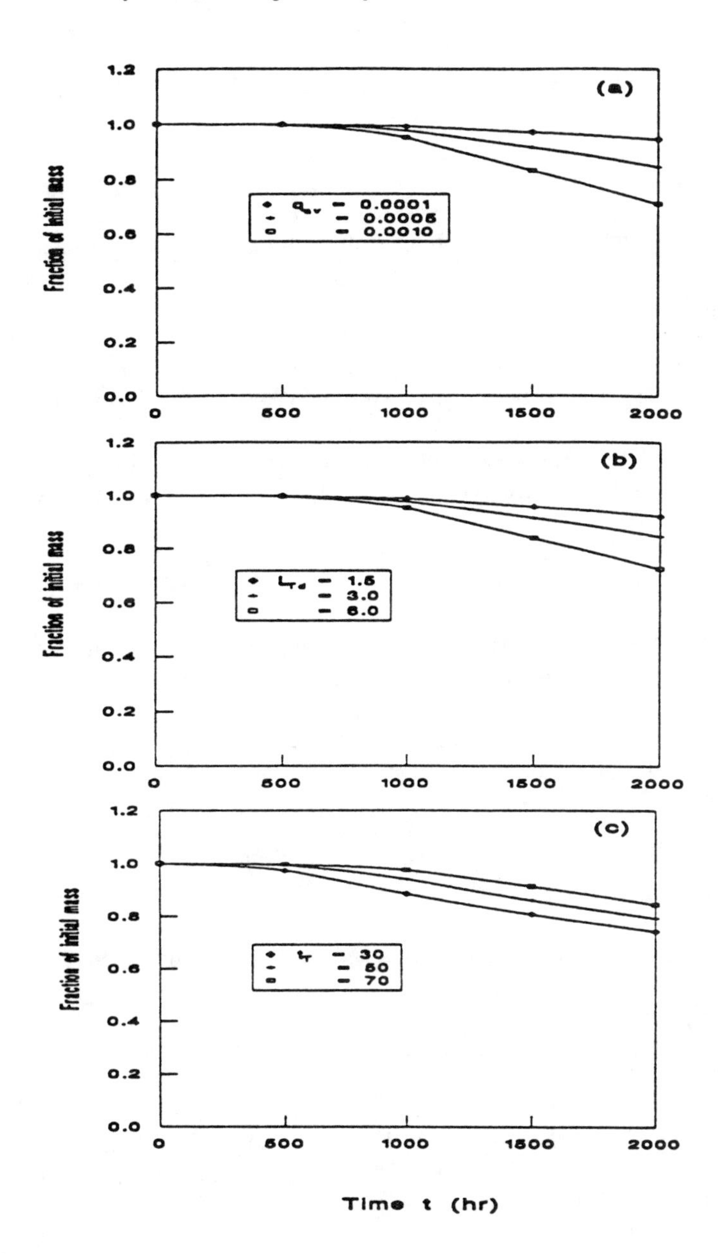

Figure 2. Effect of root uptake on hydrocarbon loss as a function of time for different values of model parameters (a) q_{av} (cm^3 cm^{-1} day^{-1}), (b) L_{Td} (cm cm^{-3}), and (c) t_T (day).

model has a certain level of uncertainty. There are situations involving complex interactive behavior of contaminants that cannot be always satisfactorily described by these parameters. A sensitivity analysis such as the one presented in this study can guide the field program to collecting appropriate data while minimizing the acquisition of irrelevant data.

ACKNOWLEDGEMENTS

This research was funded through a grant from the U.S. EPA R 8254-01-0.

REFERENCES

Bell, R.M. (1992). *Higher plant accumulation of organic pollutants from soils. EPA/600/R-92/138,* U.S. Environmental Protection Agency, Cincinnati, Ohio.

Bresler, E. (1987). "Application of a conceptual model to irrigation water requirement and salt tolerance of crops." *Soil Sci. Soc. Am. J.,* 51(3), 788-793.

Chang, Y.Y. and M.Y. Corapcioglu. 1998. "Plant-enhanced subsurface bioremediation of nonvolatile hydrocarbons." *J Env. Engr.* ASCE. 124:162-169.

Corapcioglu, M.Y. (1992). "Modeling plant uptake and bioremediation of semi-volatile hydrocarbon compounds." *Wat. Sci. Tech.*, 26(7-8), 1651-1658.

Davis, L.C., L.E. Erickson, E. Lee, J.F. Shimp, and J.C. Tracy. (1993). "Modeling the effect of plants on the bioremediation of contaminated soil and ground water." *Environ. Prog.,* 12(1), 67-74.

Jones, K.C. (1991). *Organic Contaminants in the Environment.* Elsevier Applied Science, New York, 189-206.

Matso, K. (1995). "Mother nature's pump and treat." *Civil Engineering*, October, 46-49.

Nye, P.H., and P.B. Tinker. 1977. *Solute Movement in the soil-root system.* 1st ed., Blackwell Scientific Publications. Oxford. pp. 100-150.

Shimp, J.F., J.C. Tracy, L.C. Davis, E. Lee, W. Huang, L.E. Erickson, and J.L. Schnoor. (1993). "Beneficial effects of plants in the remediation of soil and groundwater contaminated with organic materials." *CRC Critical Rev. Environ. Sci. Technol.,* 23(1), 41-77.

Taylor, H.M., and B. Klepper. (1975). "Water uptake by cotton root systems: An examination of assumptions in the single root model." *Soil Sci.,* 120 (1), 57-67.

Tracy, J.C., and M.A. Marino. (1989). "Solute movement through root-soil environment." *J. Irrig. Drain. Eng.,* ASCE. 115, 608-625.

PILOT-SCALE USE OF TREES TO ADDRESS VOC CONTAMINATION

Harry R. Compton (U.S. EPA, Edison, NJ)
Dale M. Haroski (Roy F. Weston Inc., Edison, NJ)
Steven R. Hirsh (U.S. EPA, Philadelphia, PA)
John G. Wrobel (U.S. Army Garrison, Aberdeen , MD)

ABSTRACT: High levels of volatile organic compounds (VOCs), primarily 1,1,2,2,-tetrachloroethane and trichloroethene, were detected in the groundwater at the J-Field Toxic Pits Site of Aberdeen Proving Ground (APG), Maryland. In April 1996, as part of a phytoremediation study at this site, 183 hybrid poplar trees (*Populus deltoides x trichocarpa*) were planted on a 4000-m^2 plot. The objectives were interception and containment of the flow of VOCs in the surficial aquifer and removal or destruction of VOCs in groundwater using enhanced natural mechanisms. Because phytoremediation is an emerging technology, extensive evaluation was performed to determine the fate of contaminants and optimal monitoring methods. The results of sample collections from the early-, mid- and late-growing season of the second year indicate that several study objectives have been met. Field data for plant tissues, transpiration gas, and condensate water indicate the presence of VOCs and their degradation products, implying the uptake of these compounds by the trees. Results from installed lysimeters and groundwater wells have indicated a potential drawdown effect from the trees. There is some evidence that the trees have had a positive effect on the soil nematode community. This study indicates that phytoremediation may be an effective treatment technology for VOCs in soils and groundwater.

INTRODUCTION

Aberdeen Proving Ground occupies more than 72,000 acres (3000 hectares) of land and water in southern Harford County northeast of Baltimore, Maryland, along the western shore of the Chesapeake Bay. The Army began testing ammunition and materiel, performing research and development, and operating training schools at the Aberdeen Proving Ground in 1918. J-Field, located at the southernmost end of Gunpowder Neck Pennisula, is mostly open fields, woods, and marsh. Specific activities at J-Field date back to 1918, when the Army began testing high explosives and munitions. Between 1940 and the 1970s, J-Field was used for the disposal of many types of chemical agents, high explosives and chemical wastes, which were burned and detonated in open pits and trenches. Hydrocarbon fuels, such as diesel fuel, were often used to produce more complete combustion of the waste materials. Various chlorinated solvents, in particular 1,1,2,2-tetrachloroethane, were used as decontaminating agents in the pits. The toxic burning pits area is about 150 by 240 meters in size and consists of several trenches. Significant levels of volatile organic compounds (VOCs) were detected in the groundwater in this area. VOCs in the groundwater ranged from non-detectable levels (ND) to 260 mg/L.

In general, the water table is within 0.9 to 1.2 meters of the land surface and consists of four major hydrologic units. Groundwater contaminants were detected primarily in the surficial aquifer at depths to approximately 12 meters. Aquifer testing over the last 12 years indicates that containment of the groundwater can be achieved by pumping the groundwater at a rate of 26.5 L per minute. Various soil and groundwater treatment technologies, such as soil washing, vapor extraction, and pump and treat, were considered, but deemed inappropriate based on site conditions, labor intensity, and cost. The sensitivity of the environment, high groundwater table, and non-critical time element at the APG site made phytoremediation using trees a feasible alternative for further investigation. Phytoremediation offers design flexibility, and trees have a proven ability to remove contaminants of concern.

The primary objective of the study is to determine if phytoremediation is a viable alternative for remediation of VOC contamination of shallow groundwater at the J-Field Toxic Burn Pits Operable Unit of the APG Edgewood Area National Priorities List Site. The study must show the extent of interception and containment of the surficial aquifer in the area of the toxic pits. The study must also show whether VOCs in the groundwater can be removed or destroyed through natural mechanisms associated with or enhanced by the trees planted for this study. The study has been designed to estimate the mass of VOCs removed or destroyed from the subsurface; determine the mechanisms responsible for VOC removal or destruction; monitor groundwater contaminant levels, lysimeters and tree transpiration gas and condensate; and determine aquifer drawdown within the study area and its zone of influence. Mechanisms for VOC removal or destruction include passive evaporation from groundwater through leaves without degradation, metabolism in plant tissue, incorporation into plant tissue, and degradation in soil by enhanced microbial populations or root exudates.

MATERIALS AND METHODS

Transpiration gas sampling was performed using a clear, 2 mil, 100 liter (30 x 36 inch) Tedlar® bag with dual stainless steel fittings, manufactured by SKC®, Inc. Samples were collected in Summa® Canisters and on Tenax®/CMS Tubes. Levels of volatile organics in the gas were determined using a TAGA (Trace Atmospheric Gas Analyzer) and an on-site Viking® GC/MS. Condensation formed in the Tedlar bag as a result of sealing the bag over the leaves of the tree. This water was sampled after the transpiration gas was sampled and before the bag was removed from the tree.

Tree tissue extract samples were analyzed as water samples using GC/MS, following a modification of U.S. EPA standard method 524.2. Antifoam agents were used to counteract analysis problems caused by foaming of the samples. EPA method 552 was followed for the analysis of haloacetic acids.

Stainless steel flux chambers were used to sample gases emanating from the ground surface. An open-path Fourier transform infrared (OP-FTIR) spectrometer manufactured by Environmental Technologies Group, Inc. was used to monitor ambient concentrations of the target compounds at several elevations within the phytoremediation area.

A Dynamax, Inc., Flow 32™ Sap Flow System was used for the measurement of sap flow in the tree trunks. A meteorological station located at J-Field collected data on wind speed, relative humidity, temperature, and net solar radiation during the sampling events. Trees were observed to determine the status of their health after the initial planting. Insect damage, deer damage caused by rutting, and leaf tip burn were recorded. Pruning was performed as needed.

There are currently 14 wells and four lysimeters located near the phytoremediation area. Five of the wells and the four lysimeters were installed in the surficial aquifer to obtain additional data necessary to determine the effects of the phytoremediation study on the groundwater. The placement of these wells and lysimeters was determined based on monitoring objectives, site conditions, and accessibility. The lysimeters, two at 1.2 m and two at 2.3 m below ground surface (bgs) permit coverage of the capillary zone during seasonal highs and lows in the groundwater level. Groundwater sampling for VOCs and metals is being conducted seasonally, in the spring, summer and fall.

Soil samples were collected for nematode extraction in October 1997. Three soil samples were taken around each tree at approximately 0.45-, 0.30-, and 0.15-meter increments perpendicular to the base of the tree. All three samples were combined together in a large plastic bag and sent to a subcontract lab. At the lab, the samples were split into three 20-g sub-samples, placed on Baerman funnels, and extracted for 48 hrs. The first nematodes encountered were placed into functional groups based on esophageal morphology and known feeding habits. The following trophic groups were identified: bacterivores, fungivores, herbivores, omnivores/ predators, hatchlings, and unknowns.

RESULTS AND DISCUSSION

During the May 1997 sampling period, 1,1,2,2-tetrachloroethane (1122) and trichloroethene (TCE) were detected in transpiration gas samples from several trees. The TAGA detected 1122 levels ranging from 2.0 to 170.0 parts per billion by volume (ppbv) and TCE levels at 3.4 and 13.5 ppbv. Similar results were detected in Summa canister samples. Sampling in the summer and fall, using only Summa canisters, produced similar results. Levels of 1122 and TCE increased in warmer summer months and decreased in the fall for both transpiration gas and condensate water samples, as summarized in Table 1 (gas results are for Summa canisters). Of all the trees screened for measurable levels of contaminants of concern (COC) in transpiration gas, only three consistently produced results. This may indicate a circuitous groundwater flow direction or tree roots within the zone of contamination.

Table 1. Seasonal Transpiration Gas and Water Results

	Spring		Summer		Fall	
VOC	**Gas (ppbv)**	**Water (µg/L)**	**Gas (ppbv)**	**Water (µg/L)**	**Gas (ppbv)**	**Water (µg/L)**
TCE	14.3	ND	210.0	6.3	99.0	1.8
1122	173.0	56.0	2000.0	640.0	919.0	160.0

Summa canisters were the most effective of the several methods employed to sample transpiration gas, producing consistent results between trees. The Tenax tubes produced inconsistent data, and the TAGA was not cost effective. The most cost-effective and efficient method for examining the release of COC from the trees would be to analyze the contaminants in condensate water. Recent statistical comparison of transpiration gas and condensate water shows a strong relationship between the two parameters for both 1122 and TCE ($r > 0.92$ for 1122, $r > 0.93$ for TCE). While further work is necessary to determine the validity of this relationship, it appears that condensate water results may be as effective as transpiration gas results for monitoring COC emissions from *Populus spp.*.

Flux chamber samples of soil emission gases were collected during the summer monitoring event to assess potential mass rates from the localized source of contaminants at the site. Only TCE was detected in three of the eight chambers sampled. Levels ranged from 9.4 to 38 ppbv. The locations of the three chambers were not strictly correlated with known shallow groundwater COC. While 1122 comprises 66% of the total VOC, flux chamber results indicate that contaminants with higher vapor pressures, such as TCE, probably have a mass emission from the soils. If more chambers were collected seasonally and diurnally, an estimate could be made of the volatile loss rate attributable to this mechanism.

OP-FTIR analysis was performed in three areas on the site to address concerns from the EPA Regional Biological and Technical Assistance Group (BTAG) regarding emissions from the site as a whole and the potential creation of exposure pathways. The instrument had minimum detection limits ranging from 1.6 to 53 parts per million-meter (ppm-m) for the COC. All monitoring events showed concentrations below the method detection limits (MDLs) for all of the target compounds.

Leaf and other tree tissue samples were collected for VOC analyses during the spring and summer monitoring events only. This monitoring was discontinued based on time, expense, and the lack of consistency between various laboratory analytical methods for detecting volatiles in tissue. U.S. EPA sampling and analytical methods established that while sampling may produce viable results, leaf tissue samples may not provide the most effective measure of parent compounds or their degradation products. Hybrid poplar TCE degradation products have been identified as: trichloroacetic acid (TCAA), dichloroacetic acid, and trichloroethanol (Newman *et al.*, 1997). These degradation products, especially trichloroethanol, are very difficult to detect instrumentally.

Nine samples of the leaves and buds were collected in early spring. Dichloroacetic acid was the only degradation compound detected at greater than five times the detection limit, with concentrations ranging from 0.34 to 0.88 μg/L in the extract. The water blank contained 0.49 μg/L, indicating that the dichloroacetic acid may have been a from an unknown source. 1122 was detected in one leaf tissue extract at 15 μg/kg wet weight. In the summer, three of seven leaf samples from three trees contained TCAA, at concentrations from 37 to 150 μg/kg wet weight. 1122 was detected in two of seven samples, ranging from 110 to 790 μg/kg wet weight. Concentrations of 1122 appeared to increase seasonally from spring to summer sampling periods, but these results are not definitive. Further seasonal tissue analyses may be performed if methodologies are improved.

Sap flow measurements were taken during the spring, summer, and fall sampling periods. Multiple factors may have influenced these measurements; therefore, results may not accurately represent actual flow rates. For example, moisture build up in the collars, poor electrical connections, or poor collar fit may have limited the quality of the data from the summer sampling period. Given these factors, sap flow rates appear to increase throughout the season and correspond with solar radiation data, as expected. The approximate average pumping rate per tree was 3.33 L/day in the spring, 6.78 L/day in the summer, and 48.9 L/day in the fall. Further seasonal sap flow data will be collected to confirm daily and seasonal trends.

Lysimeter readings from 2.3 meters bgs show both TCE and 1122, indicating that the tree roots are exposed to the COC. Lysimeter 1 produced readings of 5.2 μg/L for TCE, 14 μg/L for 1122, and 1.7 μg/L for chloroform. Lysimeter 2 produced readings of 29 μg/L for TCE, 36 μg/L for 1122, and ND for chloroform.

A comparison of water level contours from April 1997 and May 1997 shows a gradient of groundwater flow towards the phytoremediation test plot with a groundwater depression of approximately 0.1 meter. Apparently, the direction of groundwater flow has moved perpendicular to the land surface contour. The preliminary data strongly suggest that the trees influence typical groundwater flow. While there was no decrease in VOCs during the three 1997 seasonal sampling events, there was a significant decrease in concentrations across the lysimeters and wells transecting the area. Values of total VOCs decreased from 260,000 μg/L to non-detectable in a distance of less than 46 meters. At present, there are insufficient data to explain such a precipitous decline within the tree planting area. Additional sampling is planned to address such questions.

Nematode samples were collected in October 1997 to assess the health and functioning of the soil community. The data were compared to data collected from the site before trees were planted for the phytoremediation project. It appears that the total number of nematodes has increased with the presence of the trees. While the reasons for this increase are currently unknown, it is possible that the trees provide improved habitat for the nematodes or have mitigated the circumstances depressing the community.

CONCLUSIONS

VOCs and their degradation products were detected in transpiration gas, condensate, and leaf tissue, indicating that the trees are removing or degrading COC at the site. While the precise mechanisms for this removal or degradation are still being investigated, detection of these compounds strongly suggests that site objectives are being met. The flux chamber sampling shows measurable TCE off-gassing from the ground surface in the study area, necessitating the inclusion of a soil emissions factor in future mass balance calculations. Sap flow rates and surficial ground water levels indicate possible drawdown occurring, indicating possible containment of groundwater flow in the future. Comparisons of the current nematode population with previously collected samples show that the soil community has increased since the trees were planted, suggesting that the trees have improved the soil community.

Future Sampling. Based on the results generated from the first and second years of monitoring, the following future sampling is planned: 1) long term groundwater monitoring, elevation and chemistry; 2) installation of microwells and/or diffusion samplers to determine the concentration of COC in the rhizosphere relative to transpiration gas, condensate water and possibly tree sap results; 3) continued soil microbial studies; 4) leaf burning study to understand the potential impact of COC emitted from leaves in the study area. Resolution of this issue is necessary to address the concerns of the site owner and local community; 5) replacement of unhealthy trees with similar hybrid and/or native poplar species which may be more suited to site conditions; 6) examination of carbon dioxide levels inside the bag used for sampling transpiration gas to determine when and if stomatal closing occurs.

REFERENCE

Newman, L.A., S.E. Strand, N. Choe, J. Duffy, G. Ekuan, M. Ruszaj, B. B. Shurtleff, J. Wilmoth, P. Heilman, and M.P. Gordon. 1997. "Uptake and Biotransformation of Trichloroethylene by Hybrid Poplars." *Environ. Sci. Technol. 31*: 1062-1067.

PHYTOREMEDIATION OF DISSOLVED-PHASE TRICHLOROETHYLENE USING MATURE VEGETATION

William J. Doucette, Bruce Bugbee and Shannon Hayhurst
(Utah State University, Logan, UT)
William A. Plaehn and Douglas C. Downey (Parsons ES, Denver, CO)
Sam A Taffinder (Air Force Center Environmental Excellence, San Antonio, TX)
Robert Edwards (Booz, Allen and Hamilton, San Antonio, TX)

ABSTRACT: A field study was conducted to determine if existing, mature vegetation is involved in the uptake, volatilization and/or metabolism of trichloroethylene (TCE) from a shallow contaminated groundwater plume. Four on-site plants were sampled, three in the plume area and one outside. One additional off-site control plant was also sampled. Groundwater concentrations of TCE below the vegetation ranged from 0.4 to 90 mg/L. Flow-through glass chambers were used to collect transpiration gases for TCE analysis. Leaf, stem and root samples were collected and analyzed for TCE and 3 metabolites (2,2,2-trichloroethanol [TCEt], 2,2,2-trichloroacetic acid [TCAA] and 2,2–dichloroacetic acid [DCAA]) identified in previous laboratory studies. Measurable levels of TCE were found in 7 of 15 transpiration gas samples and 2 of 3 apparatus blanks. TCE (6 of 34 samples), TCEt (10 of 31), TCAA (22 of 27), or DCAA (3 of 27) were identified in all four plants from the site. The off-site control plant showed no measurable levels of TCE or its metabolites, suggesting that the plant thought to be outside of the plume area had been exposed to TCE. Yearly precipitation data and root distribution imply that direct groundwater use by existing vegetation is less than that obtained from precipitation.

INTRODUCTION

The potential effectiveness of phytoremediation for chlorinated solvents, such as TCE, in shallow contaminated soils and groundwater has been examined in various laboratory studies (Anderson and Walton 1995, Walton and Anderson 1990, Narayanan et al. 1996, Schnabel et al. 1997, Newman et al. 1997). In addition, several field scale phytoremediation studies using planted hybrid poplars have been recently initiated. To date, no studies have examined the potential impact of existing vegetation on TCE contaminated groundwater.

This paper presents the results of an Air Force Center for Environmental Excellence sponsored study designed to determine if existing, mature trees and shrubs are involved in the uptake, transpiration and/or metabolism of TCE from a shallow contaminated groundwater plume located at Cape Canaveral Air Station (CCAS), Florida.

SITE DESCRIPTION

The site of interest at CCAS is the Ordnance Support Facility denoted CCAS Site 1381. From 1968 to 1977, the facility housed an in-place cleaning laboratory where acids and chlorinated solvents were used to clean metals with the use of drip tanks. It is believed that the chlorinated solvent plume at CCAS Site 1381 is a result of accidental spills and poor handling practices of the drummed solvents. The Central OU resides on a shallow surficial aquifer with measured depth to groundwater between 3.5 and 4.8 feet above mean sea level, and approximately 5 feet below ground surface. Elevated levels of TCE (1 to 10 mg/L) were detected in monitoring wells at the southwestern corner of the facility, as well as *cis*-1,2-dichloroethene and vinyl chloride.

MATERIALS AND METHODS

Plant Tissue Samples. Plant tissue samples were collected from a live oak, castor bean and saw palmetto located above the TCE contaminated groundwater plume. A second live oak located several hundred feet outside of the current plume area was also sampled. Leaf samples were also collected from a third live oak located in Orlando, FL. Because of the differences in physical/chemical properties, separate collection and extraction procedures were used for TCE, TCEt, and the chlorinated acetic acids (TCAA, DCAA). All plant tissue analysis were performed using a capillary column gas chromatography (GC) with an electron capture detector (ECD).

Collection of Plant Samples for TCE. Plant tissue samples (5 grams [g] fresh mass) were placed directly in pre-weighed 40 milliliter (mL) glass vials equipped with Teflon®-lined rubber septa. After the plant tissue samples were added to the vials, the vials were re-weighed to determine the amount of plant tissue collected and then filled with purge-and-trap-grade methanol to minimize volatilization losses during transport. The vials were again weighed to determine the amount of methanol added, capped, and shipped at 4±2 degrees Celsius (°C) for analysis.

Collection of Plant Samples for TCEt, TCAA, and DCAA. Samples of plant tissue (10-30 g fresh mass) were collected for the extraction and analysis of TCEt, TCAA, and DCAA. Tissue samples were placed directly in pre-weighed wide-mouth glass vials equipped with Teflon®-lined lids. After the plant tissue samples were added to the vial, the vials were re-weighed, capped, and shipped at 4±2 °C for analysis. No methanol was added to these vials.

Extraction for TCE. TCE in plant tissue samples was extracted and analyzed using a purge and trap, gas chromatographic method. Prior to analysis, the plant tissue and methanol mixture was agitated for 24 hours in a rotary tumbler. A 250 µl aliquot of the methanol extract was removed from the sealed vial with a syringe and injected into the purge and trap vessel where it was diluted with 20 mL of deionized water.

Extraction for TCEt. Leaf tissue samples were flash frozen in liquid nitrogen and macerated in a mortar. The "woody" tissues (stems and roots) were macerated using a coffee grinder. Approximately 5 g of "processed" tissue were weighed into 60 mL Teflon® centrifuge tubes and combined with 15 mL of a 0.1 N sulfuric acid 10% NaCl solution. The centrifuge tubes were shaken for 10 min on a reciprocating shaker and then centrifuged for 10 min at 9750 rpm. The supernatant was transferred from the Teflon® centrifuge tube to a glass centrifuge tube. This extraction procedure was performed a total of 3 times, each time combining the supernatant into the same glass centrifuge tube. The combined aqueous extract was subsequently extracted 3 additional times with 7 mL of methyl tert-butyl ether (MTBE) by shaking for 5 min followed by centrifugation for 5 min at 2500 rpm. The supernatants from the triplicate MTBE extractions were combined and brought to exactly 25 mL. The extract was dried over 2 g of anhydrous sodium sulfate and then analyzed by direct injection GC/ECD.

Extraction for TCAA and DCAA. Tissue samples were "processed" as previously described for TCEt. Approximately 5 g of "processed" tissue were weighed into 60 mL Teflon® centrifuge tubes and combined with 15 mL of a 0.25 N sodium hydroxide solution. The centrifuge tubes were shaken for 10 min using a reciprocating shaker and then centrifuged for 10 min at 9750 rpm. The supernatant was transferred from the Teflon® centrifuge tube into a glass centrifuge tube. The extraction procedure was performed a total of three times, each time combining the supernatant into the same glass centrifuge tube. The combined aqueous extracts were acidified to pH 1 with 50% sulfuric acid and subsequently extracted three additional times with 7 mL of MTBE by shaking for 5 min followed by centrifugation for 5 min at 500 rpm. The supernatant from the triplicate MTBE extractions were combined and brought to exactly 25 mL. The extract was dried over 2 g of anhydrous sodium sulfate. A 2 mL aliquot of the dried extract was derivatized with diazomethane and analyzed by GC/ECD.

Plant Transpiration Samples. To determine the potential flux of TCE from leaves to the atmosphere, 3 transpiration gas samples were collected from each of the 3 test species in the contaminated area, and from a live oak plant located northwest of the main site in an area originally thought to be uncontaminated. Three chamber blanks also were collected in addition to duplicates and breakthrough traps. A schematic of the flow-through sampling apparatus is shown in Figure 1.

A glass chamber was placed over a representative section of each plant and sealed on the open end with closed-cell foam, latex silicon sealant, and electrical tape to produce a flexible, yet tight seal around the stem and chamber. The chambers were fixed in the same location during replicate sample collection. To prevent TCE emitted from the soil surface to be drawn into the sampling apparatus, compressed breathing air was used to purge the chambers. All tubing and connections prior to and after the chamber were constructed of stainless steel. Typically, samples were collected for 30 minute intervals at 200 mL/min using portable sampling pumps. This resulted in a total sample volume of about 5 L. Subsampling was necessary because of the high flow rates (5 to 10 liters per minute [L/min]) required to maintain the humidity level in the chamber equivalent to that measured outside the chamber. Rapid flow rates of air through the chamber also prevented unnatural accumulation of TCE near the plant leaves, and minimized condensation of transpired water on the inside of the chamber. Tenax® was used as the sorbent for the TCE traps because of its high sorption capacity for volatile chlorinated organics and its low affinity for water. Silica gel traps were used to determine the amount of water transpired. Humidity, light, and temperature were measured during the sampling. After sampling, the Tenax® sorbent traps were capped with stainless steel Swagelok® caps fitted with Teflon® ferrules and shipped for analysis. Samples were shipped at ambient temperature to prevent moisture from condensing in the traps.

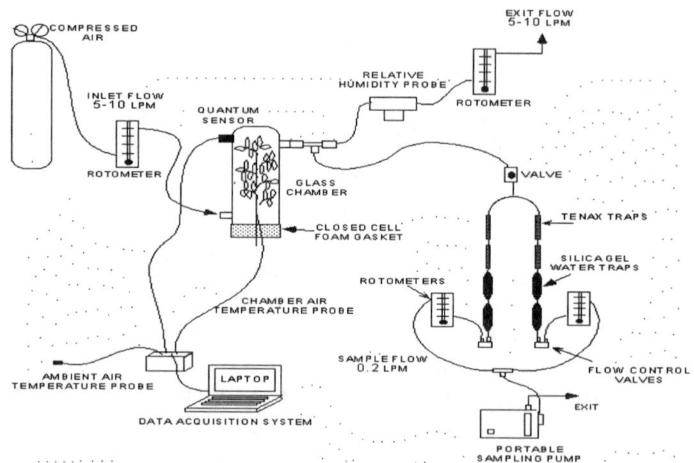

FIGURE 1. Schematic of transpiration gas sampling.

Additional Samples. Soil cores were obtained near each plant species to determine the root length and area density within the unsaturated soil column. Groundwater and soil samples were collected and analyzed for TCE and the metabolites previously mentioned. Soil gas and surface flux measurements also were collected near each plant to quantify the amount of TCE volatilizing directly from the groundwater and soil surface.

RESULTS

The flow-through glass chamber used to collect transpiration gas samples was specifically designed to provide realistic environmental conditions to the portion of the plant being sampled. Based on ambient temperature and humidity measurements, and the mass of water collected in the silica gel traps, the portion of the plant covered by the chamber transpired normally during sample collection.

Measurable levels of TCE were found in 7 of 15 transpiration gas samples analyzed by GC/MS. However, TCE was also found in 2 of the 3 apparatus blanks collected. With the small sample set and low amounts of TCE collected, no statistical difference between samples and blanks was found. Larger Tenax® traps could be used in future studies to increase the sensitivity of this measurement.

Table 1 summarizes the results of the plant tissue analysis. TCE, TCEt, TCAA, or DCAA were identified in all plant tissue types (root, stem and leaf) and in all three species including the on-site live oak believed to be outside the plume area. Metabolite concentrations were generally higher than TCE. The highest concentrations of TCE (48 μg/kg dry weight) were found in the root samples while the highest concentrations of the metabolites (TCAA at 1107 μg/kg dry weight) were found in the leaf and stem samples.

TABLE 1. Maximum concentrations (μg/kg dry weight) of TCE and its metabolites in plant tissue samples collected at CCAS June 10-11, 1997

Species	tissue	TCE*	TCEt*	TCAA*	DCAA*
castor bean	root	12 (1/3)	<MDL (0/2)	<MDL (0/2	<MDL (0/2)
	stem	<MDL (0/3)	<MDL (0/2)	207 (2/2)	<MDL (0/2)
	leaf	<MDL (O/3)	262 (2/2)	1107 (2/2)	<MDL (0/2)
saw palmetto	root	48 (2/2)	<MDL (0/2)	131 (2/2)	<MDL (0/2)
	stem	2 (1/3)	<MDL (0/2)	394 (2/2)	<MDL (0/2)
	leaf	<MDL (0/3)	86 (1/2)	125 (2/2)	<MDL (0/2)
live oak inside the plume	root	<MDL (0/2)	<MDL (0/2)	86 (2/2)	275 (2/2)
	stem	2 (2/3)	<blank* *(0/2)	36 (1/2)	<MDL (0/2)
	leaf	<MDL (0/3)	273 (4/6)	189 (3/3)	723 (1/3)
live oak outside the plume	root	<MDL (0/3)	<MDL (0/2)	63 (1/2)	<MDL (0/2)
	stem	<MDL (0/3)	<blank* *(0/2)	38 (2/3)	<MDL (0/3)
	leaf	<MDL (0/3)	295 (3/5)	132 (3/3)	<MDL (0/3)
Total	All tissue	48 (6/34)	295 (10/32)	1107 (22/27)	723 (3/27)

* Maximum concentration detected in μg/kg dry plant (number of samples >MDL/total samples)
**Value was lower than the associated methanol blank

Detection of TCE and its metabolites in the live oak thought to be outside the plume area suggested that it had been previously exposed to TCE. To further investigate this possibility, a second round of leaf samples was collected on **July 21, 1997** from the two previously sampled on-site live oaks and a third live oak located off-site in Orlando, FL. The off-site control showed no measurable levels of either TCE or its metabolites, while measurable levels of TCAA were found in both of the on-site live oaks (96-674 μg/kg dry weight). As previously suggested, accumulation of TCAA in the leaf tissue, may be the result of previous exposure to TCE contaminated groundwater as live oak trees typically retain their leaves for two growing seasons. While no information is available regarding the stability of

TCAA in leaf tissue, contaminants are known to be sequestered in the vacuoles of the leaf. Additional sampling is needed to address this hypothesis.

Yearly precipitation patterns (Mallander, 1990) and root distribution (70% in top 2 ft, 90% in top 4 ft) imply that direct groundwater use by existing vegetation is less than that obtained by precipitation. This may reduce the impact of vegetation on the removal of TCE at this site. Similar measurements at more arid sites, where direct groundwater use by vegetation is typically greater, must be made before conclusions regarding the potential usefulness of phytoremediation for chlorinated solvents can be made.

REFERENCES

Anderson, T. A., and Walton, B. T. (1995). "Comparative Fate of ^{14}C Trichloroethylene in the Root Zone of Plants from a Former Solvent Disposal Site." *Environmental Toxicology and Chemistry*, 14(12), 2041-2047.

Mallander, J. L. (1990). "Climate of the Kennedy Space Center and Vicinity." NASA Technical memorandum No. 103498.

Narayanan, M., Russell, N. K., Davis, L. C., and Erickson, L. E. "Fate and Transport of Trichloroethylene in a Chamber with Alfalfa Plants: Experimental and Modeling Studies." *1996 HSRC/WERC Joint conference*, Albuquerque, NM.

Newman, L. A., Strand, S. E., Choe, N., Duffy, J., Ekuan, G., Ruszaj, M., Shurtleff, B. B., Wilmoth, J., Heilman, P., and Gordon, M. P. (1997). "Uptake and Biotransformation of Trichloroethylene by Hybrid Poplars." *Environmental Science and Technology*, 31(4), 1062-1067.

Schnabel, W.E., A.C. Dietz, J.G. Burken, J.L. Schnoor and P.J. Alvarez. 1997. Uptake and transformation of trichloroethylene by edible garden plants. *Water Research.* 4:816-824.

Walton, B. T., and Anderson, T. A. (1990). "Microbial Degradation of Trichloroethylene in the Rhizosphere: Potential Application to Biological Remediation of Waste Sites." *Applied and Environmental Microbiology*, 56(4), 1012-1016.

EVALUATION OF TAMARISK AND EUCALYPTUS TRANSPIRATION FOR THE APPLICATION OF PHYTOREMEDIATION

Robert W. Tossell (Beak International Incorporated, Guelph, Ontario, Canada)
Kinsley Binard, ***Luisa Sangines-Uriarte***, Michael T. Rafferty (Geomatrix Consultants Inc., San Francisco, California, USA)
Neil P. Morris (Beak International Incorporated, Brampton, Ontario, Canada)

ABSTRACT: A multi-phase study is underway to determine if trees can be used as an alternative to, or in conjunction with, a slurry wall to control flow of arsenic-bearing groundwater from a Site in East Palo Alto, California to a tidal marsh in San Francisco Bay. Results of a greenhouse study indicated that Tamarisk (*Tamarix parviflora*) and Eucalyptus (*Eucalyptus camaldulensis*) can tolerate the elevated concentrations of dissolved arsenic and sodium in soil and groundwater found at the Site. The relatively high concentrations of arsenic in soil and water did not greatly affect tree water use or growth. However, the high concentrations of sodium used in this study had a negative effect on tree growth and water use. The water use and growth data from this greenhouse study, along with published water use rates of Tamarisk and Eucalyptus, were used to estimate the number of trees required to alter hydraulic gradients at a pilot-scale area of the Site.

INTRODUCTION

A greenhouse study was conducted to evaluate phytoremediation as a means of providing hydraulic control of arsenic-bearing groundwater at the Site. The greenhouse study was one component of this multi-phase feasibility study which began in 1996. It was hypothesized that hydraulic control may be obtained if plants, primarily trees, could transpire sufficient quantities of groundwater to alter the hydraulic gradient at the Site boundary, thereby limiting the migration of arsenic-bearing groundwater off-Site. This approach is being considered as both a stand-alone remedial method and as a supplementary water management method used in conjunction with other proposed remedial strategies such as containment using a slurry wall.

Objective. The objective of this study was to evaluate the effect of arsenic and sodium found in Site groundwater and soil on the growth, health and water use (transpiration) rates of candidate species (Tamarisk and Eucalyptus trees). If selected tree species were found to tolerate Site conditions and transpire significant quantities of water, these tree species would be further evaluated under field conditions as part of a pilot-scale study.

Site Description: The Site is underlain by fine-grained (silts and clays) and coarse-grained (sands and gravels) alluvial and marine sediments. The shallow groundwater zone extends to approximately 18.5 metres below ground surface (m bgs) and consists of two water-yielding units. These units are referred to as the upper shallow groundwater zone, which is found between 1.5 and 4.6 m bgs, and the lower shallow groundwater zone, which is found at 7.7 to 10.8 m bgs. The flow in the shallow groundwater zones is generally in a south-easterly direction, toward the tidal marsh. Arsenic is present in the groundwater at concentrations ranging from 0.01 to 0.2 milligrams per litre (mg/L) over most of the area, but can be as high as 200 mg/L in localized areas. Arsenic is the only chemical of concern, but other constituents exist at the Site including other metals, pesticides and chlorinated solvents. Total dissolved solids (TDS [650 mg/L to 4,800 mg/L]), sodium concentrations (150 to 2,000 mg/L) and specific conductivity of groundwater are elevated due to saltwater intrusion due to tidal effects and local groundwater extraction in the area.

MATERIALS AND METHODS

A completely randomized block (CRB) fractional factorial design, was used in this study and is summarized in Table 1. Main experimental treatments included five (5) variables including: tree type (Tamarisk and Eucalyptus); arsenic concentration (as sodium arsenate); sodium concentration (as sodium chloride); phosphate treatment (i.e. phosphate fertilizer); and soil type (treated and untreated Site soil). A control was used (water and nutrients only in untreated soil) to collect data on baseline performance of trees unaffected by treated soil, sodium and arsenic. The factorial design included two levels of each of the variables for a total of 9 treatments for each tree species. Each treatment was replicated four (4) times to measure variability within treatments and for the purpose of conducting statistical analyses. The CRB was established within an auto-ventilated greenhouse with dimensions of 12 m by 24 m. A HOBO™ temperature/relative humidity sensor and data logger were used to monitor greenhouse air temperature and relative humidity (RH).

Bare-root tamarisk saplings (approximately 0.9 to 1.2 m tall) and potted Eucalyptus seedlings (approximately 0.3 to 0.6 m tall) were planted in 13 L pots with drainage holes at the base. Each of these planting pots was placed in a larger (19 L) pot to contain the water and solutions (as outlined in Table 1) which were supplied to the trees as shown in Figure 1. A mariotte constant head reservoir (4 L) was used to supply the stock solution to the plant on-demand by maintaining 5 cm of standing water in each larger pot. The constant water head in the large pot simulated a water table.

Water use was measured every 3 to 4 days during the study by determining volume loss in the mariotte reservoir. Tree height and girth (diameter) were measured at the beginning of the study, and every 4 weeks there after until the termination of the study.

Statistical analysis included the Kolmogorov-Smirnov test to test for normality, analysis of variance (ANOVA) using a confidence interval of 95% and least squares regression to determine relationships of dependent and independent variables.

RESULTS AND DISCUSSION

Temperature and Relative Humidity: Daily average temperatures ranged from 17.7 (average daily minimum) to 40.9 (average daily maximum) degrees Celsius (°C) and averaged 26.7 °C over the entire study period. Daily RH had significant diurnal variability which ranged from 12.3% to 74.6% and averaged 45.2% over the study period.

TABLE 1: EXPERIMENTAL TREATMENTS: GREENHOUSE STUDY EAST PALO ALTO, CALIFORNIA

Treatment Number		Treatments			
Tamarisk (*Tamarix parviflora*)	Eucalyptus (*Eucalyptus camaldulensis*)	Soil Type	Sodium (mg/L)	Arsenic (mg/L)	Phosphate (mg/L_{Soil})
1	10	Treated	2000	25	None
2	11	Untreated	2000	2.5	None
3	12	Untreated	2000	25	100
4	13	Treated	2000	2.5	100
5	14	Untreated	500	25	None
6	15	Treated	500	2.5	None
7	16	Treated	500	25	100
8	17	Untreated	500	2.5	100
9	18	Untreated	None	None	100

Notes:

Treated soil: composed of 60% (by volume) untreated soil, 20% treated soil (10% cement), and 20% peat.

Untreated soil: composed of 80% (by volume) soil from untreated stockpile, 20% peat.

Phosphate Treatments 20 g of 0-20-0 (100 mg/L_{Soil} equivalent) solid fertilizer added to pots at the time of planting.

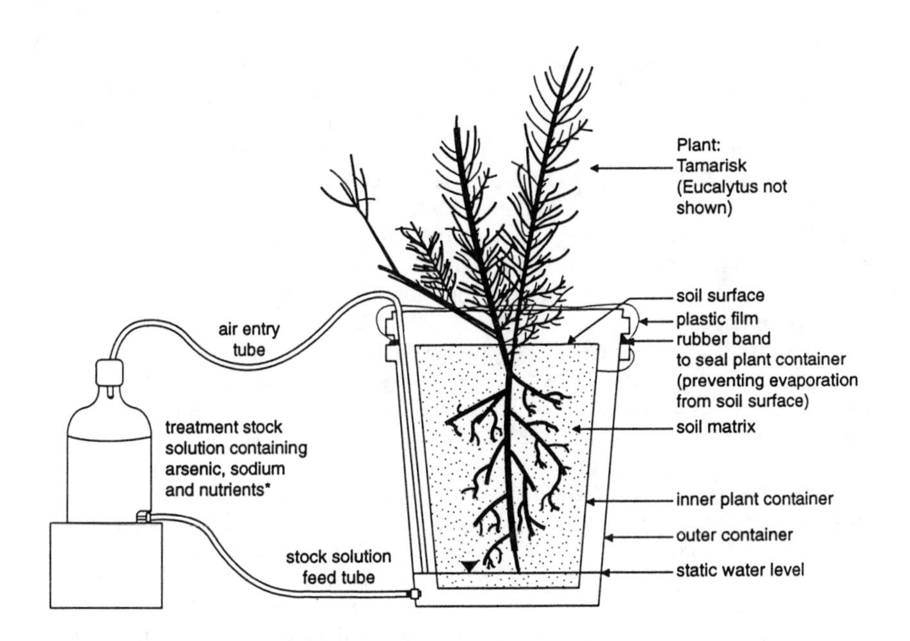

FIGURE 1. Schematic of potted plant and water supply vessel.

Water Use: Typically the rate of water use of Tamarisk and Eucalyptus increased rapidly at the beginning of the study compared to the latter portion of the study. Results of average daily water use rates (averaged by treatment and over the study duration) for Tamarisk and Eucalyptus are presented in Figure 2A. Over the course of the study, the Tamarisk used significantly more water (approximately 0.5 L/day) compared to Eucalyptus (approximately 0.3 L/day) which may be a reflection of the relative species demand for water. However, Eucalyptus water use (normalized to shoot mass at the end of the study) was substantially higher, averaging approximately 2 times the Tamarisk rate (Figure 2A). The Tamarisk rate ranged from less than 3.3 to 4.2 (g water per g dry shoot weight per day [$g\ g^{-1}\ d^{-1}$]) while Eucalyptus ranged from 5.3 to 6.8 ($g\ g^{-1}\ d^{-1}$). Hagemeyer and Waisel (1988) and Anderson (1981) reported rates of water use for Tamarisk which ranged from 8 to 28 ($g\ g^{-1}\ d^{-1}$) depending on time of day. These studies also reported greater average temperatures (30 °C) and/or a greater photoperiod, which may explain why the reported water use rates are approximately 2.5 times those found in this study. Futhermore, since tree growth was so rapid, pot size may have limited both plant growth and water use towards the end of the study.

Sodium concentration had the greatest influence on water use of both Tamarisk and Eucalyptus. Water use of both species was substantially lower for those trees exposed to high sodium treatments (2,000 mg/L). Tamarisk in treatment groups 1 to 4 (i.e. high sodium) had low rates of daily average water use (0.401 to 0.415 L/tree/day), while Tamarisks in treatments 5 to 8 (low sodium treatment) had higher daily average water use rates (0.495 to 0.579 L/tree/day). The Tamarisk control group (treatment 9) had the highest water use (0.827 L/tree/day) and represented growing conditions without additions of either sodium or arsenic. Eucalyptus in treatments 10 to 13 (high sodium treatment - 2,000 mg/L) had low rates of water use (0.228 to 0.258 L/tree/day) in contrast to Eucalyptus in treatment groups 14 to 17 (low sodium - 500 mg/L) which had higher water use rates ranging from 0.293 to 0.399 L/tree/day. As with the Tamarisks, the Eucalyptus trees in the control group (treatment 18) had the highest rates of water use (0.515 L/tree/day). Arsenic and phosphorus treatments did not significantly affect water use of Tamarisk or Eucalyptus.

Both Tamarisk and Eucalyptus exhibited an ability to transpire large quantities of water during this study. Sodium treatments had a negative effect on average daily water use of both species. Tamarisk was less affected by high concentrations of sodium than was Eucalyptus.

Height: Average final height of Tamarisk exposed to high sodium (treatments 1 to 4) ranged from 131.4 cm to 150.5 cm while the average final height of trees exposed to low sodium (treatments 5 to 8) ranged from 136.92 cm to 168.70 cm. The greatest average tree height (170.82 cm) was produced by the control group of trees (Figure 2B). The average height of treatment groups 5 (high sodium) and 9 (control) were found to be significantly greater than treatment groups 3 (high sodium and 7 (low Sodium). Although there appears to be a sodium treatment effect, due to the variability in height data, no statistically significant sodium trend was found.

Similarly, no clear trend was found for the Eucalyptus tree height, although the lowest average height was measured for trees in treatment 12 (high sodium) and

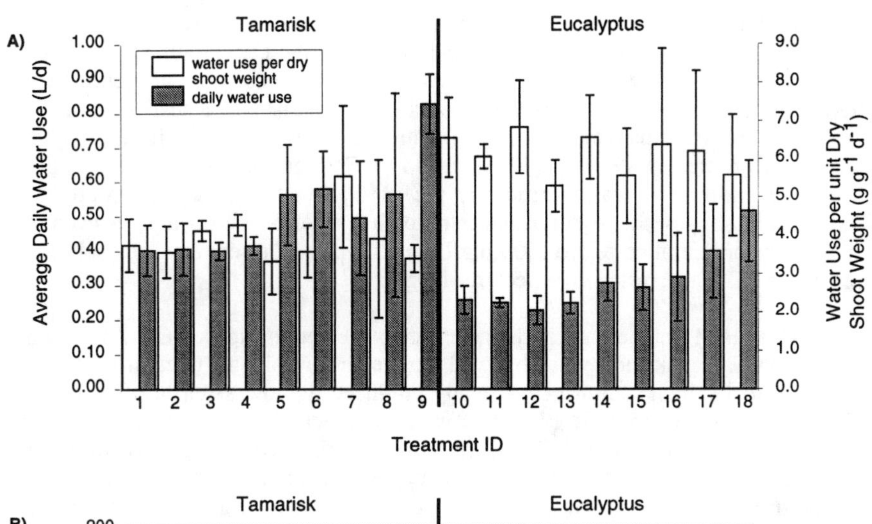

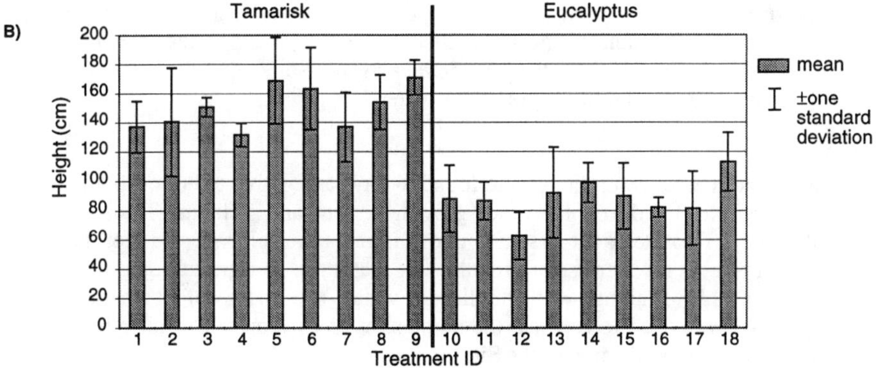

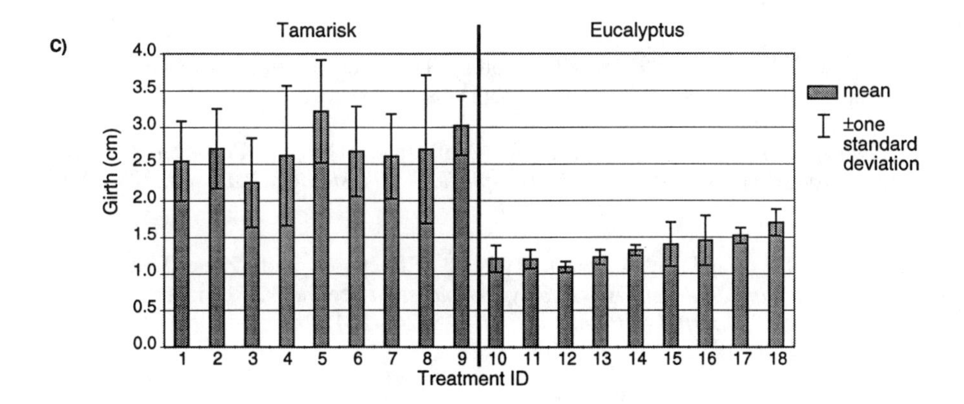

FIGURE 2. Plant response to treatments. (a) Treatment average daily water use and water use per shoot weight. (b) Tree height. (c) Tree girth (diameter).

the greatest average height of Eucalyptus (other than control trees) was treatment 14 (low sodium) (Figure 2B). The average final height of the Eucalyptus control trees was greater than the average heights of all other treatments (this difference was more pronounced for Eucalyptus than it was for Tamarisks). Statistical analysis shows that trees in treatment group 18 (control) had significantly greater heights than trees from treatment groups 12, 16 and 17. Furthermore, average tree height in treatment group 14 was significantly greater than that of treatment group 12. These results suggest that sodium concentration had an effect on Eucalyptus even at the lowest concentration tested (500 mg/L). Soil type and arsenic and phosphorus treatments did not significantly affect tree height.

Girth: Results of final tree girth (diameter) are presented in Figure 2C. As shown no trend in Tamarisk girth was observed. However, Eucalyptus trees in treatment groups 10 to 13 (high sodium) consistently produced the lowest girth. Tree girths of treatment groups 15, 16, 17 and 18 were significantly greater than treatment group 12. Treatment groups 17 and 18 were significantly greater than treatment groups 10, 11, 12 and 13. These results suggest that high sodium concentrations had a significant effect on Eucalyptus tree girth. Marcar and Termaat (1989) found that sodium chloride induced growth reductions in the same species of Eucalyptus used in this study (*Eucalyptus camaldulensis*). Soil type and arsenic and phosphorus treatment did not significantly affect tree girth.

CONCLUSIONS

Aqueous sodium concentrations were found to have a negative effect on both water use (transpiration) and in some cases growth (height and girth) of Tamarisk and Eucalyptus. Treatments of arsenic, phosphorus and soil type did not have a significant effect on water use or growth. The water use and growth data from this greenhouse study, along with published water use rates of Tamarisk and Eucalyptus, were used to estimate the number of trees required to alter hydraulic gradients at a pilot-scale area of the Site.

REFERENCES

Anderson, J.E. 1981. "Factors Controlling Transpiration and Photosynthesis in *Tamarix Chinensis* Lour." *Ecology*, 63(1) 48-56.

Hagemeyer J. and Y. Waisel. 1988. "Influence of NaCl, $Cd(NO_3)_2$ and Air Humidity on Transpiration of *Tamarix aphylla*.." *Physiologia Plantarum*, 75: 280-284.

Marcar, N.E., and A. Termaat. 1989. "Effects of Root-Zone Solutes on *Eucalyptus camaldulensis* and *Eucalyptus bicostata* seedlings: Responses to Na^+, Mg^{2+} and Cl^-." *Journal of Plant and Soil*, 125: 245-254.

PHREATOPHYTE INFLUENCE ON REDUCTIVE DECHLORINATION IN A SHALLOW AQUIFER CONTAINING TCE

Roger W. Lee (U.S. Geological Survey, Austin, Texas)
Sonya A. Jones and Eve L. Kuniansky (U.S. Geological Survey, Austin, Texas)
Gregory J. Harvey (ASC/EMR, Wright-Patterson AFB, Ohio)
Sandra M. Eberts (U.S. Geological Survey, Columbus, Ohio)

ABSTRACT: A field demonstration designed to remediate aerobic shallow ground water (less than 12 ft (3.7 m) deep) containing trichloroethene (TCE) using eastern cottonwood trees (phreatophytes) began in April 1996 at the Naval Air Station, Fort Worth, Tex. Eighteen months after planting the trees, chemical data from ground water indicated concentrations of dissolved oxygen decreased slightly beneath the trees. Changes in concentrations of TCE and its degradation products since 1996 were not measurable, indicating that tree roots had not yet affected ground-water geochemical conditions sufficiently to promote microbially mediated reductive dechlorination. However, TCE is degrading in shallow ground water beneath a mature 19-year old cottonwood tree about 200 ft (61 m) southwest of the cottonwood plantings. A microbially mediated iron-reducing environment capable of degrading TCE has developed from low molecular weight organic acids entering the aquifer from decaying roots or root exudates of the mature cottonwood tree. As the planted cottonwood trees establish more mature root systems, associated microbial activity should flourish, creating reducing environments in the shallow aquifer that could promote reductive dechlorination of the dissolved TCE, similar to the ground-water environment beneath the mature cottonwood tree.

INTRODUCTION

The surficial terrace alluvial aquifer that underlies U.S. Air Force Plant 4 (AFP4) and the adjacent Naval Air Station (NAS), Fort Worth, Tex., contains TCE. Dissolution and transport of TCE and its degradation products have occurred, creating a plume of contaminated ground water that extends beneath AFP4, parts of NAS, and Carswell Golf Course. A phytoremediation demonstration project southeast of AFP4 (Figure 1) is underway to determine if eastern cottonwood trees are effective in removal and degradation of TCE from shallow ground water southeast of the AFP4. The focus of the U.S. Geological Survey research is to determine if the trees change the chemistry and microbiology in the aquifer to the extent that degradation of the chlorinated aliphatic hydrocarbons (CAH) in the shallow ground water could result.

Site Description. The study area is on the northwest side of Fort Worth, Tex. The project area is 25 acres (10 hm^3) just northeast of the Carswell Golf Course. The geology of the site consists of Quaternary terrace alluvial deposits, that include gravel, sand, silt, and clay, overlying the fossiliferous Goodland-Walnut confining unit of Cretaceous age (Kuniansky et al, 1996). Alluvial deposits range from 6 to 15 ft (1.8 to 4.6 m) thick, with about 2 ft (0.6 m) of saturated thickness in water-bearing sand and silt just above the relatively impermeable confining unit. The land surface slopes gently southeast, and sharply southwest from Roaring Springs Road toward Farmers Branch Creek. The slope is nearly flat adjacent to Farmers Branch Creek (Figure 1).

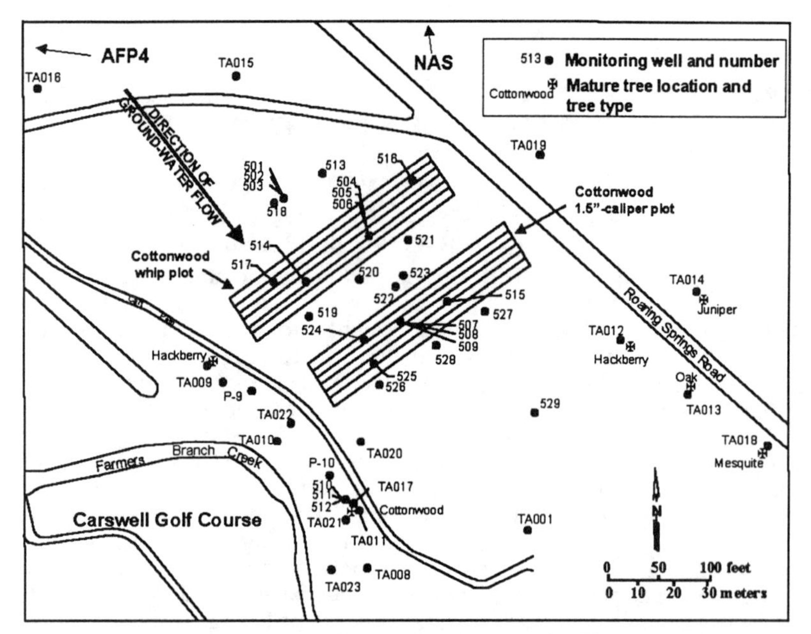

Figure 1. Well locations and tree plantings at the Phytoremediation Demonstration Site, Fort Worth, Tex.

The climate is subhumid, with generally long, hot summers (daily high temperature averaging about 96 °F (36 °C) in July) with short, mild winters (daily low temperature averaging about 34 °F (1 °C) in January). Average annual temperature is 65 °F (18 °C), and average annual precipitation is 32 in. (81 cm).

Two tree plots, approximately orthogonal to ground-water flow, were prepared and planted in April 1996. The plots are about 50 by 250 ft (15 by 76 m) each, consisting of 6 rows of 50 equally spaced plantings. The northernmost plot was planted with less costly eastern cottonwood whips (cuttings). The second plot was planted with 1- to 1.5-in.- (2.5- to 3.8-cm-) caliper (trunk diameter) eastern cottonwood trees.

Chemical Reactions. Aerobic degradation of TCE in ground water is not a factor at this site and will not be discussed. TCE is known to undergo reductive dechlorination under anaerobic conditions by the pathway

trichloroethene (TCE)→dichloroethene (DCE)+Cl →vinyl chloride (VC)+2Cl→ethene+3Cl

```
Trichloroethene                          cis-1,2-Dichloroethene
  Cl    H                                   H    H
    \  /                                     \  /
     C=C   +    H2      =>                   C=C    + HCl  (1)
    /  \                                     /  \
  Cl    Cl                                 Cl    Cl
```

Molecular hydrogen is produced as an intermediate product of anaerobic microbial metabolism (Chapelle, 1993), then consumed in reaction (1).

The efficiency of dechlorination differs depending on redox conditions present in an aquifer (most efficient to least efficient—methanogenesis, sulfate reduction, iron(III) reduction, or aerobic). Thus, an accurate determination of redox conditions in an aquifer is necessary to evaluate the potential for natural attenuation of TCE (Chapelle, 1996). Microbially mediated redox reactions of interest in the shallow aquifer at the demonstration site are

Oxidation of organic matter

$$CH_2O + O_2 => CO_2 + H_2O \tag{2}$$

Iron reduction

$$CH_2O + 4Fe(OH)_3 + 7H^+ => 4Fe^{2+} + CO_2 + 10H_2O + OH^- \tag{3}$$

Sulfate reduction

$$2CH_2O + SO_4{}^{2-} => S^{2-} + 2CO_2 + 2H_2O \tag{4}$$

Methanogenesis

$$2CH_2O => CO_2 + CH_4 \tag{5}$$

(Stumm and Morgan, 1981). All of these reactions require organic carbon (CH_2O) as an electron donor. Organic carbon generally is limited in the shallow aquifer at the site, but labile organic residues could be introduced as decaying roots of phreatophytes or exudates from root systems associated with mature trees.

Determination of the redox environment and the terminal electron-accepting process (TEAP) in an aquifer can be accomplished by several on-site measurements of ground-water chemistry. If appreciable dissolved oxygen (DO) is present in the ground water (more than 2 mg/L), reductive dechlorination is an unlikely process. Measurement of redox pairs Fe^{+2}/Fe^{+3} and $SO_4{}^{-2}/S^{-2}$ using standard methods usually distinguishes between iron(III)-reduction and sulfate-reduction processes. Detection and measurement of CH_4 usually provides an indication of methanogenesis as a TEAP.

These lines of evidence sometimes conflict, which necessitates the measurement of molecular hydrogen to determine the TEAP (Chapelle, 1996). H_2 concentrations range from 0.2 to 0.8 nM for iron(III) reduction, from 1 to 4 nM for sulfate reduction, and from 5 to 15 nM for methanogenesis.

METHODS

In order to determine the geochemical environments of ground water at the site that could affect degradation of the CAHs, the properties and constituents listed in Table 1 were determined on-site or in the laboratory, as indicated. Collection and measurement of the chemical properties and constituents provide data necessary to determine geochemical environments that can produce reductive dechlorination of CAHs.

RESULTS

Concentrations of TCE in ground water for November 1997 (Figure 2) range from less than 50 to more than 700 µg/L, with typical concentrations of 500 to 650 µg/L. The concentrations decrease from northeast to southwest toward Farmers Branch Creek in the direction of decreasing depth to water but not in the direction of ground-water flow. Concentrations of TCE are very low (less than 50 µg/L) in wells adjacent to the mature cottonwood tree about 200 ft (76 m)

TABLE 1. Analyses necessary for evaluation of natural attenuation.
(VOC, volatile organic compounds; GCMS, gas chromatography/mass spectroscopy; DOC, dissolved organic carbon)

Property or Constituent	Method	Procedure
pH	Calomel electrode	On-site method, flow-through cell, field calibration
Oxygen	Chemets[1]	On-site color match (1 to 12 mg/L); colorimetric (0 to 2 mg/L)
Fe^{+2}/Fe^{+3}	Hach[1] method	On-site colorimetric methods for both species of iron
Sulfide	Hach[1] method	On-site colorimetric analysis of sample
Carbon dioxide, methane	Gas chromatography	On-site method, headspace, thermal conductivity
Hydrogen	Gas chromatography	On-site method, headspace analysis, (0.1 to 30 nM)
VOC	GCMS	Laboratory analysis
Alkalinity	Titration	On-site microburet, fixed-endpoint titration
DOC fractions	Lab separation	0.45-μm silver filter to glass bottle, chilled to 4 °C

[1] Any use of trade, product, or firm names is for descriptive purposes only and does not imply endorsement by the U.S. Government.

southwest of the tree plantings. Additionally, concentrations of *cis*-1,2-DCE increased from less than 100 to more than 200 μg/L and VC increased from less than 1 to more than 80 μg/L at this location, indicating that degradation of CAHs occurs in the vicinity of the tree. No degradation is indicated beneath the newly planted cottonwoods.

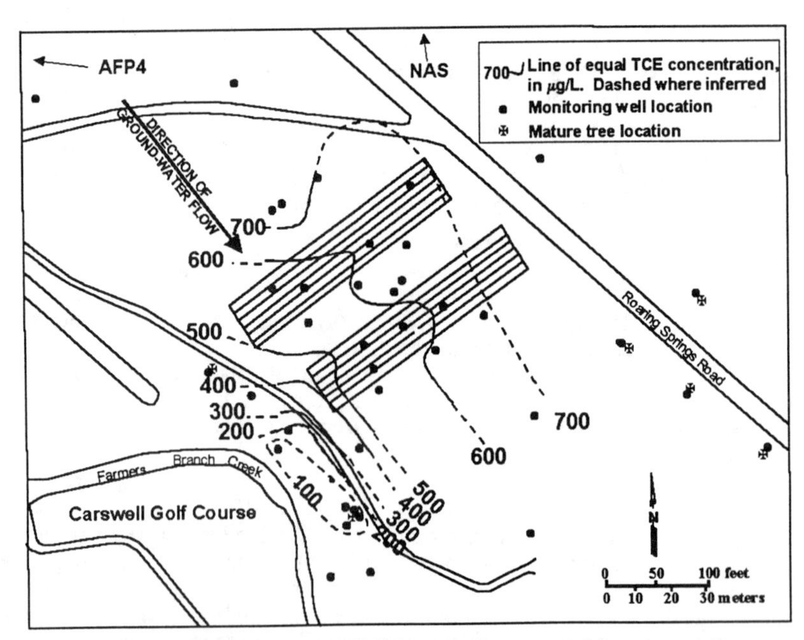

Figure 2. Distribution of TCE in shallow ground water at the Phytoremediation Demonstration Site, November 1997.

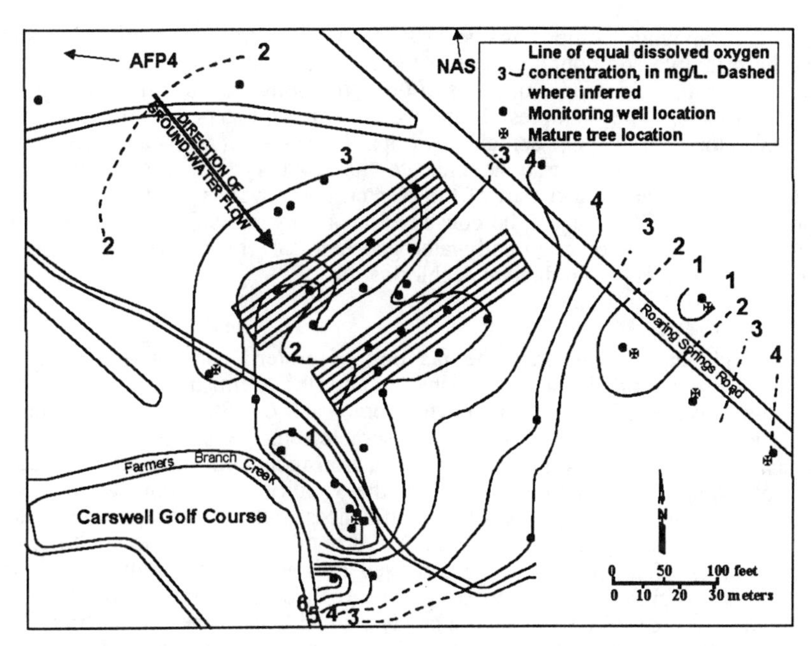

Figure 3. Distribution of DO in shallow ground water at the Phytoremediation Demonstration Site, November 1997.

Concentrations of DO in ground water for November 1997 (Figure 3) range from less than 0.1 to 8.0 mg/L, with typical concentrations of 2.5 to 3.5 mg/L. Concentrations of DO are lowest in the vicinity of the mature cottonwood tree and the area just to the northwest (Figure 3). This indicates that oxygen could be consumed by microbial oxidation of organic matter as in reaction (2). DO is less than 1 mg/L in well TA014, near a mature Juniper tree (Figure 1), indicating a similar process could be occurring in the ground water beneath this tree. DO is somewhat low at the southwestern end of the whip plot, indicating that some oxygen consumption could be occurring here as the tree roots descend to the water table. The largest concentrations of DO are at wells TA023 and TA008 (Figure 1), 8.0 and 5.0 mg/L, respectively. These large concentrations probably resulted from flooding and infiltration of streamflow from Farmers Branch Creek that occurred the week before sampling.

Dissolved iron and dissolved sulfide concentrations typically are less than 0.01 mg/L for most of the wells sampled. Well 511 (Figure 1), near the mature cottonwood tree, contains a dissolved iron concentration greater than 7 mg/L, indicating an iron(III)-reducing environment in the aquifer beneath the tree. Similarly, molecular hydrogen concentrations are typically less than 0.05 nM. However, well 511 has molecular hydrogen concentration of 0.3 nM, supporting the hypothesis of iron(III) reduction as the TEAP at this location in the aquifer.

DOC was sampled from nine wells in the vicinity of the newly planted cottonwood trees and from well 511 adjacent to the mature cottonwood tree. Concentrations range from 0.8 to 1.8 mg/L with typical concentrations of 0.8 to 1.1 mg/L. Wells 511 and 514 contain the largest concentrations of DOC, 1.7 and 1.8 mg/L, respectively. The concentrations indicate that root systems at these locations could be producing labile organic carbon that enters the ground water. Analyses of acid, base, and neutral fractions of hydrophobic and hydrophilic organic compounds composing the DOC (Leenheer and Huffman, 1979) show that the principle compound classes at the site are hydrophobic neutral and acid, and hydrophilic acid. Concentrations of these fractions are less than 0.4 mg/L for most samples. The samples from wells 511 and 514, with the largest concentrations of DOC, contain hydrophilic acids of 0.6 and 0.9 mg/L, respectively. The concentrations indicate that root systems under the mature cottonwood tree and under part of the planted cottonwoods produce hydrophilic acids (less than C5) that can be consumed as an organic substrate by microbes, thus altering the aerobic ground-water geochemical environment to anaerobic conditions conducive to reductive dechlorination.

CONCLUSIONS

Eighteen months after the planting of eastern cottonwood trees, root systems were not sufficiently established to alter the chemistry and microbiology of ground water at the Phreatophyte Demonstration Site. A nearby mature cottonwood tree has produced a measurable change in the ground-water geochemistry, causing oxygen consumption, iron reduction, methane production, and reductive dechlorination of TCE in the vicinity of its root system. Other mature trees in the area show similar influences, but the data are incomplete. Decaying root tissue and root exudates of hydrophilic acids (less than C5) appear to be sources of labile organic carbon used by microbes to produce anaerobic conditions in ground water conducive to reductive dechlorination of TCE. Maturation of root systems under the newly planted trees at the demonstration site should similarly increase the labile organic carbon content of the shallow aquifer, thus stimulating microbial activity to produce anaerobic conditions capable of reductively dechlorinating TCE.

ACKNOWLEDGMENTS

This research project is funded by the Environmental Security Technology Certification Program of the Department of Defense, the Superfund Innovative Technology Evaluation Program of the U.S. Environmental Protection Agency, and the Aeronautical Systems Center/Environmental Management Directorate at Wright-Patterson AFB.

REFERENCES

Chapelle, F.H. 1993. *Ground-Water Microbiology and Geochemistry*. John Wiley and Sons, Inc., New York, 424 p.

______1996. "Identifying Redox Conditions that Favor the Natural Attenuation of Chlorinated Ethenes in Contaminated Ground-Water Systems." *Symposium on Natural Attenuation of Chlorinated Organics in Ground Water.* September 1996, EPA/540/R–96/509, p 17–20.

Kuniansky, E.L., Jones, S.A., Brock, R.D., and Williams, M.D. 1996. *Hydrogeology at Air Force Plant 4 and Vicinity and Water Quality of the Paluxy Aquifer, Fort Worth, Texas.* U.S. Geological Survey Water-Resources Investigations Report 96–4091, 41 p.

Leenheer, J.A., and Huffman, E.W.D., Jr. 1979. *Analytical Method for Dissolved-Organic Carbon Fraction.* U.S. Geological Survey Water-Resources Investigations 79–4, 16 p.

Stumm, W., and Morgan, J. J. 1981. *Aquatic Chemistry.* John Wiley & Sons, Inc., New York, 780 p.

AUTHOR INDEX

This index contains names, affiliations, and book/page citations for all authors who contributed to the six books published in connection with the First International Conference on Remediation of Chlorinated and Recalcitrant Compounds, held in Monterey, California, in May 1998. Ordering information is provided on the back cover of this book.

The citations reference the six books as follows:

1(1): Wickramanayake, G.B., and R.E. Hinchee (Eds.). 1998. *Risk, Resource, and Regulatory Issues: Remediation of Chlorinated and Recalcitrant Compounds*. Battelle Press, Columbus, OH. 322 pp.

1(2): Wickramanayake, G.B., and R.E. Hinchee (Eds.). 1998. *Nonaqueous-Phase Liquids: Remediation of Chlorinated and Recalcitrant Compounds*. Battelle Press, Columbus, OH. 256 pp.

1(3): Wickramanayake, G.B., and R.E. Hinchee (Eds.). 1998. *Natural Attenuation: Chlorinated and Recalcitrant Compounds*. Battelle Press, Columbus, OH. 380 pp.

1(4): Wickramanayake, G.B., and R.E. Hinchee (Eds.). 1998. *Bioremediation and Phytoremediation: Chlorinated and Recalcitrant Compounds*. Battelle Press, Columbus, OH. 302 pp.

1(5): Wickramanayake, G.B., and R.E. Hinchee (Eds.). 1998. *Physical, Chemical, and Thermal Technologies: Remediation of Chlorinated and Recalcitrant Compounds*. Battelle Press, Columbus, OH. 512 pp.

1(6): Wickramanayake, G.B., and R.E. Hinchee (Eds.). 1998. *Designing and Applying Treatment Technologies: Remediation of Chlorinated and Recalcitrant Compounds*. Battelle Press, Columbus, OH. 348 pp.

Abraham, Brian M. (Target Laboratories, Inc./USA) 1(3):171

Abrajano, Teofilo (Memorial University of Newfoundland/ Canada) 1(1):67

Abriola, Linda M. (University of Michigan/USA) 1(2):91

Accashian, J.V. (Camp Dresser & McKee Inc./USA) 1(3):281

Acha, Victor (Catholic University of Louvain/Belgium) 1(4):135

Acree, Steve D. (U.S. EPA/USA) 1(6):9

Adams, Mary-Linda (Resolution Resources, Inc./USA) 1(2):155

Adams, Timothy V. (ENSR Consulting & Engineering/USA) 1(5): 97, 103

Adewuyi, Yusuf G. (North Carolina A&T State University/USA) 1(5):467

Adrian, Lorenz (Technical University Berlin/Germany) 1(4):77

Agathos, Spyros (Catholic University of Louvain/Belgium) 1(4):135

Ahn, Iksung (Cornell University/ USA) 1(4):51
Aines, Roger D. (Lawrence Livermore National Laboratory/ USA) 1(5):133, 359
Albiter, Veronica (Universidad Nacional Autónoma de México/ Mexico) 1(6):177
Allan, Jane (University of Stuttgart/ Germany) 1(2):107
Allan, Jason (Beak International Inc./USA) 1(4):227
Alleman, Bruce (Battelle/USA) 1(1):19
Allen, Tim J. (Raytheon Missile Systems Co./USA) 1(5):169
Allen-King, Richelle (Washington State University/USA) 1(6):39
Alpay, H. Erden (Ege University/ Turkey) 1(5):83
Alphenaar, Arne (Tauw Environmental Consultants/The Netherlands) 1(3):105
Ananda, M.M.A. (University of Nevada/USA) 1(1):91
Anderson, Jack (RMT, Inc./USA) 1(6):249
Anderson, Jack W. (Columbia Analytical Services/USA) 1(1):1
Andersson, Rolf (Umeå University/ Sweden) 1(6):195
Anthony, Thomas (University of Waterloo/Canada) 1(2):149
Anzia, Mary Jo (ENVIRON International Corp./USA) 1(1):199
Appel, Christine (Eaton Corp./USA) 1(2):161
Aravena, Ramon (University of Waterloo/Canada) 1(1):67
Archabal, Steven R. (Parsons Engineering Science, Inc./USA) 1(5):409
Ardito, Cynthia (Duke Engineering Services/USA) 1(1):59
Armstrong, John (Florida Dept. of Environmental Protection/USA) 1(1):229
Arnold, Robert G. (University of Arizona/USA) 1(4):21, 1(5):473
Ashburn, Arnold (Radian International LLC/USA) 1(2):187
Atalay, Ferhan Sami (Ege University/Turkey) 1(5):83
Atalay, Süheyda (Ege University/ Turkey) 1(5):83
Atamańczuk, Tomasz (Institute of Chemical Engineering/Poland) 1(5):43
Atchue, Joseph A. (Tetra Tech EM, Inc./USA) 1(1):157
Aubertin, Howie (U.S. Air Force/ USA) 1(1):205
Aurelius, Marc (Radian International LLC/USA) 1(2):187
Aveggio, John J. (SHN Consulting Engineers & Geologists Inc./ USA) 1(5):231
Aziz, Carol (Groundwater Services/ USA) 1(3):237

Babcock, James W. (Malcolm Pirnie, Inc./USA) 1(3):159
Baehr, John (U.S. Army Corps of Engineers/USA) 1(5):437
Baek, Nam H. (Occidental Chemical Corp./USA) 1(6):15
Bailey, Andrew C. (First Union Corp./USA) 1(5):147
Baker, Brady (U.S. Air Force/USA) 1(5):409
Baker, John Alan (Waste Management Technology Center/USA) 1(5):371
Baker, Ralph S. (ENSR Consulting & Engineering/USA) 1(5):175, 377, 397
Ball, Roy O. (ENVIRON International Corp./USA) 1(1):73, 85, 181, 199

Ballard, Deborah L. (Burns & McDonnell Waste Consultants, Inc./USA) 1(5):247
Ballbach, J. Daniel (Landau Associates, Inc./USA) 1(1):247
Balshaw-Biddle, Kathy (Rice University/USA) 1(5):109
Banceu, Claudia (University of Guelph/Canada) 1(5):449
Bandala, Erick R. (Instituto Mexicana de Technologia del Agua/Mexico) 1(4):7, 1(6):177
Bandy, Brian (Watkins-Johnson Co./USA) 1(5):231
Barczewski, Baldur (Universität Stuttgart/Germany) 1(5):199
Barker, James F. (University of Waterloo/Canada) 1(3):327, 1(6):289
Bartow, Gregory W. (California Regional Water Quality Control Board/USA) 1(1):151
Basel, Michael D. (Montgomery Watson/USA) 1(3):243
Bass, David H. (Fluor Daniel GTI/USA) 1(5):285
Bauer, Susanne (Fraunhofer IGB/Germany) 1(4):181
Beam, Paul (U.S. DOE/USA) 1(1):229
Beaubien, André (Gama Innovation Inc./Canada) 1(4):91
Beaudry, William T. (U.S. Army/USA) 1(6):265
Becker, Jennifer G. (Northwestern University/USA) 1(3):93
Becker, Mark (Dames & Moore/USA) 1(3):231
Becvar, Erica S.K. (Applied Research Associates/USA) 1(4):121
Bedient, Philip B. (Rice University/USA) 1(3):219
Bender, Judith (Clark Atlanta University/USA) 1(4):233
Beneteau, Kevin (University of Waterloo/Canada) 1(1):67
Bent, Tim (Bridgestone-Firestone, Inc./USA) 1(3):81
Bentley, Harold (HydroGeoChem/USA) 1(4):21
Berardesco, Gina (Northwestern University/USA) 1(3):93
Bergersen, Ove (SINTEF/Norway) 1(6):237
Bergsman, Theresa M. (Battelle/USA) 1(5):63
Betterton, Eric A. (University of Arizona/USA) 1(5):473
Betts, Paul C. (U.S. Air Force/USA) 1(2):187, 1(6):115
Bhattacharyya, Dibakar (University of Kentucky/USA) 1(5):323
Bice, Nancy T. (Geomatrix Consultants, Inc./USA) 1(6):145
Binard, Kinsley (Geomatrix Consultants Inc./USA) 1(4):257
Bisch, Gerhard (Universität Stuttgart/Germany) 1(4):129
Blickle, Frederick W. (Conestoga-Rovers & Associates/USA) 1(3):139
Böckle, Karin (DVGW-Technology Center/Germany) 1(4):65
Bonneur, Ad (Tauw Milieu b.v./The Netherlands) 1(6):201
Bosma, Tom N.P. (TNO/The Netherlands) 1(3):7
Bostick, Kent (Jacobs Engineering/USA) 1(5):265
Bourquin, Al W. (Camp Dresser & McKee Inc./USA) 1(3):281
Bracco, Angelo A. (GE Corporate R&D Center/USA) 1(3):177, 231
Brame, Scott (Clemson University/USA) 1(2):205
Brandon, William (U.S. EPA/USA) 1(3):193
Brandt, Randy (Dames & Moore, Inc./USA) 1(1):271

Brannan, James (Clemson University/USA) 1(2):125
Braun, Jürgen (Universtität Stuttgart/Germany) 1(2):107, 1(5):199
Breck, Michael (Booz Allen, & Hamilton/USA) 1(1):241
Brennan, Michael J. (GE Corporate R&D Center/USA) 1(3):133
Bresette, D.M. (Hargis & Associates, Inc./USA) 1(5):169
Bressan, Mario (Università "G. D'Annunzio"/Italy) 1(5):433
Brockman, Fred J. (Battelle/USA) 1(1):79
Broderick, John (California EPA/ USA) 1(1):193
Brodman, Bruce W. (U.S. Army/ USA) 1(6):249
Brouillard, Lee A. (Duke Engineering and Services/USA) 1(1):59
Brown, Kandi L. (IT Corp./USA) 1(3):333
Brown, Richard A. (Fluor Daniel GTI/USA) 1(5):285, 1(6):39, 301
Brown, Steve G. (CH2M HILL/USA) 1(6):115
Bruce, Cristin L. (Arizona State University/USA) 1(5):293
Bruce, Kevin R. (ARCADIS Geraghty & Miller/USA) 1(5):69
Bruner, D. Roger (Foth & Van Dyke and Associates/USA) 1(4):27
Brusseau, Mark L. (University of Arizona/USA) 1(5):155
Bryniok, Dieter (Fraunhofer IGB/ Germany) 1(4):181
Buermann, Werner (University of Karlsruhe/Germany) 1(2):43
Buffalo, Daniel (Tetra Tech EM, Inc./USA) 1(1):157
Bugbee, Bruce (Utah State University/USA) 1(4):251
Bunce, Nigel J. (University of Guelph/Canada) 1(5):449
Burdick, Jeffrey S. (Geraghty & Miller, Inc./USA) 1(3):81
Burris, David R. (U.S. Air Force/ USA) 1(5):341
Burt, Ronald Allen (Eckenfelder Inc./USA) 1(2):119, 175
Butler, Barbara J. (University of Waterloo/Canada) 1(1):67, 1(3):327, 1(6):163
Büyüksönmez, Fatih (University of Idaho/USA) 1(6):277
Buzzelli, Maurizio (Ambiente SpA/ Italy) 1(6):309

Campbell, Brett E. (Geosafe Corp./ USA) 1(6):231
Cantaloub, Michael G. (Oregon State University/USA) 1(2):137
Cantrell, Kirk J. (Battelle/USA) 1(5):335
Carey, Grant (Environmental Software Solutions/Canada) 1(1):25, 1(3):127, 213
Carl, F.G. (TerraTherm Environmental Services/USA) 1(5):25
Carman, Jeff (Jacobs Engineering/ USA) 1(5):265
Caron, Denise (U.S. Air Force/USA) 1(3):333
Cawein, Christopher C. (Pima County Solid Waste Management/ USA) 1(3):159
Cervini-Silva, Javiera (University of Illinois/USA) 1(5):329
Chaloupka, Katherina J. (Tetra Tech EM, Inc./USA) 1(1):157
Chambliss, Glenn (University of Wisconsin/USA) 1(6):249
Chang, Daniel P.Y. (University of California/USA) 1(3):341
Chang, Soon W. (Oregon State University/USA) 1(3):51
Chang, Yoon-Yang (Korea Institute of Science and Technology/ Republic of Korea) 1(4):239

Chen, Dingwang (National University of Singapore/Singapore) 1(5):443
Chen, Jin-Song (Dynamac Corp./USA) 1(3):205
Cheng, Shu-Fen (National Taiwan University/Taiwan ROC) 1(5):299
Cherry, John A. (University of Waterloo/Canada) 1(2):55
Chiang, Sheau-Yun (Georgia Institute of Technology/USA) 1(3):87
Chiarappa, Marina L. (Lawrence Livermore National Laboratory/USA) 1(5):133
Chiarenzelli, Jeffrey R. (State University of New York/USA) 1(6):189
Chilakapati, Ashok (Battelle/USA) 1(1):79
Choi, Sue Hyung (Kwangju Institute of Science & Technology/Republic of Korea) 1(4):161
Christ, John A. (U.S. Air Force/USA) 1(1):235, 1(6):33
Christensen, Thomas H. (Technical University of Denmark/Denmark) 1(2):193, 1(5):37
Christians, Greg L. (Eckenfelder Inc./USA) 1(2):175
Church, Clinton D. (Oregon Graduate Institute/USA) 1(3):327
Cirpka, Olaf (Universität Stuttgart/Germany) 1(4):129
Cizerle, Kimberly D. (ENVIRON Corp./USA) 1(1):163
Clark, Boyce (Electrokinetics, Inc./USA) 1(5):461
Clark, Wendy (Laguna Construction Co./USA) 1(6):59
Clarke, H. Steve (Chemical Waste Management of the Northwest/USA) 1(5):371
Clausen, Christian A. (University of Central Florida/USA) 1(6):71
Clayton, Wilson S. (Fluor Daniel GTI/USA) 1(5):285, 383, 389
Cleary, John P. (TH Agriculture & Nutrition Co., Inc./USA) 1(6):21
Clouse, Matt (Oregon Dept. of Environmental Quality/USA) 1(1):277
Clover, Charles C. (Louisiana State University/USA) 1(3):75
Coates, John T. (Clemson University/USA) 1(2):205
Colberg, Robert R. (Environmental Consulting & Technology, Inc./USA) 1(5):147
Cole, Jason D. (Oregon State University/USA) 1(6):121
Compton, Harry R. (U.S. EPA/USA) 1(4):143, 245
Constant, W. David (Louisiana State University/USA) 1(3):69
Corapcioglu, M. Yavuz (Texas A&M University/USA) 1(2):79, 1(4):239
Cousino, Matthew A. (ENSR Corp./USA) 1(2):19
Cox, Chris D. (University of Tennessee/USA) 1(4):221
Cox, Evan E. (Beak International Inc./Canada) 1(1):67, 1(3):275, 309, 1(4):227
Craig, Harry D. (U.S. EPA/USA) 1(1):13
Crawford, Ronald L. (University of Idaho/USA) 1(6):277
Crouse, Daniel G. (Roy F. Weston, Inc./USA) 1(4):143
Crowe, R. Ken (BDM Management Services Co./USA) 1(1):19
Culbreth, Mark A. (Environmental Consulting & Technology, Inc./USA) 1(5):147
Cumming, Lydia (Battelle/USA) 1(5):425, 1(6):91, 97
Cummings, Charles (Battelle/USA) 1(5):415

Cura, Jerome J. (Menzie-Cura & Associates, Inc./USA) 1(1):133
Curtis, Mike (CH2M HILL/USA) 1(2):187

Dablow, Jay (Fluor Daniel GTI/ USA) 1(5):109
Dacyk, Peter I. (Parsons Engineering Science, Inc./USA) 1(4):215, 1(5):279
Daftary, David (IT Corp./USA) 1(3):333
D'Alessandro, Nicola (Università "G. D'Annunzio"/Italy) 1(5):433
Daley, Paul (Lawrence Livermore National Laboratory/USA) 1(3):57
Damera, Raveendra (General Physics Corp./USA) 1(1):51
Davey, Christine (Clemson University/USA) 1(2):113
Davis, Eva L. (U.S. EPA/USA) 1(5):49, 115
Day, Steven (Conestoga-Rovers & Associates/Canada) 1(3):127
Day, Steven R. (Inquip Associates/ USA) 1(5):19
Dean, Wm. Gordon (WRS Infrastructure & Environment/USA) 1(5):211
deFlaun, Mary F. (Envirogen, Inc./ USA) 1(6):15
Degher, Alexandra B. (Oregon State University/USA) 1(6):183
Delfino, Thomas A. (Geomatrix Consultants, Inc./USA) 1(1):39, 1(4):103
Dempster, Helen D. (University of Toronto/Canada) 1(3):133
Deng, Baolin (New Mexico Tech/ USA) 1(5):341
Desai, Naren (U.S. Army/USA) 1(1):51
Devlin, R. (University of Waterloo/ Canada) 1(6):289
de Rouffignac, E.P. (Shell E&P Technology Co./USA) 1(5):25
De Voe, Corinne (University of California/USA) 1(2):61
de Wit, Johannes C.M (Tauw Milieu b.v./The Netherlands) 1(5):139
Dibley, Michael J. (Lawrence Livermore National Laboratory/ USA) 1(5):359
Diekert, Gabriele (University of Stuttgart/Germany) 1(4):155
Dietz, Linda (U.S. EPA/USA) 1(3):257
DiGiulio, Dominic C. (U.S. EPA/ USA) 1(5):155
Di Palma, Luca (Università di Roma "La Sapienza"/Italy) 1(4):71
Dober, Maria (Environment Canada/ Canada) 1(1):217
Dooher, Brendan P. (University of California/USA) 1(1):145
Dooley, Maureen A. (ABB Environmental Services, Inc./USA) 1(6):1
Dott, Wolfgang (RWTH Aachen/ Germany) 1(4):33
Doucette, William J. (Utah State University/USA) 1(4):251
Downey, Douglas C. (Parsons Engineering Science, Inc./USA) 1(4):251
Downey, Jerome (Hazen Research, Inc./USA) 1(5):7
Dreiling, Douglas (Burns & McDonnell Waste Consultants, Inc./USA) 1(5):247
Drescher, Eric (Battelle/USA) 1(5):237, 415, 425, 1(6):91, 97
Drescher, Melody (Battelle/USA) 1(5):425, 1(6):97
Drew, Malcolm (Texas A&M University/USA) 1(4):239
Dumdei, Bruce (ENSR Consulting and Engineering/USA) 1(5):193
Duner, Marianne D. (RMT, Inc./ USA) 1(6):53

Dwight, Donald M. (Metcalf & Eddy/USA) 1(5):253, 259
Dworatzek, Sandra (Water Technology International Corp./ Canada) 1(5):347

Eaker, Craig L. (Southern California Edison Co./USA) 1(5):133
Ebbs, Bill (ISK Biosciences/USA) 1(3):153
Eberts, Sandra (U.S. Geological Survey/USA) 1(4):263
Edwards, Robert (Booz Allen and Hamilton/USA) 1(4):251
Eguchi, Masahiro (Organo Corp./ Japan) 1(4):187
Ehlenbeck, Donald R. (Environmental Consulting & Technology, Inc./USA) 1(5):147
Eidså, G. (SINTEF/Norway) 1(6):237
Elgal, Galoust M. (U.S. Air Force/ USA) 1(5):187
Ellsworth, Jim (Environment Canada/Canada) 1(1):217
Ely, Roger L. (University of Idaho/ USA) 1(2):181, 1(3):39, 45
Enfield, Carl G. (U.S. EPA/USA) 1(2):37, 1(5):37
English, Joseph (EG&G Environmental/USA) 1(5):253, 259
Estuesta, Paul (Minnesota Pollution Control Agency/USA) 1(3):193
Evans, John C. (Battelle/USA) 1(5):335
Eweis, Juana B. (University of California/USA) 1(3):341

Fabrega, Jose (Purdue University/ USA) 1(1):111
Falta, Ronald W. (Clemson University/USA) 1(2):67, 85, 125, 205
Faris, Bart H. (New Mexico Environment Dept./USA) 1(3):263
Farnham, Irene (Harry Reid Center for Environmental Studies/USA) 1(1):99
Farrell, James (University of Arizona/USA) 1(6):85
Fate, Richard (Sandia National Laboratories/USA) 1(1):59
Feenstra, Stanley (Applied Groundwater Research, Ltd./Canada) 1(2):55
Ferguson, Ginger K. (Black & Veatch Special Projects Corp./ USA) 1(1):13
Ferrey, Mark (Minnesota Pollution Control Agency/USA) 1(3):193
Field, J.A. (Oregon State University/ USA) 1(2):199
Figura, Mike (U.S. Navy/USA) 1(3):231
Findley, Joseph (U.S. Army/USA) 1(5):437
Fisher, Arthur (U.S. Navy/USA) 1(3):315, 1(4):121
Fisher, R. Todd (Groundwater Services, Inc./USA) 1(6):47
Fishman, Michael (Dynamac Corp./ USA) 1(2):37
Flathman, Paul E. (OHM Remediation Services Corp./USA) 1(6):271
Forbes, Jeffrey (Daniel B. Stephens & Associates, Inc./USA) 1(3):183
Foster, G. David (NYS Dept. of Environmental Conservation/ USA) 1(2):25
Fox, Tad C. (Battelle/USA) 1(1):127, 1(6):157
Franz, Thomas (Beatty Franz Associates/Canada) 1(3):231
Frape, Shaun (University of Waterloo/Canada) 1(1):67
Freed, Darryl (Plasma Environmental Technologies, Inc./Canada) 1(5):69
Frez, William A. (Earth Tech, Inc./ USA) 1(1):117

Friday, Greg (University of Waterloo/Canada) 1(6):163
Fruchter, John S. (Battelle/USA) 1(5):335
Fujita, Masanori (Osaka University/Japan) 1(4):109
Fulton, John W. (ENSR Corp./USA) 1(2):19
Funk, John G. (Rust Environment & Infrastructure/USA) 1(5):89
Futamata, Hiroyuki (Marine Biotechnology Institute Co., Ltd./ Japan) 1(3):99

Gaillot, Gary (IT Corp./USA) 1(3):287
Gale, Robert J. (Louisiana State University/USA) 1(5):461
Gallinatti, John D. (Geomatrix Consultants, Inc./USA) 1(4):103
Gan, D. Robert (Gannett Fleming, Inc./USA) 1(6):283
Ganzel, Larry A. (ERM–North Central, Inc./USA) 1(5):365
Gaudette, Carol (U.S. Air Force/ USA) 1(6):59
Gavaskar, Arun R. (Battelle/USA) 1(5):237, 415, 425, 1(6):91, 97, 157, 169
Gehring, Kristine (University of Wisconsin/USA) 1(6):249
Gehringer, Peter (Austrian Research Centre/Austria) 1(5):455
Geiger, Cherie L. (University of Central Florida/USA) 1(6):71
Gerdes, Kevin S. (University of California/USA) 1(2):31
Gerritse, Jan (TNO/The Netherlands) 1(3):7
Gervason, Ron (California Regional Water Quality Control Board/ USA) 1(1):283
Ghosh, Mriganka M. (University of Tennessee/USA) 1(6):225
Giammar, Daniel (Battelle/USA) 1(5):237, 1(6):91
Gierke, William (Golder Associates, Inc./USA) 1(3):205
Gilbert, Craig D. (Oregon Graduate Institute/USA) 1(5):293
Gillespie, Rick D. (Battelle/USA) 1(1):19
Gillham, Robert W. (University of Waterloo/Canada) 1(6):65, 163
Gilpin, Luke (Lockheed Martin/USA) 1(5):409
Ginn, Jon S. (U.S. Air Force/USA) 1(1):19, 1(2):143, 211
Gogos, A. Stacey (Environmental Planning & Management, Inc./ USA) 1(1):169
Goltz, Mark N. (U.S. Air Force/ USA) 1(1):235, 1(6):33
Gonzales, James (U.S. Air Force/ USA) 1(3):237, 1(5):409
Gordon, E. Kinzie (Parsons Engineering Science, Inc./USA) 1(5):409
Görisch, Helmut (Technical University Berlin/Germany) 1(4):77
Gossett, James M. (Cornell University/USA) 1(4):121
Gotpagar, Jayant K. (University of Kentucky/USA) 1(5):323
Gottipati, Sarayu (Oregon State University/USA) 1(2):137
Govind, Rakesh (University of Cincinnati/USA) 1(4):55
Granade, Steve (U.S. Navy/USA) 1(6):27
Granzow, Silke (University of Stuttgart/Germany) 1(4):155
Graves, Duane (IT Corp./USA) 1(3):147, 287
Green, Roger B. (Waste Management, Inc./USA) 1(6):271
Groher, Daniel M. (ENSR/USA) 1(3):171, 1(5):175
Grotefendt, Amy J. (EnviroIssues/ USA) 1(1):247
Grulke, Eric A. (University of Kentucky/USA) 1(5):323

Gu, Man Bock (Kwangju Institute of Science & Technology/Republic of Korea) 1(4):161
Guarnaccia, Joseph F. (Ciba Specialty Chemicals Corp./USA) 1(2):37
Guest, Peter R. (Parsons Engineering Science, Inc./USA) 1(5):409
Guiot, Serge R. (Biotechnology Research Institute/Canada) 1(3):21, 1(4):91, 1(6):207
Guo, Jicheng (University of Cincinnati/USA) 1(4):55
Gupta, Neeraj (Battelle/USA) 1(1):127, 1(6):91, 157, 169
Guswa, John H. (HSI GeoTrans, Inc./USA) 1(2):1

Haas, Patrick E. (U.S. Air Force/ USA) 1(3):237, 1(6):47
Hafenmaier, Mark F. (The River Corp., Inc./USA) 1(2):161
Hageman, Clarissa J. (University of Idaho/USA) 1(3):39
Haglund, Peter (Umeå University/ Sweden) 1(1):7
Hahn, Melinda W. (ENVIRON International Corp./USA) 1(1):73, 85, 181
Haling, Greg (Metcalf & Eddy/USA) 1(6):59
Hampton, Mark L. (U.S. Army/ USA) 1(1):123, 1(6):259
Hankins, Deborah A. (Intersil, Inc./ USA) 1(6):145
Hansen, John (IT Corp./USA) 1(3):147
Happel, Anne M. (Lawrence Livermore National Laboratory/ USA) 1(1):145
Harayama, Shigeaki (Marine Biotechnology Institute Co., Ltd./ Japan) 1(3):99
Harju, John A. (Gas Research Institute/USA) 1(4):193
Harkness, Mark R. (GE Corporate R&D Center/USA) 1(3):177, 231
Haroski, Dale M. (Roy F. Weston Inc./USA) 1(4):245
Harrar, Bill (Geological Survey of Denmark and Greenland/ Denmark) 1(1):45
Harris, Steven M. (Conestoga-Rovers & Associates/Canada) 1(3):127
Harrison, Vicky L. (North Carolina A&T State University/USA) 1(5):467
Hartley, James (CH2M HILL/USA) 1(4):175
Hartman, Blayne (Transglobal Environmental Geochemistry/ USA) 1(1):117
Hartwig, Dale S. (AECL/Canada) 1(6):77
Harvey, Gregory J. (U.S. Air Force/ USA) 1(4):263
Harvey, Steven P. (U.S. Army/USA) 1(6):265
Hater, Gary R. (Waste Management, Inc./USA) 1(6):271
Havermans, Willem J.M. (Ingenieursbureau Oranjewoud b.v./The Netherlands) 1(3):111
Havlicek, Stephen C. (Coast-to-Coast Analytical Specialists/ USA) 1(3):153
Hawari, Jalal (National Research Council Canada/Canada) 1(6):207
Hayes, Thomas D. (Gas Research Institute/USA) 1(4):193
Hayes Martin, Elizabeth M. (ENSR/USA) 1(3):171
Hayhurst, Shannon (Utah State University/USA) 1(4):251
Hazen, Terry C. (University of California/USA) 1(4):193
Helmig, Rainer (Technische Universität Braunschweig/ Germany) 1(2):107

Henderson, Douglas A. (Troutman Sanders LLP/USA) 1(1):253
Henning, Leo G. (Kansas Dept. of Health & Environment/USA) 1(5):247
Heredy, Laszlo (Plasma Environmental Technologies, Inc./ Canada) 1(5):69
Heron, Gorm (Technical University of Denmark/Denmark) 1(2):193, 1(5):37, 49
Heron, Tom (Nellemann, Nielsen, & Rauschenberger/Denmark) 1(2):193
Herridge, Brian (Resolution Resources, Inc./USA) 1(2):155
Hess, Thomas F. (University of Idaho/USA) 1(6):277
Hewitt, Alan (U.S. Army/USA) 1(5):377
Hicken, Steve (U.S. Air Force/USA) 1(2):211, 1(6):115
Hicks, James E. (Battelle/USA) 1(1):127, 1(6):127, 169
Hiett, Richard (California Regional Water Quality Control Board/ USA) 1(1):151
Hightower, Mike (Sandia National Laboratories/USA) 1(1):229
Hill, Michael (General Physics Corp./USA) 1(1):51
Hill, Stephen (California Regional Water Quality Control Board/ USA) 1(1):283
Hinchee, Robert E. (Parsons Engineering Science/USA) 1(3):299
Hirsch, Jack M. (TerraTherm Environmental Services, Inc./ USA) 1(5):25
Hirsh, Steven R. (U.S. EPA/USA) 1(4):245
Hise, Wally (Radian International LLC/USA) 1(1):205
Hocking, Grant (Golder Sierra LLC/USA) 1(6):103, 151
Hoekstra, Roy (Bechtel Environmental, Inc./USA) 1(5):217
Hoeppel, Ronald E. (U.S. Navy/ USA) 1(3):315
Hoffman, Adam H. (Shepherd Miller, Inc./USA) 1(5):181
Holcomb, David H. (Focus Environmental, Inc./USA) 1(6):21
Holder, Anthony W. (Rice University/USA) 1(3):219
Hollander, David J. (Northwestern University/USA) 1(3):117
Honniball, James H. (Geomatrix Consultants, Inc./USA) 1(4):103
Hopkins, Gary D. (Stanford University/USA) 1(1):235
Hopkins, Omar (Oregon State University/USA) 1(2):137
Hu, Shaodong (New Mexico Tech/ USA) 1(5):341
Huang, Tiehong L. (University of Cincinnati/USA) 1(4):167
Huber, William A. (Dames & Moore/USA) 1(3):231
Hudak, Curtis (Foth & Van Dyke and Associates/USA) 1(4):27
Hughes, Joseph B. (Rice University/ USA) 1(3):219, 1(6):47
Hughes, William D. (Parsons Engineering Science, Inc./USA) 1(4):215, 1(5):279
Humphrey, M.D. (Oregon State University/USA) 1(2):199
Hund, Gretchen (Battelle/USA) 1(1):211
Hussain, S. Tariq (Dames & Moore/ USA) 1(1):271
Hutchens, Ronald E. (Environ International Corp./USA) 1(5):365
Hyman, Michael R. (Oregon State University/USA) 1(3):51, 321

Ike, Michihiko (Osaka University/ Japan) 1(4):109
Ingle, David S. (U.S. DOE/USA) 1(1):229

Ingle, James D. (Oregon State University/USA) 1(3):57
Ishii, Hirokazu (Japan Sewage Works Agency/Japan) 1(4):167
Isosaari, Pirjo (National Public Health Institute/Finland) 1(6):213
Istok, Jonathan D. (Oregon State University/USA) 1(2):137, 199
Itamura, Michael T. (Sandia National Laboratories/USA) 1(5):57
Iversen, Gary M. (University of South Carolina/USA) 1(2):73

Jackson, Richard E. (Duke Engineering & Associates/USA) 1(2):143
Jackson, W. Andrew (Louisiana State University/USA) 1(3):63, 69, 75, 1(5):461
Jadalla, Riyad A. (North Carolina A&T State University/USA) 1(5):467
Jaffé, Peter (Princeton University/ USA) 1(4):83
Jafvert, Chad T. (Purdue University/ USA) 1(1):111
Janosy, Robert J. (Battelle/USA) 1(6):127
Jedral, Wojceich (University of Guelph/Canada) 1(5):449
Jenkins, Thomas F. (U.S. Army Corps of Engineers/USA) 1(1):13
Jensen, Betty K. (USA) 1(1):223
Jensen, Richard A. (Hofstra University/USA) 1(1):223
Jerger, Douglas E. (OHM Remediation Services Corp./ USA) 1(3):33, 1(6):27, 271
Jerome, Karen M. (Westinghouse Savannah River Co/USA) 1(2):113, 1(5):353
Jiménez, Blanca E. (Universidad Nacional Autónoma de México/ Mexico) 1(4):7
Jin, Minquan (Duke Engineering & Associates/USA) 1(2):143
Jitnuyanont, Pardi (Oregon State University/USA) 1(4):149
Johns, Robert (Tetra Tech Inc./ USA) 1(1):193
Johnson, Jaret C. (ABB Environmental Services, Inc./ USA) 1(6):1
Johnson, Jr., John H. (Troutman Sanders LLP/USA) 1(1):253
Johnson, Paul C. (Arizona State University/USA) 1(5):293
Johnson, Randall G. (University of Idaho/USA) 1(3):39
Johnson, Richard L. (Oregon Graduate Institute/USA) 1(5):293
Johnson, Timothy L. (Oregon Graduate Institute/USA) 1(5):371
Johnson, Waynon (U.S. EPA/USA) 1(4):143
Jones, Jennifer M. (Columbia Analytical Services/USA) 1(1):1
Jones, Sonya A. (U.S. Geological Survey/USA) 1(4):263
Jørgensen, Kirsten S. (Finnish Environment Agency/Finland) 1(4):45
Josef, Reinhold (University of Stuttgart/Germany) 1(5):199
Jowett, Robin (Waterloo Barrier Inc./Canada) 1(6):77
Jurgens, Robert (Kansas Dept. of Health & Environment/USA) 1(5):247
Jurka, Val (Lucent Technologies, Inc./USA) 1(5):97, 1(5):193

Kampbell, Donald H. (U.S. EPA-NRMRL/USA) 1(3):123
Kanaras, Louis (U.S. Army Environmental Center/USA) 1(5):75
Kaslusky, Scott (University of California/USA) 1(2):31
Kastenberg, William E. (University of California/USA)1(1):145

Kawakami, Brett T. (Stanford University/USA) 1(6):33
Kawala, Zdzislaw (Institute of Chemical Engineering/Poland) 1(5):43
Kean, Judie (Florida Dept. of Environmental Protection/USA) 1(3):147
Keasling, Jay (University of California/USA) 1(4):97
Keizer, Meindert (Wageningen Agricultural University/The Netherlands) 1(6):201
Kellar, Edward M. (QST Environmental Inc./USA) 1(5):437
Keller, Arturo A. (University of California/USA) 1(2):13, 131
Kelley, Mark E. (Battelle/USA) 1(3):315
Kerfoot, William B. (K-V Associates, Inc./USA) 1(5):271
Khan, Tariq A. (Groundwater Services, Inc./USA) 1(3):237
Kim, Byung C. (Battelle/USA) 1(5):415
Kim, Eunki (Inha University/ Republic of Korea) 1(4):51
Kim, In Soo (Kwangju Institute of Science & Technology/Republic of Korea) 1(4):39, 167
Kim, Jeongkon (Hydro Geo Logic/USA) 1(2):79
Kim, Kyoung-Woong (Kwangju Institute of Science & Technology/Republic of Korea) 1(6):243
Kim, Soon-Oh (Kwangju Institute of Science & Technology/Republic of Korea) 1(6):243
Kim, Young J. (Kwangju Institute of Science & Technology/Republic of Korea) 1(4):39
Kinzel, Julia (Darmstadt University of Technology/Germany) 1(2):43
Kline, Travis R. (TechLaw/USA) 1(1):117
Kmetz, Tom (Bechtel Savannah River Inc./USA) 1(5):161
Knackmuss, Hans-Joachim (Fraunhofer IGB/Germany) 1(4):181
Knauss, Kevin G. (Lawrence Livermore National Laboratory/ USA) 1(5):133, 359
Kneidel, Amy L. (The Ohio State University/USA) 1(4):205
Kornguth, Steven E. (University of Wisconsin/USA) 1(6):249
Koschitzky, Hans-Peter (University of Stuttgart/Germany) 1(5):199
Koziollek, Petra (Fraunhofer IGB/ Germany) 1(4):181
Kraft, Daniel (Booz-Allen & Hamilton, Inc./USA) 1(5):409
Kram, Mark (U.S. Navy/USA) 1(2):131
Krijger, P. (Ecotechniek Bodem b.v./ The Netherlands) 1(5):1
Kringstad, Alfhild (SINTEF/Norway) 1(6):237
Kulik, Ernest J. (Eaton Corp./USA) 1(2):161
Kuniansky, Eve L. (U.S. Geological Survey/USA) 1(4):263
Kupferle, Margaret J. (University of Cincinnati/USA) 1(4):167
Kwon, Paul (Oregon State University/USA) 1(3):321

LaChance, John C. (ENSR Consulting & Engineering/USA) 1(5):377, 397
Laine, M. Minna (Finnish Environment Institute/Finland) 1(4):45
Lakin, D.D. (U.S. Army Corps of Engineers/USA) 1(3):281
LaPlante, Laurie L. (University of Utah/USA) 1(5):223
Larsen, Thomas H. (Hedeselskabet Danish Land Development/ Denmark) 1(2):193

Larson, Richard A. (University of Illinois/USA) 1(5):329
La Torre, Keith A. (University of Tennessee/USA) 1(6):225
Lau, Nancy (University of Central Florida/USA) 1(6):71
Lau, Peter C.K. (National Research Council Canada/Canada) 1(6):207
Layton, Alice C. (The University of Tennessee/USA) 1(6):225
Leahy, Maureen C. (Fluor Daniel GTI, Inc./USA) 1(3):165
Lee, C.M. (Clemson University/ USA) 1(2):205
Lee, David R. (AECL/Canada) 1(6):77
Lee, Ken W. (Kwangju Institute of Science & Technology/Republic of Korea) 1(4):161
Lee, Linda S. (Purdue University/ USA) 1(1):111
Lee, Roger W. (U.S. Geological Survey/USA) 1(4):263
Lee, Seung-Bong (University of Washington/USA) 1(4):199
Lee, Tae Ho (Osaka University/ Japan) 1(4):109
Lees, Raymond (Fluor Daniel GTI/USA) 1(6):301
Legrand, Robert (Radian International LLC/USA) 1(4):193
Lehmicke, Leo Gregory (Beak International Inc./USA) 1(3):309, 1(4):227
Leif, Roald N. (Lawrence Livermore National Laboratory/USA) 1(5):133, 359
Leigh, Daniel P. (IT Corp./USA) 1(3):33, 1(6):27
Leland, David F. (California Regional Water Quality Control Board/USA) 1(1):151
Lelinski, Dariusz (University of Utah/USA) 1(5):223
Leone, Gerald A. (USA Waste Services Co./USA) 1(3):139
Leonhart, Leo S. (Hargis & Associates, Inc./USA) 1(5):169
Levin, Richard S. (QST Environmental/USA) 1(5):437
Lewis, Mark (General Physics Corp./ USA) 1(1):51
Lewis, Ronald F. (U.S. EPA/USA) 1(6):1
Li, Henian (The University of Manitoba/Canada) 1(2):97
Li, Hui (Purdue University/USA) 1(1):111
Liberatore, Lolita (Università "G. D'Annunzio"/Italy) 1(5):433
Lion, Leonard W. (Cornell University/USA) 1(4):51
Lipkowski, Jack (University of Guelph/Canada) 1(5):449
Liu, Anbo (Cornell University/USA) 1(4):51
Liu, Zhijie (University of Arizona/ USA) 1(5):473
Londergan, John T. (Duke Engineering & Associates/USA) 1(2):143
Looney, Brian (Westinghouse Savannah River Co/USA) 1(2):113, 1(5):353
Lorah, Michelle M. (U.S. Geological Survey/USA) 1(3):27
Lovelace, Jr., Kenneth A. (U.S. EPA/USA) 1(1):265
Lowry, Gregory V. (Stanford University/USA) 1(5):311
Lubbers, Renate (TAUW Milieu b.v./The Netherlands) 1(5):139
Lucas, Mark (CH2M HILL/USA) 1(3):257
Ludwig, Ralph D. (Water Technology International Corp./Canada) 1(5):347
Lundstedt, Staffan (Umeå University/Sweden) 1(1):7
Lute, James R. (Fluor Daniel GTI, Inc./USA) 1(6):295

Luthy, Richard G. (Carnegie Mellon University/USA) 1(6):39
Lutz, Edward (DuPont/USA) 1(6):15

Macauley, Douglas D. (Reynolds Metals, Inc./USA) 1(3):139
MacDonald, Jacqueline A. (National Research Council/USA) 1(6):39
MacKenzie, Patricia D. (General Electric Corporate R&D/USA) 1(3):133
Magar, Victor S. (Battelle/USA) 1(3):315, 1(4):121
Mahaffey, William R. (Pelorus Environmental & Biotechnology Corp./USA) 1(6):15
Maier, Walter (University of Minnesota/USA) 1(4):83
Major, David W. (Beak International Inc./Canada) 1(1):67, 1(3):309, 1(4):227, 1(6):145
Major, Leslie (Pacific Groundwater Group/USA) 1(6):65
Makdisi, Richard S. (Stellar Environmental Solutions/USA) 1(1):283
Mandalas, Glenn Christopher (U.S. Air Force/USA) 1(1):235
Manning, John F. (Argonne National Laboratory/USA) 1(6):259
Mantovani, Pier F. (Metcalf & Eddy/USA) 1(5):253, 259
Manz, Robert D. (Environmental Soil Management, Inc./USA) 1(5):13
Marchesi, Primo (American Color and Chemical/USA) 1(6):21
Marcoux, Sébastien (McGill University/Canada) 1(4):91
Marcus, Donald (EMCON/USA) 1(6):85
Margrave, Todd (U.S. Navy/USA) 1(6):27
Markos, Andrew G. (Black & Veatch Special Projects Corp./USA) 1(1):13
Marley, Michael C. (XDD/USA) 1(3):269
Martian, Pete (Lockheed Martin Idaho Technologies Co./USA) 1(3):299
Marvin, Bruce K. (Fluor Daniel GTI, Inc./USA) 1(5):383
Mason, William (Oregon Dept. of Environmental Quality/USA) 1(1):277
Maull, Elizabeth A. (U.S. Air Force/USA) 1(1):157
May, Lawrence D. (Hazen Research, Inc./USA) 1(5):7
May, Ralph J. (General Electric Co. R&D/USA) 1(2):1
Maynard, J. Barry (University of Cincinnati/USA) 1(3):33
McAlary, Todd (Beak International Inc./Canada) 1(4):227, 1(6):9
McBean, Edward A. (Conestoga-Rovers & Associates/Canada) 1(1):25, 1(3):213
McCall, Sarah (Battelle/USA) 1(5):237
McCann, Michael J. (Oregon Dept. of Environmental Quality/USA) 1(1):277
McCarter, Robert L. (University of California/USA) 1(5):121
McCarty, Perry L. (Stanford University/USA) 1(1):235, 1(6):33
McDermott, Raymond (U.S. Army/USA) 1(1):51
McFarland, Michael J. (Utah State University/USA) 1(6):109
McGill, Ken (CH2M HILL/USA) 1(3):257
McInnis, Dean L. (Toxicological & Environmental Assoc./USA) 1(3):69

McKay, Daniel J. (U.S. Army Corps of Engineers/USA) 1(5):377, 397
McMaster, Michaye (Beak International Inc./Canada) 1(3):309, 1(6):289
McNab, Jr., Walter W. (Lawrence Livermore National Laboratory/ USA) 1(5):305, 311
McQuillan, Dennis M. (New Mexico Environment Dept./USA) 1(3):263
Mechaber, Richard A. (GeoInsight, Inc./USA) 1(3):275
Meehan, Jennifer (University of Idaho/USA) 1(3):45
Meinardus, Hans W. (Duke Engineering and Services/USA) 1(2):143
Merica, Simona G. (University of Guelph/Canada) 1(5):449
Merli, Carlo (Università di Roma "La Sapienza"/Italy) 1(4):71
Meurens, Marc (Catholic University of Louvain/Belgium) 1(4):135
Mew, Daniel A. (Lawrence Livermore National Laboratory/ USA) 1(5):359
Miller, Jan D. (University of Utah/ USA) 1(5):223
Miller, Mark (Daniel B. Stephens & Associates, Inc./USA) 1(3):183
Miller, Thomas F. (Roy F. Weston, Inc./USA) 1(4):143
Milloy, Claire (Deep River Science Academy/Canada) 1(6):77
Mims, Diane K. (Versar Inc./USA) 1(1):151
Minier, Jeffrie (Daniel B. Stephens & Associates, Inc./USA) 1(3):183
Mitroshkov, Alex V. (Battelle/USA) 1(5):335
Miyake, Yusaku (Organo Corp./ Japan) 1(4):187
Miyamura, Akira (EBARA Corp./ Japan) 1(6):195
Mobarry, Bruce K. (University of Idaho/USA) 1(3):39
Mohn, Michael F. (Roy F. Weston, Inc./USA) 1(4):143
Monzyk, Bruce (Battelle/USA) 1(5):237
Moon, Seung-Hyun (Kwangju Institute of Science & Technology/ Republic of Korea) 1(4):161, 1(6):243
Moore, Kari D. (Hazen Research, Inc./USA) 1(5):7
Moos, Larry (Argonne National Laboratory/USA) 1(5):19
Morecraft, Andrew J. (Radian International LLC/USA) 1(4):193
Morkin, Mary (University of Waterloo/Canada) 1(6):289
Morris, Neil P. (Beak International Inc./Canada) 1(4):257
Morvillo, Antonino (Università di Padova/Italy) 1(5):433
Mosher, Jackie (CDM Federal Programs Corp./USA) 1(3):281
Muhlbaier, David R. (Soil Remediation Barriers Co./USA) 1(6):133
Mullen, Jo Jean (U.S. Air Force/ USA) 1(2):7
Munakata, Naoko (Stanford University/USA) 1(5):305
Munster, Clyde (Texas A&M University/USA) 1(4):239
Murali, Dev M. (General Physics Corp./USA) 1(1):51
Murdoch, Larry (Clemson University/USA) 1(1):45
Murphy, Brian K. (U.S. Army/ USA) 1(2):7
Murphy, J. Richard (Conestoga-Rovers & Associates/ Canada)1(3):213
Murray, Christopher J. (Battelle/ USA) 1(1):79
Murray, Willard A. (ABB Environmental Services, Inc./USA) 1(6):1

Muzzin, Paolo F. (Ambiente SpA/ Italy) 1(6):309
Myoga, Haruki (Organo Corp./ Japan) 1(4):187

Nalipinski, Michael (U.S. EPA/ USA) 1(3):193
Nambi, Indumathi M. (Clarkson University/USA) 1(2):49
Nardini, Christine A. (Environmental Soil Management, Inc./ USA) 1(5):13
Naus, Jerry (O'Connor Associates Environmental Inc./Canada) 1(3):249
Naveau, Henry (Catholic University of Louvain/Belgium) 1(4):135
Naymik, Thomas G. (Battelle/USA) 1(1):127
Neff, Jerry (Battelle/USA) 1(1):139
Nelson, Christopher H. (Fluor Daniel GTI/USA) 1(5):383, 1(6):295
Neville, Scott (Aerojet General Corp./ USA) 1(1):39, 1(3):309, 1(4):227
Newell, Charles J. (Groundwater Services, Inc./USA) 1(3):237, 1(6):47
Newmark, Robin L. (Lawrence Livermore National Laboratory/ USA) 1(5):133
Nicell, Jim (McGill University/ Canada) 1(4):91
Nielsen, R. Brent (University of California/USA) 1(4):97
Nienkerk, Monte M. (Clayton Group Services/USA) 1(1):175
Noakes, John E. (University of Georgia/USA) 1(4):233
Norris, Robert D. (Eckenfelder, Inc./ USA) 1(2):119, 175, 1(3):269, 1(4):13
Nzengung, Valentine A. (University of Georgia/USA) 1(4):233

Öberg, Lars G. (Umeå University/ Sweden) 1(1):7, 1(6):195
O'Connor, Michael J. (O'Connor Associates Environmental Inc./ Canada) 1(3):249
Octaviano, Juan A. (Instituto Mexicano de Tecnología del Agua/Mexico) 1(6):177
Odish, Kareem (University of Auckland/New Zealand) 1(3):13
O'Dwyer, Deirdre (Tetra Tech EM Inc./USA) 1(6):157, 169
Odziemkowski, Merek K. (University of Waterloo/Canada) 1(6):163
O'Hannessin, Stephanie (EnviroMetal Technologies, Inc./ Canada) 1(6):59, 139
Olie, Jaap J. (Delft Geotechnics/The Netherlands) 1(3):1
Olsen, Lisa D. (U.S. Geological Survey/USA) 1(3):27
Ong, Say-Kee (Iowa State University/USA) 1(5):415
O'Niell, Walter L. (University of Georgia/USA) 1(4):233
O'Reilly, Kirk (Chevron Research and Technology Co./USA) 1(3):321
Orient, Jeffrey P. (Tetra Tech NUS Inc./USA) 1(3):231
O'Shaughnessy, James C. (Worcester Polytechnic Institute/ USA) 1(5):13
Ospina, Rafael I. (Golder Sierra LLC/USA) 1(6):103, 151
O'Steen, William N. (U.S. EPA/ USA) 1(1):259
O'Sullivan, Dennis (U.S. Air Force/ USA) 1(6):127
Otten, Almar (Tauw Miliew e.v./The Netherlands) 1(3):105, 1(5):139, 1(6):201
Ozsu-Acar, Elif (Electrokinetics, Inc./USA) 1(5):461

Pantazidou, Marina (Carnegie Mellon University/USA) 1(3):199
Pardue, John H. (Louisiana State University/USA) 1(3):63, 69, 75, 1(5):461
Park, Jong-Sup (Kwangju Institute of Science & Technology/ Republic of Korea) 1(4):39
Parvatiyar, Madan (University of Cincinnati/USA) 1(4):55
Paszczynski, Andrzej (University of Idaho/USA) 1(6):277
Patrick, Guy (Golder Associates Ltd./Canada) 1(2):149
Patrick, J. Dudley (U.S. Navy/USA) 1(5):217
Patton, Jim (University of Washington/USA) 1(4):199
Pavlostathis, Spyros G. (Georgia Institute of Technology/USA) 1(4):115
Payne, Tamra H. (Clemson University/USA) 1(2):125
Pemberton, Bradley E. (Gregg In Situ Co./USA) 1(5):161
Pennell, Kurt D. (Georgia Institute of Technology/USA) 1(2):91, 1(4):115
Perina, Tomas (IT Corp./USA) 1(3):333
Perry, Christopher (Battelle/USA) 1(2):155
Petrucci, Elisabeth (Università di Roma "La Sapienza"/Italy) 1(4):71
Peurrung, Loni (Battelle/USA) 1(5):63
Phillips, Peter C. (Clark Atlanta University/USA) 1(4):233
Pintenich, Jeffrey (Eckenfelder, Inc./USA) 1(2):175
Pirkle, William A. (University of South Carolina/USA) 1(2):73
Pitts, David M. (Focus Environmental, Inc./USA) 1(6):21
Plaehn, William A. (Parsons Engineering Science, Inc./USA) 1(4):251
Pocsy, Ferenc (Hungaroplazma KFT/Hungary) 1(5):69
Pollack, Albert J. (Battelle/USA) 1(1):19
Pon, George W. (Oregon State University/USA) 1(3):57
Pope, Jeffery L. (Clayton Group Services/USA) 1(1):175
Pound, Michael J. (U.S. Navy/USA) 1(3):293
Powell, Thomas D. (Battelle/USA) 1(5):63
Powers, Gregory M. (RCS Corp./ USA) 1(6):133
Powers, Susan E. (Clarkson University/USA) 1(2):49
Pruijn, Marc (Heidemij Realisatie/ The Netherlands) 1(3):7
Puls, Robert W. (U.S. EPA/USA) 1(5):317

Qazi, Mahmood (Concurrent Technologies Corp./USA) 1(5):75

Rabbi, M. Fazzle (Louisiana State University/USA) 1(5):461
Rafalovich, Alex (Metcalf & Eddy, Inc./USA) 1(6):59
Rafferty, Mike T. (Geomatrix Consultants Inc./USA) 1(4):257
Ramanand, Karnam (IT Corp./ USA) 1(3):287
Rappe, Christoffer (Umeå University/Sweden) 1(1):7, 1(6):195
Rathfelder, Klaus M. (University of Michigan/USA) 1(2):91
Ravi, Varadhan (Dynamac Corp./ USA) 1(3):205, 1(5):155
Rawson, James R.Y. (General Electric Co. R&D/USA) 1(2):1
Ray, Ajay K. (National University of Singapore/Singapore) 1(5):443

Raycroft, Carl (Environ International Corp./USA) 1(5):365
Reeter, Charles (U.S. Navy/USA) 1(6):157, 169
Reinhard, Martin (Stanford University/USA) 1(5):305, 311
Reinhart, Debra R. (University of Central Florida/USA) 1(6):71
Reitsma, Stanley (ENSR Consulting & Engineering/Canada) 1(5):377, 397
Rhykerd, Robert L. (Texas A&M University/USA) 1(4):239
Rice, Barry (Roy F. Weston, Inc./USA) 1(5):127
Rich, Corey A. (Tetra Tech NUS Inc./USA) 1(3):231
Richgels, Christopher M. (County of Sacramento/USA) 1(4):175
Riha, Brian D. (Savannah River Technology Center/USA) 1(5):161
Rittmann, Bruce E. (Northwestern University/USA) 1(3):93, 117
Ro, Kyoung S. (Louisiana State University/USA) 1(6):253
Robb, Joseph (ENSR/USA) 1(3):171
Roberts, David B. (Oregon State University/USA) 1(6):121
Roberts, Edward (Conestoga-Rovers & Associates/Canada) 1(3):127
Roberts, Janet S. (Battelle/USA) 1(5):63
Roberts, Paul V. (Stanford University/USA) 1(5):305
Roberts, Susan V. (Parsons Engineering Science, Inc./USA) 1(2):7
Robinson, Kevin G. (University of Tennessee/USA) 1(4):221
Roeder, Eberhard (Clemson University/USA) 1(2):67, 85, 205
Romer, James (EMCON/USA) 1(6):139
Røneid, Turid (SINTEF/Norway) 1(6):237
Rong, Yue (California Regional Water Quality Control Board/USA) 1(1):105
Roovers, Claude (Tauw Environmental Consultants/The Netherlands) 1(3):105
Rosansky, Stephen H. (Battelle/USA) 1(5):415
Rose, Tom (General Physics Corp./USA) 1(1):51
Ross, Christopher G.A. (Hargis & Associates, Inc./USA) 1(5):169
Rossabi, Joseph (Savannah River Technology Center/USA) 1(2):125, 1(5):161
Rovers, Frank A. (Conestoga-Rovers & Associates/Canada) 1(1):25, 1(3):127, 213
Rudnick, Barbara (U.S. EPA/USA) 1(3):257
Ruiz, Nancy E. (University of Central Florida/USA) 1(6):71
Rulison, Christopher (Krüss USA/USA) 1(2):73
Russell, Richard (U.S. EPA/USA) 1(1):193
Rzeczkowska, Alina (Water Technology International Corp./Canada) 1(5):347

Salinas, Alejandro (Universidad Nacional Autónoma de México/Mexico) 1(4):7
Salvo, Joseph J. (General Electric Co./USA) 1(6):9
Samorano, Daniel (City of Tucson/USA) 1(4):21
Sánchez, Morella (University of Idaho/USA) 1(2):181
Sangines-Uriarte, Luisa (Geomatrix Consultants Inc./USA) 1(4):257
Sanseverino, John (University of Tennessee/USA) 1(4):221
Sasaki, Shoichi (Organo Corp./Japan) 1(4):187

Sass, Bruce M. (Battelle/USA) 1(5):237, 425, 1(6):91, 97, 157, 169
Satrom, Jon (U.S. Air Force/USA) 1(1):193
Saunders, F. Michael (Georgia Institute of Technology/USA) 1(3):87
Sawyer, T.E. (Oregon State University/USA) 1(2):199
Sayavedra-Soto, Luis (Oregon State University/USA) 1(4):149
Sayles, Gregory D. (U.S. EPA/USA) 1(4):167
Schalla, Ronald (Battelle/USA) 1(5):63
Scheibe, Timothy D. (Battelle/USA) 1(1):79
Schilling, Keith E. (Montgomery Watson/USA) 1(3):243
Schirmer, Kristin (University of Waterloo/Canada) 1(3):327
Schirmer, Mario (University of Waterloo/Canada) 1(3):327
Schlegel, Thomas (Antipollution Techniques Entreprise/France) 1(6):219
Schlott, David A. (The ERM Group/USA) 1(1):199
Schmid, Henry (Oregon Dept. of Transportation/USA) 1(6):183
Schneider, James R. (CH2M HILL/USA) 1(6):115
Scholz-Muramatsu, Heidrun (University of Stuttgart/Germany) 1(4):155
Schouten, C.J.J.M. (Mateboer Milieutechniek b.v./The Netherlands) 1(5):271
Schroeder, Edward D. (University of California, Davis/USA) 1(3):341
Schroth, Martin H. (Oregon State University/USA) 1(2):199
Schwall, Jim D. (Hargis & Associates, Inc./USA) 1(5):169
Schwartz, Franklin (The Ohio State University/USA) 1(2):217, 1(5):403
Scogin, Gail (U.S. EPA/USA) 1(4):143
Scow, Kate M. (University of California/USA) 1(3):341
Scrudato, Ronald J. (State University of New York/USA) 1(6):189
Seagren, Eric A. (University of Maryland/USA) 1(3):117
Seiden, Steven (North Carolina State University/USA) 1(6):253
Semprini, Lewis (Oregon State University/USA) 1(2):137, 1(3):39, 57, 1(4):149, 1(6):27
Serie, Patricia J. (EnviroIssues/USA) 1(1):247
Sevcik, Angela E. (The ERM Group/USA) 1(1):73, 85
Sewell, Guy W. (U.S. EPA/USA) 1(4):121, 1(6):15
Shalabi, George (Olin Corp./USA) 1(6):249
Sheahan, Joseph W. (Groundwater Solutions/USA) 1(1):181
Sheldon, Jack K. (Montgomery Watson/USA) 1(3):243
Sheremata, Tamara (National Research Council Canada/Canada) 1(3):21
Sherwood, Lollar B. (University of Toronto/Canada) 1(3):133
Shi, Jianyou (The Ohio State University/USA) 1(2):217
Shi, Zhou (University of Tennessee/USA) 1(6):225
Shikaze, Steve G. (National Water Research Institute/Canada) 1(6):77
Shim, Hojae (The Ohio State University/USA) 1(4):205
Shipley, Kevin B. (XCG Consultants Ltd./Canada) 1(2):169
Shuler, M.L. (Cornell University/USA) 1(4):51

Sibbett, Bruce (IT Corp./USA) 1(3):333
Siegrist, Robert (Colorado School of Mines/USA) 1(1):45
Sinclair, Nate (U.S. Navy/USA) 1(2):155
Singh, Anita (Lockheed-Martin Env. Systems & Technologies Co./ USA) 1(1):91
Singh, Ashok K. (University of Nevada/USA) 1(1):91, 99
Singhal, Naresh (University of Auckland/New Zealand) 1(3):13, 1(4):83
Sisk, Wayne E. (U.S. Army Environmental Center/USA) 1(1):123
Skeen, Rodney S. (Battelle/USA) 1(6):27
Skladany, George J. (Fluor Daniel GTI/USA) 1(3):165, 1(5):383, 1(6):295
Slack, William W. (FRx Inc./USA) 1(1):45
Slater, Greg F. (University of Toronto/Canada) 1(3):133
Small, Mitchell J. (Carnegie Mellon University/USA) 1(3):199
Smets, Barth F. (University of Connecticut/USA) 1(3):117
Smiecinski, Amy. (Harry Reid Center for Environmental Studies/ USA) 1(1):99
Sminchak, Joel R. (Battelle/USA) 1(6):157
Smith, Ann P. (Groundwater Services, Inc./USA) 1(3):237
Smith, Gregory John (ENSR Consulting & Engineering/USA) 1(5):97, 103, 193
Smith, Stewart (HydroGeoChem/ USA) 1(4):21
Smith, William (Environmental Alliance, Inc./USA) 1(1):187
Smyth, David J.A. (University of Waterloo/Canada) 1(6):77
Snyder, Frances A. (Battelle/USA) 1(6):157
Song, Eun (Kwangju Institute of Science & Technology/Republic of Korea) 1(4):161
Song, Qi (University of Cincinnati/ USA) 1(4):55
Song, Stephen (ENVIRON Corp./USA) 1(1):163
Sorenson, Jr., Kent S. (Parsons Infrastructure & Technology Group, Inc./USA) 1(3):299
Sørheim, Roald (Centre for Soil and Environmental Research/Norway) 1(6):237
Spadaro, Jack T. (AGRA Earth & Environmental, Inc./USA) 1(6):183
Spencer, Linda L. (California Regional Water Quality Control Board/USA) 1(1):151
Spivack, Jay L. (GE Corporate R&D Center/USA) 1(3):133
Spuij, Frank (TAUW Milieu/The Netherlands) 1(5):139
Stahl, David A. (Northwestern University/USA) 1(3):93, 117
Stanforth, Michael Trent (Excel Environmental Associates PLLC/ USA) 1(2):161
Stansbury, John (ISK Biosciences/ USA) 1(3):153
Steeneken, Marieke J.J. (Ecoloss/ The Netherlands) 1(3):1
Stefanovic, Marko (East Central University/USA) 1(3):123
Stegemeier, G.L. (GLS Engineering/USA) 1(5):25
Steimle, Rich (U.S. EPA/USA) 1(1):229
Steiof, Martin (TU Berlin/Germany) 1(4):33
Stensel, H. David (University of Washington/USA) 1(4):199

Stephens, Daniel B. (Daniel B. Stephens & Associates, Inc./USA) 1(3):183
Stewart, Craig A. (Geomatrix Consultants, Inc./USA) 1(1):39
Stewart, Lloyd D. (Praxis Environmental Technologies Inc./USA) 1(2):211
Stiber, Neil (Carnegie Mellon University/USA) 1(3):199
Stolzenburg, Thomas R. (RMT, Inc./USA) 1(6):53
Strand, Stuart E. (University of Washington/USA) 1(4):199
Streile, Gary P. (Battelle/USA) 1(1):79
Stucki, Gerhard (Ciba Specialty Chemicals Inc./Switzerland) 1(4):1
Studer, James E. (Duke Engineering & Services/USA) 1(1):59
Su, Benjamin Y. (GEI Consultants, Inc./USA) 1(3):275
Su, Chunming (U.S. EPA/USA) 1(5):317
Sullivan, Kevin M. (Fluor Daniel GTI, Inc./USA) 1(5):383
Sung, Kijune (Texas A&M University/USA) 1(4):239
Suthersan, Suthan (Geraghty & Miller, Inc./USA) 1(3):81
Swanson, Baird H. (New Mexico Environment Dept./USA) 1(3):263
Szafraniec, Linda L. (U.S. Army/ USA) 1(6):265
Szecsody, James E. (Battelle/USA) 1(5):335
Szerdy, Frank S. (Geomatrix Consultants, Inc./USA) 1(6):145
Szewzyk, Ulrich (Technical University Berlin/Germany) 1(4):77

Taat, Jan (Delft Geotechnics/The Netherlands) 1(3):7
Tabak, Henry H. (U.S. EPA/USA) 1(4):55
Taffinder, Sam (U.S. Air Force/ USA) 1(2):143, 1(4):251
Tang, Jinshang (HydroGeoChem/ USA) 1(4):21
Taniguchi, Shin (EBARA Corp./ Japan) 1(6):195
Tanilmis, Tamer (Ege University/ Turkey) 1(5):83
Tartakovsky, Boris (National Research Council Canada/ Canada) 1(6):207
Taylor, Michael (Rust Environment & Infrastructure/USA) 1(5):89
Taylor, Tammy P. (Georgia Institute of Technology/USA) 1(2):91
Tenenti, Stefano (Ambiente SpA/ Italy) 1(6):309
Thompson, Christopher J. (Battelle/ USA) 1(5):335
Thorne, Phillip (U.S. Army Corps of Engineers/USA) 1(1):13
Thornton, Edward C. (Battelle/ USA) 1(5):335
Thüer, Markus (Ciba Specialty Chemicals Inc./Switzerland) 1(4):1
Timmerman, Craig L. (Geosafe Corp./USA) 1(6):231
Tituskin, Sue (IT Corp./USA) 1(3):287
Toda, Hisayuki (EBARA Corp./ Japan) 1(6):195
Tolbert, James N. (Earth Tech, Inc./ USA) 1(1):117
Tonnaer, H. (Tauw Environmental Consultants/The Netherlands) 1(3):105
Torres, Luis Gilberto (Universidad Nacional Autónoma de México/ Mexico) 1(4):7, 1(6):177
Tossell, Robert W. (Beak International Inc./Canada) 1(4):257
Touati, Abderrahame (ARCADIS Geraghty & Miller/USA) 1(5):69

Trail, Kristi L. (ERM-Southwest, Inc./USA) 1(3):69
Tratnyek, Paul G. (Oregon Graduate Institute/USA) 1(5):371
Trewartha, Mark A. (SECOR International Inc./USA) 1(1):31
Trizinsky, Melinda (Clean Sites Inc./USA) 1(1):229
Troxler, William L. (Focus Environmental, Inc./USA) 1(6):21
Tsentas, Constantine (Dames & Moore/USA) 1(3):231
Tuck, David M. (Westinghouse Savannah River Co./USA) 1(2):73
Tuhkanen, Tuula (University of Kuopio/Finland) 1(6):213
Tyner, Larry (IT Corp./USA) 1(3):333

Uchida, Ryuji (EBARA Research/Japan) 1(6):195
Udell, Kent S. (University of California/USA) 1(2):31, 61, 1(5):57, 121
Unversagt, Louis (Olin Corp./USA) 1(6):249

van Aalst-van Leeuwen, Martine A. (TNO/The Netherlands) 1(3):7
van Bavel, Bert (Umeå University/Sweden) 1(1):7, 1(6):195
Vancheeswaran, Sanjay (Oregon State University/USA) 1(3):57
Vanderglas, Brian R. (Parsons Engineering Science Inc./USA) 1(2):7
van Engen-Beukeboom, V.C.M. (Bureau Bodemsanering Dienst Water en Milieu/The Netherlands) 1(5):271
Van Geel, Paul J. (Carleton University/Canada) 1(1):25, 1(3):213
van Heiningen, Erwin (TNO/The Netherlands) 1(3):7
van Laarhoven, Jeroen (Wageningen Agricultural University/The Netherlands) 1(6):201
Vanpelt, Robert S. (Bechtel Savannah River, Inc./USA) 1(5):161
Van Zutphen, Marcus (Netherlands Institute of Applied Geoscience TNO/The Netherlands) 1(5):37
Vartiainen, Terttu (National Public Health Institute/Finland) 1(6):213
Verhagen, Ingrid J. (Minnesota Pollution Control Agency/USA) 1(4):27
Vessely, Mark J. (Parsons Engineering Science, Inc./USA) 1(5):409
Vilardi, Christine L. (STV Inc./USA) 1(1):169
Villanueve, Thomas. (Tetra Tech Inc./USA) 1(1):193
Vinegar, Harold J. (Shell E&P Technology/USA) 1(5):25
Vis, P.I.M. (Ecotechniek b.v./The Netherlands) 1(5):1
Vogan, John L. (EnviroMetal Technologies, Inc./Canada) 1(6):145, 163
Vogel, Catherine (U.S. Air Force/USA) 1(1):19, 1(4):121
Vogel, Timothy M. (Rhodia Eco Services/France) 1(6):219

Waddill, Dan W. (Anderson and Associates, Consulting Engineers/USA) 1(3):225
Wade, Roy (U.S. Army Corps of Engineers/USA) 1(6):253
Wang, Lili (Battelle/USA) 1(5):237
Wang, Xialoan (Emory University/USA) 1(3):87
Ward, David B. (Jacobs Engineering/USA) 1(5):265
Ware, Leslie (Anniston Army Depot/USA) 1(5):437

Warner, Scott D. (Geomatrix Consultants, Inc./USA) 1(1):39, 1(5):371, 1(6):145
Warren, C.J. (University of Waterloo/Canada) 1(6):65
Warren, Stephen (U.S. DOE/USA) 1(1):241
Washburn, Steve (ENVIRON Corp./ USA) 1(1):163
Watanabe, Kazuya (Marine Biotechnology Institute/Japan) 1(3):99
Watts, Phillip (Earth Tech, Inc./ USA) 1(1):193
Watts, Richard J. (Washington State University/USA) 1(6):277
Webb, Eileen L. (AGRA Earth & Environmental, Inc./USA) 1(6):183
Weesner, Brent (ConsuTec, Inc./ USA) 1(6):9, 15
Weigand, M. Alexandra (IT Corp./ USA) 1(3):287
Weik, Lois J. (Dames & Moore/ USA) 1(5):205
Weissenborn, Richard C. (Malcolm Pirnie, Inc./USA) 1(3):159
Wells, Samuel L. (Golder Sierra LLC/USA) 1(6):103, 151
Werner, Peter (Technical University of Dresden/Germany) 1(4):65
Weststrate, Frans A. (Delft Geotechnics/The Netherlands) 1(3):1
Wetzstein, Doug (Minnesota Pollution Control Agency/USA) 1(4):27
Weytingh, Koen R. (Ingenieursbureau Oranjewoud b.v./The Netherlands) 1(3):111
Whiting, Timothy (Fuss & O'Neill/ USA) 1(1):45
Wible, Lyman (RMT Corp./USA) 1(6):249
Wickramanayake, Godage B. (Battelle/USA) 1(3):315
Widdowson, Mark A. (Virginia Tech/USA) 1(3):225
Wiedemeier, Todd H. (Parsons Engineering Science, Inc./USA) 1(3):293
Williams, David R. (O'Connor Associates Environmental Inc./ Canada) 1(3):249
Williamson, Kenneth J. (Oregon State University/USA) 1(3):51, 57, 321, 1(6):121, 183
Williamson, Travis K.J. (Battelle/USA) 1(5):425, 1(6):97
Wilson, David J. (Eckenfelder, Inc./ USA) 1(2):119, 1(3):269, 1(4):13
Wilson, James (Geo-Cleanse International, Inc./USA) 1(5):353, 437
Windfuhr, Claudia (University of Stuttgart/Germany) 1(4):155
Witt-Smith, Carol A. (U.S. EPA/USA) 1(1):13
Woo, Hae-Jin (University of Tennessee/USA) 1(4):221
Wood, Lynn (U.S. EPA/USA) 1(2):37
Woodbury, Allan D. (University of Manitoba/Canada) 1(2):97
Woodhull, Patrick M. (OHM Remediation Services Corp./ USA) 1(6):271
Woods, Sandra L. (Oregon State University/USA) 1(6):121
Wray, Darrin J. (U.S. Air Force/ USA) 1(6):109
Wright, C.W. (Clemson University/ USA) 1(2):205
Wright, Charles L. (University of Tennessee/USA) 1(4):221
Wright, Cliff (Gannett Fleming, Inc./ USA) 1(6):283
Wrobel, John G. (U.S. Army/USA) 1(4):245
Wu, Shian-Chee (National Taiwan University/Taiwan ROC) 1(5):299

Wunderlich, Michele (State University of New York/USA) 1(6):189
Wynn, Jennifer C. (University of Georgia/USA) 1(3):87

Yamane, Carol L. (Geomatrix Consultants, Inc./USA) 1(6):145
Yan, Y. Eugene (The Ohio State University/USA) 1(5):403
Yan, Zhanghua (University of Kuopio/Finland) 1(6):213
Yang, Shang-Tian (The Ohio State University/USA) 1(4):205
Yeh, Daniel H. (Georgia Institute of Technology/USA) 1(4):115
Yemut, Emad (California EPA/USA) 1(1):193
Yong, Raymond N. (McGill University/Canada) 1(3):21
Yoon, Woong-Sang (Battelle/USA) 1(6):169

Zettler, Berthold (TU Berlin/Germany) 1(4):33
Zhang, Jian (IT Corp./USA) 1(3):287
Zhang, S. Allan (O'Connor Associates Environmental Inc./Canada) 1(3):249

KEYWORD INDEX

This index contains keyword terms assigned to the articles in the six books published in connection with the First International Conference on Remediation of Chlorinated and Recalcitrant Compounds, held in Monterey, California, in May 1998. Ordering information is provided on the back cover of this book.

In assigning the terms that appear in this index, no attempt was made to reference all subjects addressed. Instead, terms were assigned to each article to reflect the primary topics covered by that article. Authors' suggestions were taken into consideration and expanded or revised as necessary to produce a cohesive topic listing. The citations reference the six books as follows:

1(1): Wickramanayake, G.B., and R.E. Hinchee (Eds.). 1998. *Risk, Resource, and Regulatory Issues: Remediation of Chlorinated and Recalcitrant Compounds.* Battelle Press, Columbus, OH. 322 pp.

1(2): Wickramanayake, G.B., and R.E. Hinchee (Eds.). 1998. *Nonaqueous-Phase Liquids: Remediation of Chlorinated and Recalcitrant Compounds.* Battelle Press, Columbus, OH. 256 pp.

1(3): Wickramanayake, G.B., and R.E. Hinchee (Eds.). 1998. *Natural Attenuation: Chlorinated and Recalcitrant Compounds.* Battelle Press, Columbus, OH. 380 pp.

1(4): Wickramanayake, G.B., and R.E. Hinchee (Eds.). 1998. *Bioremediation and Phytoremediation: Chlorinated and Recalcitrant Compounds.* Battelle Press, Columbus, OH. 302 pp.

1(5): Wickramanayake, G.B., and R.E. Hinchee (Eds.). 1998. *Physical, Chemical, and Thermal Technologies: Remediation of Chlorinated and Recalcitrant Compounds.* Battelle Press, Columbus, OH. 512 pp.

1(6): Wickramanayake, G.B., and R.E. Hinchee (Eds.). 1998. *Designing and Applying Treatment Technologies: Remediation of Chlorinated and Recalcitrant Compounds.* Battelle Press, Columbus, OH. 348 pp.

A

accelerated cleanup (groundwater) **1(1):**253
activated charcoal (filter) **1(5):**187
adsorption **1(4):**91, **1(5):**237
air sparging **1(1):**205, **1(4):**51, **1(5):**247, 265, 279, 285, 293
air stripping **1(5):**187, 271, 205, 247, 415, 467, **1(6):**309
analytical methods **1(1):**13, **1(2):**125, **1(3):**111
aquifer modification **1(5):**193
arsenic **1(4):**257

B

barium **1(3):**33
barometric pumping **1(5):**161
base catalyzed degradation **1(6):**195
bedrock **1(2):**7, **1(5):**205
    fractured **1(2):**7, **1(5):**205, **1(6):**85
beneficial use (groundwater) **1(1):**151
benzene, toluene, ethylbenzene, and xylenes (BTEX) **1(3):**333, **1(4):**215
bioaugmentation **1(4):**149

BIOCHLOR **1(3)**:237
biodegradation **1(1)**:67, **1(3)**:21, 27, 93, 213, 243, 321, 341, **1(4)**:13, 45, 51, 71, 161, 233, **1(5)**:97, **1(6)**:9, 27, 47, 265
aerobic **1(1)**:19, **1(3)**:327, **1(4)**:1, 27, **1(6)**:27, 121, 225, 289
anaerobic **1(4)**:1, 71, **1(6)**:121
cometabolic **1(1)**:235, **1(3)**:39, 57, 263, 321, **1(4)**:149, 155, 175, 181, 187, 193, 199, 205, 215, 221, 227, **1(6)**:1, 27, 33, 183
enhanced **1(3)**:81, 99, **1(4)**:7, 143, 149, 155, 167, **1(6)**:33, 53, 177, 207, 271, 277
herbicides **1(6)**:183
pesticides **1(6)**:177
sequential anaerobic/aerobic **1(4)**:109, **1(6)**:1, 121, 219, 207, 271
biofilm **1(3)**:39
biofilter **1(3)**:341, **1(4)**:221
biofouling **1(6)**:163
bioluminescent bacteria **1(4)**:221
BIOPLUME II **1(3)**:219
bioreactor **1(4)**:77, 109, 135, 199, 205, **1(6)**:183, 265
BIOREDOX **1(3)**:213
bioslurry **1(6)**:259
Biosolve™ **1(6)**:27
biotransformation **1(3)**:51
bioventing **1(1)**:19, **1(4)**:227
BTEX, *see* benzene, toluene, ethylbenzene, and xylenes

C

carbon tetrachloride (CCl_4) **1(3)**:165, 263, **1(5)**:83
catalytic combustion **1(5)**:83
catalytic hydrodehalogenation **1(5)**:305, 311
cation exchange **1(1)**:111, **1(5)**:237
CERCLA, *see* Comprehensive Environmental Response, Compensation, and Liability Act
CFC-113, *see* trichlorotrifluoroethane *and* refrigerants
chemometric **1(1)**:39
chlorinated aromatics (*see also* chlorobenzenes *and* chlorophenols) **1(4)**:33, **1(5)**:443
chlorine dioxide **1(5)**:193
chlorobenzenes (*see also* chlorinated aromatics *and* chlorophenols) **1(1)**:19, **1(3)**:63, 75, **1(4)**:115, 167, **1(5)**:449, **1(6)**:201
chloroethanes **1(3)**:315
chloroethenes **1(3)**:81, 105, 249, 315, **1(4)**:13, 21, 103, 121, 129, 155, 175, **1(5)**:161, 169, 187, 285, 433, **1(6)**:1, 15, 27, 47, 289
chloroform **1(3)**:309
chloromethanes **1(5)**:299
chlorophenols (*see also* chlorinated aromatics *and* chlorobenzenes) **1(4)**:7, 45, 143, 161, **1(5)**:121, 383, 433, **1(6)**:121, 201, 243, 295
cleanup levels (alternate) **1(1)**:157
cleanup objectives **1(1)**:175, **1(3)**:249
community initiative/involvement **1(1)**:211, 217, 229, 247
composting **1(1)**:123, **1(4)**:45, **1(6)**:109, 253, 271
Comprehensive Environmental Response, Compensation, and Liability Act (CERCLA) **1(1)**:253
cone penetrometer technology **1(6)**:127
containment **1(4)**:245, **1(5)**:217, **1(6)**:115, 133
cosolvent **1(2)**:67, 85, 113, 205
flooding **1(2)**:67, 85, 205

cost
- analysis **1(5):**409, **1(6):**109
- avoidance **1(1):**157
- comparison **1(5):**231
- efficiency **1(5):**365
- estimation **1(1):**199, **1(5):**187
- life-cycle **1(1):**205

cost-limited **1(1):**277
creosote **1(2):**149, **1(5):**133

D

data analysis **1(1):**91
data quality objective (DQO) (*see also* geostatistics *and* statistics) **1(1):**73, 99
DCA, *see* dichloroethane
DCE, *see* dichloroethene
deep well injection **1(5):**365
dehalogenation (*see also* reductive dehalogenation) **1(4):**135, **1(5):**299, **1(6):**71, 85, 237
- abiotic **1(5):**335, 347, **1(6):**289
- microbial **1(6):**237
- zero-valent metals **1(6):**71

demilitarization **1(5):**75, **1(6):**265
dense, nonaqueous-phase liquid (DNAPL) **1(1):**181, **1(2):**1, 13, 19, 25, 31, 37, 49, 61, 67, 73, 85, 107, 113, 131, 155, 161, 187, 211, **1(5):**175, 353, 425, 437
- detection **1(2):**149, 155
- dissolution **1(2):**49
- fractured media transport **1(2):**1, 19
- quantification **1(2):**19
- recovery **1(2):**187
- remediation **1(2):**91, 161, 193, 199, 211
- steam-enhanced extraction **1(2):**61
- surfactant-enhanced remediation **1(2):**91, 199
- thermally enhanced remediation **1(2):**13, 193, 211
- transport modeling **1(2):**25

density **1(2):**67, 85, 181
depressurization **1(5):**57
detection limits **1(1):**99
dichloroethane (DCA) **1(3):**7, 205, 249, 309, **1(4):**1
dichloroethene (DCE) **1(3):**147, 159, 165, 219, 281, **1(4):**129, 181, **1(5):**169, 205, 211, 217, **1(6):**301
dinitrotoluene (DNT) **1(6):**249
dioxin **1(1):**1, 223, **1(5):**7, 69, **1(6):**21, 195, 201, 213
dithionite **1(5):**329, 335, 371
DNAPL, *see* dense nonaqueous-phase liquid
DNT, *see* dinitrotoluene
DQO, *see* data quality objective
dual-phase extraction (*see also* multi-phase extraction) **1(5):**169

E

ecological risk (*see also* human health risk *and* risk assessment) **1(1):**133, 139
electrochemical/electrolytic reduction **1(5):**473
electrochemical peroxidation **1(6):**189
electrokinetics **1(5):**461
electron transfer **1(5):**329
electroosmosis **1(6):**243
enhanced biotransformation **1(5):**97
enhanced dissolution **1(5):**193
environmental isotopes **1(1):**67, **1(3):**133
enzymes **1(6):**53
EPA **1(1):**265
EPA Region 4 **1(1):**259
ethene **1(4):**181
expert system **1(3):**199
explosives **1(1):**123, **1(6):**109, 249, 253, 259, 271

F

fate and transport **1(6):**39
FBR, *see* fluidized-bed reactor
Fenton chemicals/chemistry **1(5):**353, **1(6):**189
field sampling **1(3):**177
fluidized-bed reactor (FBR) **1(4):**91
fractured media **1(2):**1, 7, 19
funnel and gate (*see also* permeable barrier *and* reactive barrier) **1(6):**139, 157, 289

G

GAC, *see* granular activated carbon
geochemical indicators **1(3):**257
geochemistry **1(1):**39
geostatistics (*see also* data quality objective *and* statistics) **1(1):**79, 85, 91
granular activated carbon (GAC) **1(4):**65, 91
groundwater
  beneficial use **1(1):**151
  containment **1(5):**199
  designation **1(1):**151
  technology comparison **1(6):**39

H

half-lives **1(6):**59
heavy metals **1(3):**63
herbicides **1(6):**21, 183
horizontal subsurface barriers **1(6):**133
horizontal wells **1(5):**181, **1(6):**9
hot air injection **1(5):**19
hot water injection **1(5):**115
human health risk (*see also* ecological risk *and* risk assessment) **1(1):**145, 151, 163
hydraulic
  barrier **1(6):**65, 309
  control **1(4):**257
  fracturing **1(5):**109, **1(6):**85, 103, 151
hydrocyclone stripping **1(5):**223
hydrogen addition **1(5):**305, 311, **1(6):**47
hydrolysis **1(6):**265
hydrous pyrolysis/oxidation **1(5):**133

I

incineration **1(6):**201
indoor air pathway **1(1):**163
inhibitors/inhibition **1(3):**63, **1(5):**305, 341
in situ vitrification **1(6):**231
institutional controls **1(6):**283
instruments **1(1):**45, **1(2):**131
interfacial tension **1(2):**73
intrinsic bioremediation, *see* natural attenuation
investigation strategy **1(3):**105
iron dechlorination **1(5):**329

K

kinetics **1(3):**39, **1(4):**55, 71, 83, 97, **1(5):**271, 317, 323, 359, 403, 443, 473, **1(6):**177, 259
kriging **1(1):**85, 91, 199

L

landfarming **1(4):**143
landfill **1(1):**25, **1(2):**7, **1(3):**275, **1(4):**21, 27, 175, **1(6):**201
laser-induced fluorescence (LIF) **1(2):**131
life-cycle cost **1(1):**205
light, nonaqueous-phase liquid (LNAPL) **1(2):**79
lindane (*see also* pesticides) **1(3):**153
LNAPL, *see* light, nonaqueous-phase liquid
long-term monitoring (LTM) **1(1):**193
LTM, *see* long-term monitoring

M

manganese **1(3):**193
marine sediments **1(6):**237
metal mixtures **1(6):**91
methanotrophic treatment **1(4):**193
methyl tertiary butyl ether (MTBE) **1(1):**105, 145, **1(3):**321, 327, 333, 341, **1(5):**293, 467
methylene chloride **1(1):**175, **1(5):**89
microbial mats **1(4):**233
microbial respiration **1(3):**87
microcosm **1(3):**27, 45
microwave radiation **1(5):**43
migration **1(1):**163, 169, **1(2):**31, 73, 79, **1(3):**315, **1(4):**175, **1(6):**115
mineral precipitation **1(6):**163
mineralization **1(6):**277
modeling **1(5):**293, **1(6):**169
    bioremediation **1(1):**25, 79, **1(3):**45, 237, **1(4):**13, 39, 239
    contaminant concentration/amount **1(2):**1, 19, 49, 55, 119, **1(3):**219, 299, **1(4):**199
    kinetics **1(5):**323, 389
    microbial transport **1(3):**13
    Monte Carlo simulation **1(1):**105
    numerical **1(2):**85, 91, **1(5):**57, 253, **1(6):**9
    site closure/screening **1(1):**283, **1(3):**237
    transport **1(1):**39, 105, 111, 127, **1(2):**25, 37, 49, 79, 97, 107, 217, **1(3):**13, 333, **1(5):**155, 265, **1(6):**27, 157
monitoring **1(1):**13, 19, 45, 51, 59, **1(2):**131, **1(3):**81, 117, 133, 147, 171, 177, 275, **1(4):**135, **1(6):**169
monitoring (*cont'd*)
    biological **1(1):**1
    remote **1(1):**19
MTBE, *see* methyl tertiary butyl ether
multi-level sampling **1(1):**31
multi-phase extraction (*see also* dual-phase extraction) **1(5):**103, 147, 175
mustard (sulfur) biodegradation **1(6):**265

N

NAPL, *see* nonaqueous-phase liquid
natural attenuation **1(1):**79, 193, **1(2):**149, 175, **1(3):**1, 7, 33, 45, 69, 75, 81, 105, 123, 127, 139, 147, 153, 159, 165, 171, 177, 183, 193, 199, 205, 213, 225, 231, 237, 243, 249, 257, 263, 269, 275, 281, 287, 293, 299, 309, 315, 327, 333, **1(6):**27, 283
    BIOCHLOR **1(3):**237
    chloroethenes **1(3):**105
    inhibition **1(3):**63
    microcosms **1(3):**69
    modeling **1(3):** 127, 199, 205, 213, 225, 231
    monitoring **1(3):**117, 123, 133, 177, 299
    oxic conditions **1(3):**257
    site characterization **1(2):**149, **1(3):**111, 159, 165, 171, 193, 243
nonaqueous-phase liquid (NAPL) **1(2):**55, 125, 137, 205
    detection **1(2):**125, 137
    quantification **1(2):**55
nutrient enhancement **1(3):**81, **1(4):**55, 103, 121, 193, **1(6):**1, 9, 15, 33, 183, 219
nutrient/metabolite delivery **1(5):**461

O

off-gas treatment **1(2):**187, **1(4):**199, **1(5):**1, 409, 415, **1(6):**309
oil well stimulation technology **1(5):**193
oils **1(5):**115
operation and maintenance **1(1):**205
organic matter **1(3):**21
organo-silicon compounds **1(3):**57
oxidation **1(3):**51, **1(5):**193, 353, 437
  catalyzed **1(5):**415, 433
  chemical **1(2):**181, **1(5):**211, 353, 365, 371, 377, 383, 389, 397, 403, 425, 433, 437, **1(6):**189, 295
  kinetics **1(5):**359
  thermal **1(5):**409
  ultraviolet **1(2):**169, **1(5):**365, 415
oxygen release compounds **1(4):**13
ozone **1(5):**271, 247, 383, 389, **1(6):**295
ozone/electron beam irradiation **1(5):**455

P

PAHs, *see* polycyclic aromatic hydrocarbons
palladium **1(5):**305, 311
partnering **1(1):**217
partnership, public/private **1(1):**229
passive extraction **1(5):**161
PCBs, *see* polychlorinated biphenyls
PCE, *see* tetrachloroethene
peat **1(3):**1
PER, *see* tetrachloroethene
perched zone remediation **1(5):**181, 231
perchloroethene, *see* tetrachloroethene
perchloroethylene, *see* tetrachloroethene
permanganate **1(5):**403
permeability **1(2):**43
permeable barrier (*see also* funnel and gate *and* reactive barrier) **1(2):**175, **1(6):**65, 71, 77, 91, 97, 109, 121, 127, 139, 145, 157, 163, 169, 289
permeable cover **1(1):**25
permitting **1(6):**231
peroxide **1(5):**365, 433, 437,**1(6):**189
pesticides **1(5):**1, **1(6):**21, 39, 177, 201
petrochemicals **1(5):**187
photocatalytic degradation **1(5):**443
photochemical degradation **1(6):**213
photolysis **1(5):**467, **1(6):**225
physical properties **1(3):**33, **1(5):**49, 115, 193, **1(6):**97
phytoremediation **1(4):**239, 245, 251, 257, 263
plasma remediation **1(5):**69, 75
plume delineation **1(1):**51, 85, **1(2):**125
plume stabilization **1(5):**217
polychlorinated biphenyls (PCBs) **1(2):**1, 25, **1(5):**1, 7, 13, 25, 69, **1(6):**39, 189, 207, 219, 225, 231, 237, 243
polycyclic aromatic hydrocarbons (PAHs) **1(1):**1, 7, 139, **1(4):**39, 51, 55, **1(5):**13, 383, **1(6):**53, 231, 295
potassium iodide **1(2):**181
potassium permanganate **1(5):**377, 397, 425
presumptive response/remedies **1(1):**253
properties
  physical **1(3):**33, **1(5):**49, 115, 193, **1(6):**97
  transport **1(1):**139
Pseudomonas **1(4):**7, **1(6):**177, 249
public
  acceptance **1(1):**247
  perception **1(1):**211, 223
public/private partnership **1(1):**229

pump-and-treat **1(1):**205, **1(2):**187, **1(5):**147, 181, 187, 199, 205, 211, 217, 223, 231, 365, 415, **1(6):**201, 309

R

RBCA, *see* Risk-Based Corrective Action
reactant sand-fracking (RSF) **1(6):**85
reactive barrier/wall/curtain (*see also* funnel and gate *and* permeable barrier) **1(6):**59, 77, 103, 45, 151, 163
  iron **1(6):**103
real time assay **1(3):**123
recirculation wells **1(4):**21, **1(5):**253, 259, 265, **1(6):**33
redox potential **1(3):**193
reduction, electrochemical **1(5):**449
reductive dehalogenation **1(3):**63, 75, 159, 183, 225, 269, 287, 299, 315, **1(4):**65, 77, 91, 97, 115, 121, **1(5):**19, 317, 329, **1(6):**47, 109
  abiotic **1(5):**323, 341, **1(6):**103, 151
  chemical **1(5):**371
  microbial **1(3):**21, 57, 93, 263, **1(4):**115, 129, 155, 263, **1(6):**15, 207
  ultraviolet **1(6):**225
refrigerants **1(1):**169, **1(3):**1
regeneration (of reactive media) **1(6):**91
regulatory issues **1(1):**253, **1(3):**293
remedial decision making **1(1):**133
remedial objectives **1(1):**151
remediation endpoint **1(3):**249
Reporter Gene System (RGS) **1(1):**1
RGS, *see* Reporter Gene System
risk assessment (*see also* ecological risk *and* human health risk) **1(1):**73, 133, 145, 163, 175, 187
Risk-Based Corrective Action (RBCA) **1(1):**73, 117, 169, 175, 181
risk management **1(1):**145
RSF, *see* reactant sand-fracking
ruthenium catalyst **1(5):**433

S

sampling **1(1):**7, 31, 51, **1(3):**111
saturated zone **1(5):**63
seismic reflection **1(2):**155
sensors (optical) **1(4):**135
SEQUENCE **1(3):**127
sequential anaerobic/aerobic biodegradation **1(4):**109, **1(6):**1, 121, 207, 219, 271
sequential reduction/oxidation **1(5):**371
sewage sludge **1(3):**1
shear stress **1(3):**13, **1(4):**161
sidewall sensors **1(1):**45
simultaneous abiotic/biotic **1(6):**277
site characterization **1(1):**1, 7, 31, 39, 51, 85, **1(2):**113, 125, 137, 143, 149, 155, **1(3):**111, 133, 159, 165, 171, 193, 243, **1(6):**115, 127
site closure **1(1):**117, 175, 181, 283, **1(3):**263
sodium bicarbonate **1(6):**195
sodium hydrosulfite **1(5):**347
software, interactive **1(1):**235
soil
  heating **1(5):**37, 63, 109
  vacuum extraction **1(5):**155
  vapor extraction (SVE) **1(1):**187, 205, 271, 283, **1(2):**7, 143, 175, **1(4):**21, **1(5):**19, 37, 43, 139, 147, 161, 211, 247, 409, **1(6):**283, 309
  venting **1(5):**181
  washing **1(6):**213
solubility **1(3):**33
source removal **1(6):**283

source term/zone **1(1):**127, **1(2):**149, **1(3):**183, **1(6):**115
Spizizen medium **1(6):**249
stable carbon isotopes **1(3):**133
statistics (*see also* data quality objective *and* geostatistics) **1(1):**73, 91, 99, **1(3):**205
steam injection **1(2):**13, 31, 61, 211, **1(5):**89, 103, 133, **1(6):**21
steam stripping **1(2):**187, **1(5):**97, 121, 127
strontium-90 **1(6):**77
surface flux measurements **1(1):**117
surfactant-enhanced solubilization **1(2):**91
surfactants **1(2):**91, 113, 143, 199, **1(4):**39, 51, 115, **1(5):**199, 237, 449, **1(6):**213
SVE, *see* soil vapor extraction
synergistic microbial processes **1(3):**51
system scale-up **1(6):**27

T

TCA, *see* trichloroethane
TCB, *see* trichlorobenzene
TCE, *see* trichloroethene
technical impracticability guidance **1(1):**259, 265
technology
  acceptance **1(1):**211
  combination **1(2):**175, **1(3):**269, **1(4):**21, **1(5):**97, 247, **1(6):**289, 301
  comparison **1(1):**277, **1(2):**175, **1(5):**231, **1(6):**39
    groundwater **1(6):**39
  evaluation/selection **1(1):**217, 229, 241, **1(3):**139, 293, **1(5):**285, **1(6):**301
temperature effects **1(5):**49, 317
tetrachloroethane **1(6):**59
tetrachloroethene (PCE, PER) **1(1):**67, **1(2):**7, 13, 61, 193, **1(3):**1, 69, 159, 165, 183, 231, 275, 287, **1(4):**1, 65, 91, 97, 109, 129, 233, **1(5):**43, 139, 147, 271, 455, **1(6):**1, 127, 277
tetrachloroethylene, *see* tetrachloroethene
thermal blanket/well **1(5):**25
thermal desorption **1(5):**1, 7, 13, **1(6):**21
thermal remediation **1(2):**13, 31, 61, 193, **1(5):**1, 7, 13, 25, 37, 49, 57, 63, 69, 75, 103, 109, 115, 121, 133, 409, **1(6):**21, 231
thermally enhanced extraction **1(2):**211
TNT, *see* trinitrotoluene
Toxic Substances Control Act (TSCA) permitting **1(6):**231
trace elements **1(1):**99
tracer test **1(2):**119, 125, 137, 143, 217, **1(3):**327, **1(6):**97, 157
transpiration **1(4):**251
transport properties **1(1):**139
treatability test/study **1(3):**165, **1(4):**187, **1(5):**7, **1(6):**53
trenching **1(6):**139
trichlorobenzene (TCB) **1(4):**77
trichloroethane (TCA) **1(3):**231, **1(4):**149, 227, **1(5):**205
trichloroethene (TCE) **1(1):**79, 157, 193, 235, 271, **1(2):**7, 97, 143, 169, 181, **1(3):**21, 27, 39, 45, 57, 99, 123, 133, 147, 159, 165, 171, 183, 199, 219, 231, 243, 257, 275, 281, 287, 299, **1(4):**21, 97, 109, 187, 193, 199, 205, 215, 221, 227, 233, 251, 263, **1(5):**37, 43, 57, 89, 97, 103, 155, 169, 181, 193, 199, 211, 217, 231, 253, 259, 317, 323, 335, 341, 359, 377, 397, 461, **1(6):**33, 59, 65, 71, 91, 109, 115, 151, 189, 283, 309
trichloroethylene, *see* trichloroethene

trichlorotrifluoroethane (CFC-113) (*see also* refrigerants) **1(3):**1
trinitrotoluene (TNT) **1(1):**123, **1(6):**259, 271
TSCA, *see* Toxic Substances Control Act

U

ultrasound energy **1(6):**71
ultraviolet **1(2):**169, **1(5):**365, 415

V

vacuum extraction **1(5):**147
vapor extraction **1(1):**59, **1(5):**103
VC, *see* vinyl chloride
vibration **1(2):**43
vinyl chloride (VC) **1(3):**147, 159, 281, **1(4):**27, 103
viscosity **1(5):**115
VOCs, *see* volatile organic compounds
volatile organic compounds (VOCs) **1(1):**151, 271, **1(3):**139, **1(4):**21, 175, 215, 227, 245, **1(5):**127, 279, 365, 467
volatilization **1(5):**89

W

wetland **1(3):**27, 87

Z

zero-valent metals **1(5):**299, 317, 323, 329, 341, **1(6):**65, 71, 85, 91, 103, 109, 139, 145, 151, 157, 163, 169, 289
zoning **1(3):**111